Cell Death
Mechanisms of Acute and Lethal Cell Injury

Volume I

Cell Death
Mechanisms of Acute and Lethal Cell Injury

Volume I

Edited by

Wolfgang J. Mergner, M.D., PH.D.

Professor
Department of Pathology
University of Maryland School of Medicine
Baltimore, Maryland
    and
Distinguished Scientist
American Registry of Pathology
Washington, DC

Raymond T. Jones, PH.D.

Associate Professor
Department of Pathology
University of Maryland School of Medicine
Baltimore, Maryland

Benjamin F. Trump, M.D.

Professor and Chairman
Department of Pathology
University of Maryland School of Medicine
Baltimore, Maryland

FIELD & WOOD
Medical Publishers, Inc.

Distributed by W. W. Norton & Company, Inc.

500 Fifth Avenue, New York, NY 10110

Distributed by:

W. W. Norton & Company, Inc.
500 Fifth Avenue
New York, New York 10110

ISSN 1048-8499

ISBN 0-938607-39-1

Printed in the United States of America

Printing: 1 2 3 4 5 6 7 8 9 10

Contents

Contributors

John B. Classen, M.D., Staff Fellow, Laboratory of Immunology, National Institute of Allergy and Infectious Diseases, National Institutes of Health, Bethesda, Maryland

Michael Costa, M.D., Fellow, Department of Pathology, The George Washington University School of Medicine, Washington, D.C.

Gertrud M. Hänsch, PH.D., Associate Professor, Institute for Immunology and Serology, University of Heidelberg, Heidelberg, Federal Republic of Germany

Raymond T. Jones, PH.D., Associate Professor, Department of Pathology, University of Maryland School of Medicine, Baltimore, Maryland

Howard J. Leventhal, M.S., Pathologist Assistant, Department of Pathology, Sinai Hospital, Baltimore, Maryland

Louis Marzella, M.D. PH.D., Associate Professor, Department of Pathology and Maryland Institute for Emergency Medical Services Systems, University of Maryland School of Medicine, Baltimore, Maryland

Wolfgang J. Mergner, M.D., PH.D, Professor, Department of Pathology, University of Maryland School of Medicine, Baltimore, Maryland and Distinguished Scientist, American Registry of Pathology, Washington, D.C.

Margo M. Moore, PH.D., Research Scientist, Pulmonary Research Laboratory, St. Paul's Hospital, University of British Columbia, Vancouver, Canada

Pierluigi Nicotera, M.D., PH.D., Department of Toxicology, Karolinska Institute, Stockholm, Sweden

Sten Orrenius, M.D., Professor and Chairman, Department of Toxicology, Karolinska Institute, Stockholm, Sweden

John C. Papadimitriou, M.D., Attending Pathologist, Department of Pathology, University of Maryland Hospital, Baltimore, Maryland

Mohammed M. Sayeed, PH.D., Associate Professor, Department of Physiology, Loyola University, Stritch School of Medicine, Maywood, Illinois

Moon L. Shin, M.D., Professor, Department of Pathology, University of Maryland School of Medicine, Baltimore, Maryland

Yasaman Shirazi, PH.D., Research Associate, Department of Pathology, University of Maryland School of Medicine, Baltimore, Maryland

Benjamin F. Trump, M.D., Professor and Chairman, Department of Pathology, University of Maryland School of Medicine, Baltimore, Maryland

Qian-Chun Yu, M.D., PH.D., Fellow, Department of Cancer Biology, Harvard University School of Public Health, Boston, Massachusetts

Heinz-Gerd Zimmer, M.D., Professor of Physiology, Physiology Institute, University of Munich, Munich, Federal Republic of Germany

Introduction

Wolfgang J. Mergner
Raymond T. Jones
Benjamin F. Trump

THIS IS A BOOK ABOUT MEDICINE AND pathology as much as about the biology of the cell. Human disease is the reference of the discussion of cell injury and cell death. It is our intention to remind the reader of: myocardial infarct, ischemic renal necrosis, the shock liver, cyanide poisoning, immune injury, poisoning by heavy metals, and many other events. These events should then be considered from the view of the cell. "Cell death," "loss of viability," "necrosis," "acute lethal cell injury," and "the point of no return" are terms that frequently appear to describe what happens to cells in our body when they are fatally wounded. We want to focus on the question: Why do cells die?

Cell injury and damage have eluded precise definitions, although the data accumulated are numerous. One important motivation in our work has been the trust that if we could understand the nature of cell injury and death, we could also modify the cell reaction and push the barrier of viability beyond the currently accepted finality. We have undertaken the task of editing this book on cell death in order to contribute to the discussion and to help investigators to redefine their own questions. In the quest for the true nature of cell reactions to injury we have attempted to stay away from dogmatism. The truth has many faces.

Viability

The test of viability is the maintenance of viability after removal of toxic, ischemic, or physical injury. For example, at the moment when a human kidney implanted into a recipient produces urine or when a transplanted heart resumes its regular pumping action the organ as a whole has proven to be viable, although some of its cells may still die. In contrast, when a patient suffers a heart attack, it is soon too late to salvage myocardium, because angioplasty, coronary bypass, or enzyme treatment did not arrive in time and could not relieve the critical ischemia. The ischemia then leads to permanent loss of myocardium. At this point, the tissues of the heart respond with an inflammatory reaction, granulation tissue forms, and the removal of necrotic, nonviable heart tissue ensues, to be replaced by fibrous tissue. These examples leave no doubt about viability and loss of viability.

Viability of Cells

The problem appears considerably more complex when we study the cellular processes that lead to loss of viability. It becomes apparent that there are no direct indicators of cellular viability. How should we define the problem if reflow in ischemia actually produces adverse conditions that lead to cell killing instead of to cell survival, for example, by alterations of the slow calcium channels, absence of chemical energy, or by production of free radicals? Investigators of cell death have therefore turned to symbols, or to indicators that can signal whether cells are alive or nonviable, fully realizing that the symbols are meant to reflect an event that cannot be seen or proven directly. The symbol, therefore, may not be the true indicator of cell viability or cell death, but may offer an approximation to such event.

The flocculent densities seen in the matrix of mitochondria of dead cells are such symbols. Are they really an early sign of cell death? Another symbol is the formation of plasma membrane blebs, protrusions filled with amorphous, partly electron-dense material. Others have suggested the presence of plasma membrane interruptions as the final sign of cell death. Coarse breaks in the plasma membrane that allow large molecules to penetrate are detected by dye exclusion tests, by leakage of cytoplasmic enzymes such as lactate dehydrogenase into the supernatant, or by uptake of large molecules such as horse radish peroxidase or albumin into the cell space. Others have defined cell death as being without respiration, without electron transport, and without the ability to blue or redden tetrazolium dyes. Cell death has been defined as a stage without cellular activity such as muscle contraction, cell division, or dye transport. The duration or severity of the same injury leading to the inversion points of each of these tests varies considerably. This confirms our thinking that no one has found the "heart beat" or the "brain waves" of cells that could tell us if they are viable or not.

Definitions

At this point we want to define the terminology and concepts that have been utilized in the contributions of this volume. We focus on *acute cell injury*, not on chronic cell injury. In acute cell injury the cells have no time for adaptive behavior. Instead, the reactions to injury proceed at considerable speed toward the *point of no return*. By definition, the point of no return is reached when no intervention is able to restore viability. Following the point of no return, the cellular reactions only lead to the breakup of organized structure and function. This phase is called *necrosis*. A frequent term found in the literature of cell injury is *autolysis*. Autolysis, translated, means self-digestion or self-dissolution. When an organ was removed experimentally and allowed to die, this experimental procedure was called *in vitro ischemia*, or autolysis. Autolysis is contrasted to regional ischemia with reduced blood supply, which was also called *in vitro ischemia*. The value of one system over the other was identified as being more pure (*in vitro* ischemia) or more lifelike (*in vivo* ischemia). This contention, however, obscures the fact that reactions in most forms of cell injury are autolytic. With the exception of mechanical disruption or some other forms of physical injury, autolytic processes lead to cell death in most forms of injury: ischemic and anoxic injury, many forms of cellular toxicity, complement injury, after the initial phase of insertion of the membrane attack complex, and others.

Cell injury can be transient or permanent. If cell injury is transient, cells do not necessarily die, but may survive and recover or may survive in a suppressed form. This lack of total restoration of cell functions is produced by *sublethal injury*. An example of sublethal injury is the persistent depression of total adenine nucleotides in transiently ischemic heart cells. Sublethal injury may have considerable importance in medicine related to reduced detoxifying functions by the liver in shock, depressed protein synthesis, and a negative protein balance. A discussion of sublethal cell injury is not included in the deliberations of this book, but the concept is implied in all interpretations (see Figure I-1).

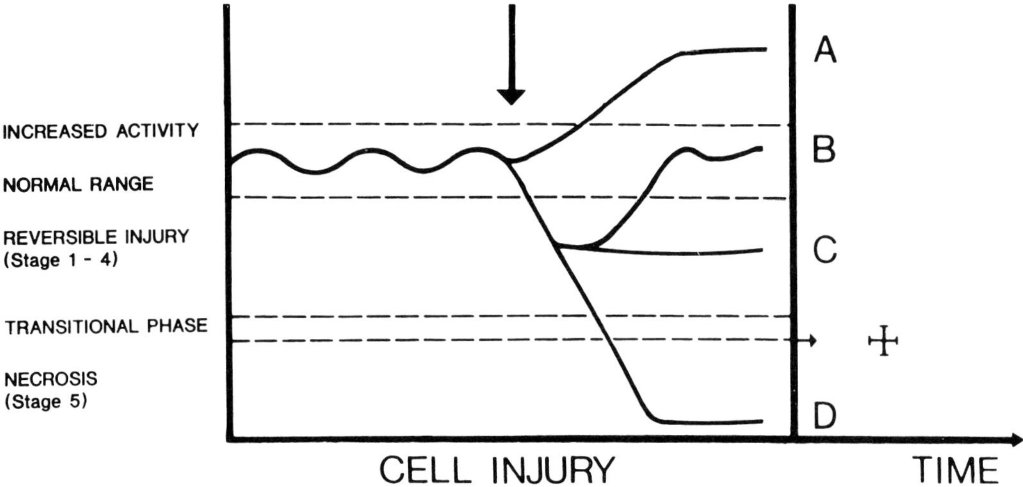

Figure I-1. Prior to cell injury, cells oscillate around a steady state that we call *Normal Range*. Following the injury (arrow), cells may adapt in the following ways: (A) If the cell injury is light and does not hamper cellular activity the steady state of the cellular potential may increase; (B) if the injury is transient without harming the cell the cell may quickly recover and resume normal activity; (C) if the cells experiences structural or functional derangement they may remain depressed; (D) finally, cells may quickly pass beyond the point or the phase of no return and enter the phase of necrosis.

Cellular Reactions to Injury

The very idea of cell damage seems to imply the creation of disorder. One thinks that during the response to cell injury structures disintegrate, chemically ordered reactions are disrupted, and catabolic mechanisms induce lysis that results in random disintegration and cell necrosis. The observer of morphological cell reactions to injury, however, becomes aware of specific sequences or morphological changes that cells undergo. A repeated and even predictable pattern of morphological sequence evolves and can be described. These changes were first described by Trump (1969). The stages are outlined in Table I-1, and some examples are shown in Figures I-2 and I-3.

Cellular Biochemistry and Cellular Physiology

Functional studies also suggest that biochemical and physiological alterations following cell injury do not occur at random but appear to be regulated and predictable under similar conditions (Table I-2). The reactions of cells to injury are probably an extension of normal cellular responses to environmental stimuli. In lethal cell injury such reactions might appear as an abortive attempt to preserve viability. The degree of freedom for the reactive response in cellular injury may be limited by the functional parameters of cellular structure and enzyme parameters. The study of glucose and fat metabolism during hypoxia and ischemia have shown the oscillating nature of the energy metabolism as ischemic time or hypoxic time progresses, all of which may be related to pH and electrolytes, restrictions of enzymes, alterations in cellular regulators such as cAMP, as well as to the availability of internal substrate in the form of glycogen catabolism.

Cells as a Transient Unit

Cells and the organization of tissues, including collagen and bone, are not static but

TABLE I-1

Morphological Stages of Acute Cell Injury

Stages	Cell Membrane	Mitochondria	Nucleus	Lysosomes	Endoplasmic Reticulum
1b	Surface vacuoles	Matrix granules ↓	Chromatin clumping	Aggregate content	Normal
2	Villous contours	Matrix granules ↓	Chromatin clumping	Clear matrix	Fragmentation & dilatation
3	Lamellar wrapping	Matrix condensed	Chromatin clumping	Clear matrix	Polysome dissociation
4	Lamellar wrapping	Condensed & swollen	Chromatin clumping	Clear matrix	Polysome dissociation
4a	Lamellar wrapping	High amplitude swelling	Chromatin clumping	Clear matrix	Polysome dissociation
4b	Lamellar wrapping	Small flocculent densities	Chromatin clumping	Loss of cationic material	Polysome dissociation
5	Defects in continuity	Large flocculent densities	Fragmentation & lysis	Loss of cationic material	Fragmentation

Trump et al., 1978.

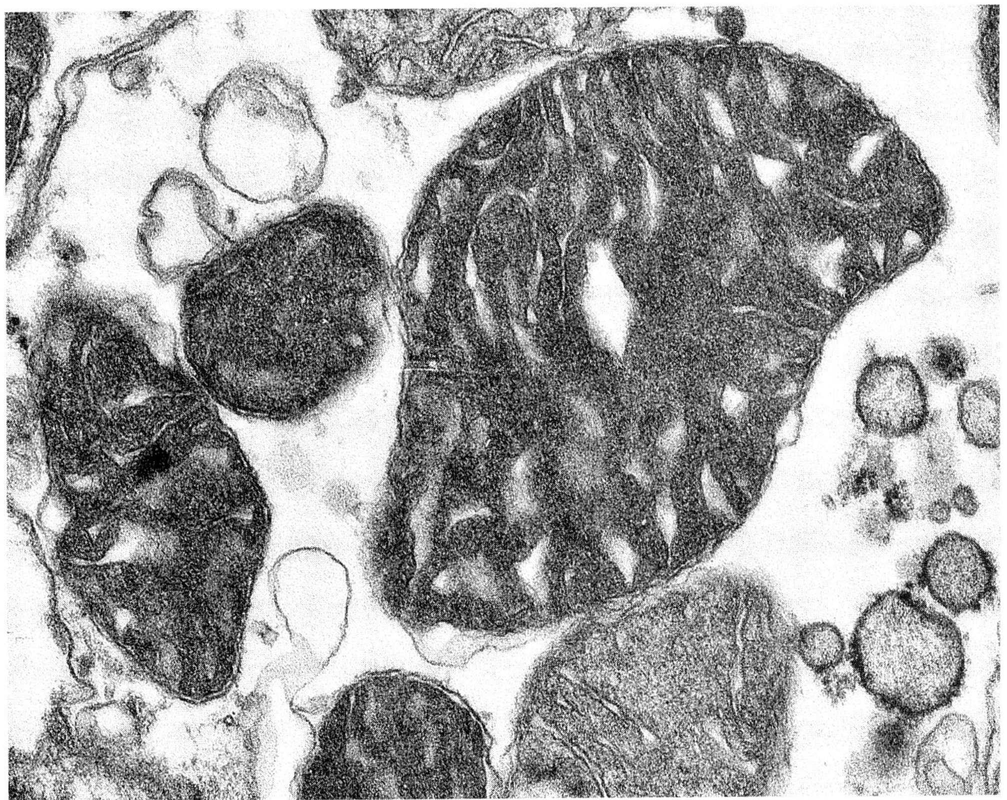

Figure I-2.   Condensed mitochondrial matrix space, indicators of stage 4 of cell injury.

constantly changing. Newly formed intestinal mucosa cells or liver parenchymal cells replace altered mucosa cells and degenerated parenchymal cells. New reserve cells fuse with the syncytium of muscle cells, epithelial cells desquamate, to be replaced by a new generation of epithelial cells. Organelles also turn over, and lipids, fatty acids, proteins, and metabolites exchange, are resynthesized and are disposed of. Microtubles and elements of the cytoskeleton are constantly polymerized and depolymerized. This persistent turnover, which takes place at variable rates for each structure, has been termed a *flow-through system*. Living organisms and living cells are such a steady-state or flow-through system. The very essence of life, from the viewpoint of a single cell, is the relative constancy of shape and the rapid fluctuation of the components. Changes also occur, concurring with the requirements of the system and in response to outside stimuli. Compressed to a short formula, one may say that, for the whole organism as well as for single cells, an essential characteristic of life is active maintenance and incessant interaction. Cell injury could then simply occur by depriving the cell of the means of maintenance and the means of responding to stimuli. In ischemia, energy-dependent maintenance is inhibited, and accelerated catabolism may destroy organelles or cells within a short time. If anabolism is not in balance with catabolism, atrophy, decay, and cell death will result.

CELL DEATH

How, then, can we define cell death? Cell death may occur when catabolism overtakes maintenance of structure and function. It

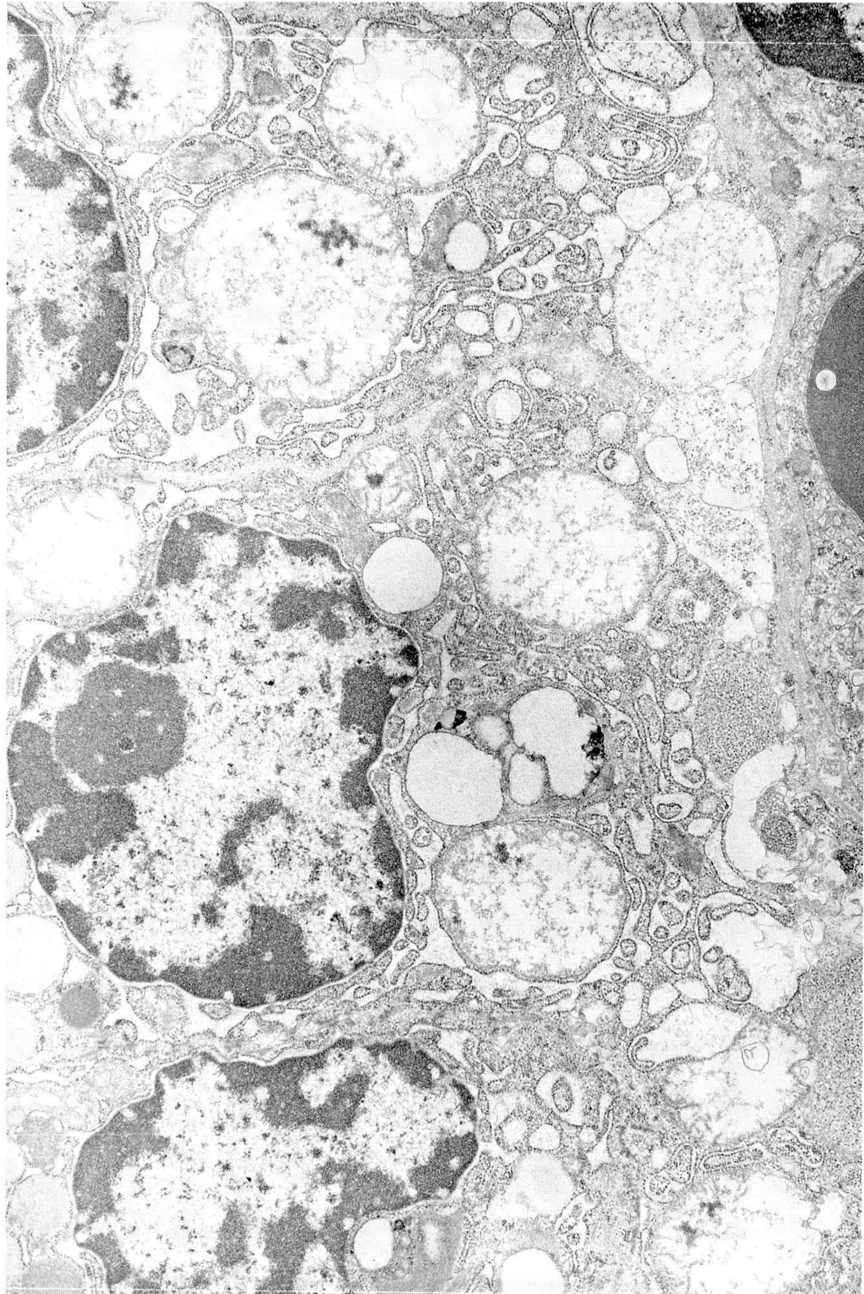

Figure I-3. Necrotic cell, indicated by flocculent densities and cell membrane alterations (state 5 of cell injury).

may be that certain structures and functions are quite essential to guarantee life, such as the plasma membrane and the anaerobic and aerobic energy metabolism; however, there is no doubt that persistence of viability requires preservation of all cell functions.

We propose that cell death may not be a fixed event. Cells with a certain set of metabolic deficiencies may not survive in one situation although they may recover in another situation. For example, by reducing the onslaught of calcium in the reflow situa-

TABLE I-2

Cell Biochemistry and Physiology Alterations Following Cell Injury

Stages	Na^+	K^+	Ca^{2+}	Mg^{2+}	ATP	Cardiolip.	Succ Dh	α-keto glutarate
1b	↑	↓↓	NC	NC	75%↓	NC	NC	NC
2	↑	↓↓↓	NC	NC	80%↓	NC	NC	NC
3	↑	↓↓↓	NC	NC	90%↓	↓	NC	↓
4	↑	↓↓↓	↓	NC	95%↓	↓	NC	↓
5	↓↑	↓↓↓	↓	↓	99%↓	↓↓	NC	↓

tion or by protecting the calcium channels, we may offer cells a chance to regain control of calcium metabolism and to recover. Indeed, the greatest hope for studies on cell injury is to find conditions under which cells can recover. Both internal and external membranes are thought to play a leading role in cellular autolysis and decay. The rapid turnover of membrane components and the necessity for constant maintenance are thought to be related to the great vulnerability of membranes. It is not yet known whether active lysis of membrane components or inability to resynthesize and detoxify cellular components dominates the process of membrane decomposition (Figure I-4).

CONCLUSION

Research in medicine serves to expand means of patient care and the cure of diseases. The focus of the reports in this book is on cell damage and reactions of cells to injury. This may leave an impression of extreme one-sidedness. Therefore, it may be necessary to consider what studies of cellular alteration teach us in regard to reactions in tissues, organs, or the whole organism (Figure I-5).

Cells and Tissues

An essential characteristic of cells is their tendency to form tissues. A tissue is an ar-rangement of sessile cells. A tissue is characterized by its specific physiologic function such as endocrine, respiratory, immune, excretory, or pump and circulatory function. In order to perform such a specialized function and to survive, a tissue forms a symbiotic relationship of cells, with specialization of cells and selective restriction of the multipotential genetic possibilities. For success, a tissue has to develop a tertiary organization of cells and their intercellular space. The organization into tissues is both externally determined by the formation of an exterior scaffolding and cell connections, and internally determined by intracellular fibers that foster the anchoring of connecting sites between cells. In spite of our considerable knowledge of the cytoskeleton, the organizing principal of tissue organization is largely unknown. Cells in tissues do not act as isolated units but in cooperation, communicating through gap junctions and various local hormonal substances. An outstanding example of such cooperation in the synchronous beating of bronchial cilia or of the beating heart.

Subcellular Changes and Tissue Injury

The thought of interrelating cellular and subcellular events in their effects on tissue organization has guided us in designing the layout of the chapters presented and the ones still to come. The study of cellular reactions cannot alone describe human disease adequately. Cell changes become disease-

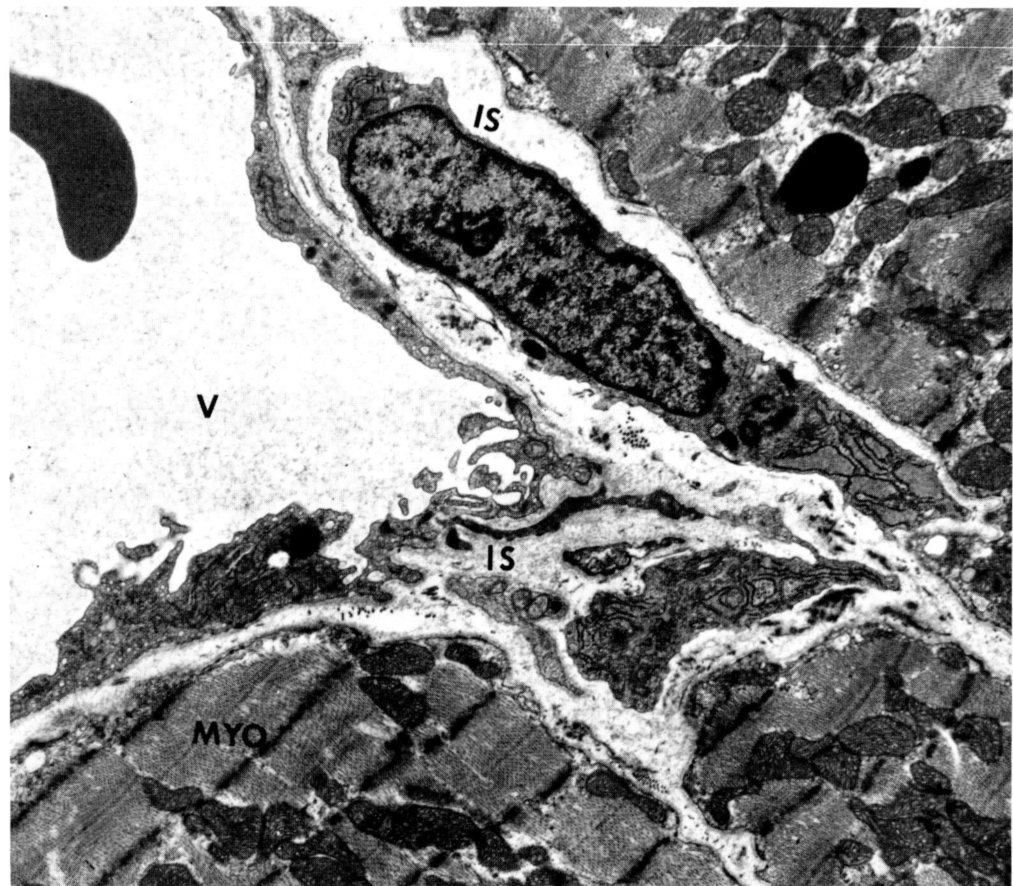

Figure I-4. Model of membrane alteration in autolysis. This figure shows (A) the unmixing of components in the bilayers of the cell membrane; (B) the insertion of free fatty acids, which cannot be transferred onto lysophosphatides; and (C) the possible role of free radicals that cannot be cleaned off the lipid bilayer; (D) the role of lysophosphatides and the dissolution of the cytoskeleton are depicted (interupted arrow); allowing (E) the conglomeration of membrane proteins (medium arrow).

producing entities if they significantly change the function and/or shape of organs or of tissues and their components. This means that the quantity of cellular involvement in tissues by a disease process is a distinct factor for the impact of injury for the organism. The pathophysiological impact of selected cell injury on disease processes may be illustrated by Disseminated Intravascular Coagulation (DIC). The function of the circulation is altered in DIC, and such alteration promotes parenchymal ischemia following the primary cell damage of endothelial cells, thrombus formation, and blood vessel occlusion.

Most tissue injuries invoke a general and nonspecific reaction from the immune system, the coagulation system, and of inflammatory cells. Inflammation is frequently followed by activation of fibroblasts and neovascularization. Altered oxygen and carbon dioxide exchange in shock lung is caused by the combined action of humoral, cellular, and local cell systems, resulting in significant structural changes. Altered hemostasis in aplastic anemia is the outcome of the failure of humoral and cellular components. The pathophysiologists have shown again and again the interrelationship and interdependent involvement of many

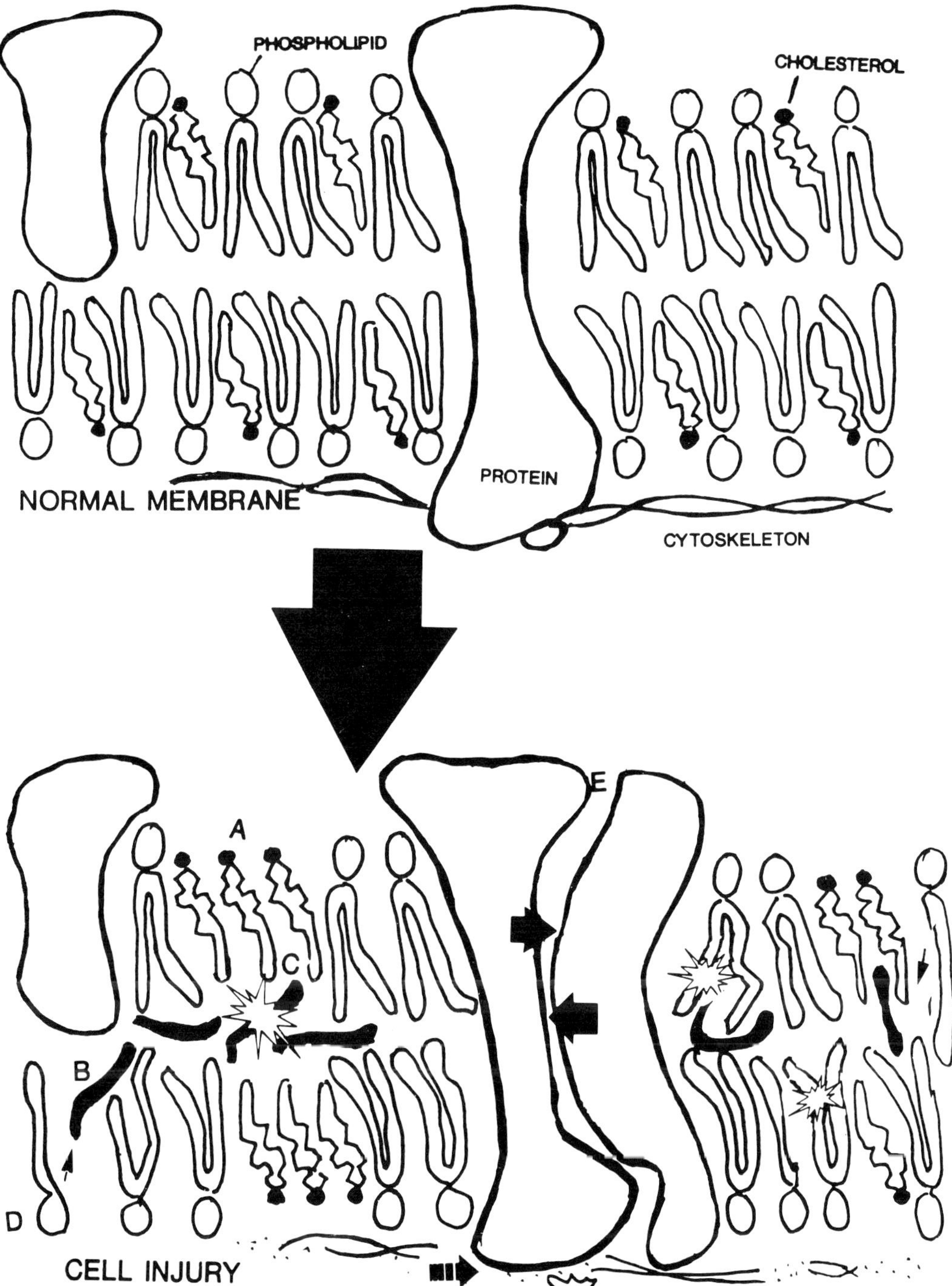

Figure I-5. Myocardium with venule indicating the tissue spaces: the cell space, the interstitial space, and the vascular space.

organ systems and tissue components. Recognizing this interdependence helps to understand mechanisms of disease, its symptoms, causes, and possible therapy.

Cellular pathology is only understandable as part of the systemic pathophysiology. Modern pathophysiology and cellular pathology are highly interrelated sciences that seek a road map through the maze of interactions generated by cell reactions of various tissues and cell systems. Understanding these relationships is basic to patient care, pharmacological treatment, patient monitoring, and diagnostic studies. Cells are frequently as much bystanders as active participants in the progression of disease processes.

ACKNOWLEDGMENT

The editors wish to sincerely and gratefully acknowledge the role that Mr. Wayne J. Ivusich had in the development of this book. In his role as Editorial Assistant, Mr. Ivusich significantly contributed his knowledge, creativity, and expertise as the Information Systems Manager for the Division of Anatomic Pathology, University of Maryland School of Medicine, in the formulation, documentation, organization, and production of this book. We thank him for his past, current, and future efforts.

FURTHER READING

Achs MJ, Garfinkel D. Simulation of the detailed regulation of glycolytic oscillation in a heart supernatant preparation. *Comp Biomed Res.* 1968;2:92.

Achs MJ, Garfinkel D. Computer simulation of energy metabolism in anoxic perfused rat heart. *Am J Physiol.* 1977;232:R164.

Achs MJ, Garfinkel D. Metabolisms of the acutely ischemic dog heart, I: construction of a computer model. *Am J Physiol.* 1979;236:R21.

Anderson KR, Popple A, Parker DJ, Sayer R, Trickey RJ, Davies MJ. An experimental assessment of macroscopic enzyme techniques for autopsy demonstration of myocardial infarction, *J Pathol.* 1979;127,321.

Braasch WS, Gudbjarnason S, Puri PS, Ravens KG, Bing RJ. Early changes in energy metabolism in the myocardium following acute coronary artery occlusion in anesthesized dogs. *Circ Res.* 1968;23:429.

Cowley RA, Mergner WJ, Fischer RS, Jones RT, Trump BF. The subcellular pathology of shock in trauma patients: studies using the immediate autopsy. *Am Surgeon.* 1979;45:255.

Dhalla NS, Das PK, Sharma GP. Subcellular basis of cardiac contractile failure. *J Mol Cell Cardiol.* 1978;10:363.

Fomenko PI, Rebror LB. Post mortem inhibition of RNA synthesis in the liver and skeletal muscle of rats. *Voprosy Meditsinskoi Khimii (Moscow).* 1979;25:408.

Garfinkel D. Simulation of the Krebs Cycle and closely related metabolism in perfused rat liver, II: properties of the model. *Comp Biomed Res.* 1971;4:18.

Garfinkel D. Lactate permeation. *Circ Res.* 1976; 38(Suppl. 1):13

Garfinkel D, Achs MJ. Metabolism of the acutely ischemic dog heart, II: Interpretation of a model. *Am J Physiol.* 1979;236:R31.

Jones RT, Garcia JH, Mergner WJ, Pendergrass RE, Valigorsky JM, Trump BF. Effects of shock on the pancreatic acinar cell: cellular and subcellular effects in humans. *Arch Pathol.* 1975;99:634–644.

Kalimo H, Garcia JH, Kamijyo Y, et al. Cellular and subcellular alterations of human CNS: studies utilizing *in situ* perfusion-fixation at immediate autopsy. *Arch Pathol.* 1974;97:352.

Kohn MC, Garfinkel D. Computer simulation of ischemic rat heart purine metabolism, I: model construction. *Am J Physiol.* 1977a;232:R158.

Kohn MC, Garfinkel D. Computer simulation of ischemic rat heart purine metabolism, II: model behavior. *Am J Physiol.* 1977b;232:H386.

Krause EG, Wollenberger A. Uber die Aktivierung der Phosphorylase und die Glycolyserate im akut Anoxischen Hundeherz. *Biochem Z.* 1965;342:171.

Krause EG, Wollenberger A. Pszillierende glycolyse im ischamischen herzmuskel. *Stud Biophysics.* 1966;1:79.

Kubler W. Tierexperimentelle Untersuchungen zum Myokardstoffwechsel im Angina-Pectoris Anfall und beim Herzinfarkt. *Bibliotheca Cardiologica.* 1971;1(suppl):87.

Kubler W, Spiekerman PG. Regulation of glycolysis in the ischemic and anoxic myocardium. *J Mol Cell Cardiol.* 1970;1:351.

Marzella LL, Yu Q, Mergner WJ, Trump BF. Unbuffered osmium staining of cell organelles: alterations induced by cell injury. *Virch Arch.* 1984; 45:273.

Mergner WJ, Chang SH, Marzella L, Kahng MW, Trump BF. Studies on the pathogenesis of ischemic cell injury, VIII: ATPase of rat kidney mitochondria. *Lab Invest.* 1979;40:686–694.

Mergner WJ, Chang SH, Trump BF. Studies on the pathogenesis of ischemic cell injury, V: morphologic changes of the pars convoluta (P1 and P2) of the proximal tubule of rat kidney made ischemic *in vitro. Virch Arch.* 1976;21:211.

Mergner WJ, Marzella L, Mergner C, Kahng MW, Smith MW, Trump BF. Studies on the pathogenesis of ischemic cell injury, VII: proton gradient and respiration of renal tissue cubes, renal mitochondria, and submitochondrial particles following ischemic cell injury. *Beitrage zur Pathologie.* 1977;161:260.

Mergner WJ, Smith MW, Sahaphong S, Trump BF. Studies on the pathogenesis of ischemic cell injury, X: accumulation of calcium by isolated mitochondria of ischemic rat kidney cortex. *Virch Arch.* 1977;26:1.

Mergner WJ, Smith MW, Trump BF. Studies on the pathogenesis of ischemic cell injury, IV: alteration of ionic permeability of mitochondria from ischemic rat kidney. *Exp Mol Pathol.* 1977a;26:1.

Mergner WJ, Smith MA, Trump BF. Studies on the pathogenesis of ischemic cell injury, XI: P/O ratio and acceptor control. *Virch Arch.* 1977b;26:17.

Opie LH, Owen P, Mansford KRL. Metabolic adjustments to acute heart work: observations in the isolated perfused rat heart. *Cardio Res.* 1971;1(suppl.):87.

Rovetto MJ, Lamberton WF, Neely JR. Mechanisms of glycolytic inhibition. *Circ Res.* 1975;37:742.

Syrakos TP, Wight DG, McMaster P, Marni A, Alfani D. Damage to the biliary tract during preservation. *Transplantation.* 1979;28:166.

Trump BF, Ginn FL. The pathogenesis of subcellular reaction to lethal injury. In: Bajusz E and Jasmin G (eds) *Methods and Achievements in Experimental Pathology.* Volume IV. Basel: Karger, pp 1–29, 1969.

Trump BF, Berezesky IK, Chang SH, Pendergrass RE, Mergner WJ. The role of ion shifts in cell injury. *Scan Electr Microsc.* 1979;3:1.

Trump BF, Berezesky IK, Collan Y, Kahng MW, Mergner WJ. Recent studies on the pathophysiology of ischemic cell injury. *Beitr Pathol.* 1976; 158:363.

Trump BF, Berezesky IK, Pendergrass RE, Chang SH, Bulger RE, Mergnew WJ. Problems of x-ray microanalysis of diffusible elements in scanning electron microscopy of biological thin section: studies of pathologically altered cells. *Scan Electr Microsc.* 1978;2:1027.

Trump BF, Laiho KU, Mergner WJ, Arstila AU. Studies on the subcellular pathophysiology of acute lethal cell injury. *Beitrage zur Pathologie.* 1974; 152:243.

Trump BF, Berezesky IK, Laiho KU, Osornia AR, Mergner WJ, Smith MW. The role of calcium in cell injury: a review. *Scan Electr Microsc.* 1980; 2:437.

Trump BF, Mergner WJ, Kahng MW, Saladino AJ. Studies on the subcellular pathophysiology of ischemia. *Circ Res.* 1976;17(suppl.I):17.

Trump BF, Valigorsky JM, Dees JH, et al. Cellular changes in human disease: a new method of pathological analysis. *Hum Pathol.* 1973;4:89.

Trump BF, Valigorsky JM, Jones RT, Mergner WJ, Garcia JH, Cowley RA. The application of electron microscopy and cellular biochemistry to the autopsy. *Hum Pathol.* 1975;6:499.

Vandor E, Varga T. Investigation of the short-time autolysis of rat hearts by means of SDS, polyacrylamide gel electrophoresis and electron microscopy. *Zeitschrift fur Rechtsmedizin.* 1979;83:225.

Williamson JR. Glycolytic control mechanisms, II: kinetics of intermediate changes during the aerobic–anoxic transition in perfused rat heart. *J Biol Chem.* 1966;241:5026.

Wollenberger A, Krause E. Metabolic control characteristic of the acutely ischemic myocardium. *Am J Cardiol.* 1968;22:349.

PART I

Induction of Cell Injury

1

Membranes in Cell Injury

Wolfgang J. Mergner
Raymond T. Jones

INTRODUCTION

The Relevance of the Lipid Bilayer Model

Membranes have been considered simple lipid bilayer envelopes surrounding cell compartments. Experimental examples are "dark membranes" and "liposomes." Impermeability to the flow of soluble molecules is the characteristic feature of these two model systems.[1] They could be experimentally modified by inserting proteins resulting in special functions and models of cellular activities. ATP synthesis by proton gradients is an example for this molecular modification. Protein molecules appear to be dissolved in the lipid bilayer. As a result of these studies on model membranes it is clear that cell membranes control permeability. Membranes also separate subcellular compartments. It is important to recognize that all biological membranes in eukaryotic cells share a common structure.[2] Even in light of new discoveries, we can still understand membranes as assemblies of lipid and protein molecules held together by noncovalent interaction.

But cell membranes are more than a bilayer system with dissolved proteins. More recent studies on membranes have revealed that the once relatively simple membrane model has become more complicated due to the discovery of the membrane structural components necessary for the maintenance of cell shape, or cell-to-cell connection. Membranes were found to undergo constant renovation and detoxification. Membranes accomplish selected transport into and out of cells, sustain biosynthesis, link receptors for targeted signals, and transmit action potentials through excitable membranes.[3-6] Cytoskeletal components interlink with the membrane in a reversible fashion and guide transport of receptors, membrane movements, and membrane cycling. Cell membranes are dynamic fluid structures that allow molecules and lipids to move about freely. Membranes are asymmetrical, with both surfaces differing greatly in structure and composition.

Membranes and Injury

The membrane theory of cell damage proposes that both internal and external membranes are targets of a suspected, but largely unknown, destructive process. Such membrane disturbances are critical for the survival of cells since they cause the disintegration of multiple membrane-related functions, as listed in Table 1-1. Although the consequences of injury are also destructive for all subcellular membrane systems, it seems that the loss of control over the internal environment outweighs all subsequent alterations. This loss of control occurs if the plasma membrane is altered.

Since we are going to review the basic

TABLE 1-1
Some Consequences of Membrane Alteration

Altered barrier function	Loss of electrochemical gradient
	Loss of volume control
	Loss of calcium-linked internal message
	Loss of pH regulation
	Loss of substrate transport
Altered electrolyte regulation	Activation of catabolic enzymes
	Depression of synthetic enzymes
	Loss of membrane potential
Disintegration of membrane-associated structures	Loss of cell-to-cell connection
	Loss of metabolic cooperation
	Alteration of receptor-mediated cell regulation
Altered cytoskeleton	Loss of mechanical membrane support
	Formation of blebs
Altered organelle membrane	Loss of mitochondrial ATP synthesis
	Loss of protein synthesis
	Loss of detoxification

concepts that are integrated into the membrane theory of cell injury and cell death, we are faced with numerous pathogenic mechanisms. These mechanisms are thought to initiate membrane injury and may send cells to a common pathway of deterioration. Leading among these causes are ischemia, toxins, and immune injury. In many systems the mechanisms that actually promote injury are not entirely clear. Ischemia and many toxins affect the plasma membrane only secondarily. Since there are only a few examples of direct and exclusive membrane attacks, these few examples deserve careful examination to elucidate the isolated role of external membranes in cell viability. Some toxic compounds forming ion selective channels are also examples of induced membrane lesions; for example, valinomycin and gramicidin. Membrane lysis by the complement system, however, is an outstanding example of cell membrane–dependent injury whose major feature is increased permeability. In complement-induced permeability, the cell destruction depends on the assembly and activation of membrane attack complexes.[7,8] These membrane attack complexes form aqueous pores through the membrane and make the membrane leaky. This disrupts the balance of water in cells. Water is drawn into cells through osmosis, and cells swell and burst. Cell-mediated immunity is another example of attack on the plasma membrane of target cells by T-lymphocytes. The T-lymphocytes exist in three classes: cytotoxic T-cells, helper T-cells, and suppressor T-cells. Some cytotoxic T-cells attack their target cells in a manner similar to complement. Following binding to the target cells, the cytotoxic T-cells are stimulated to release a pore-forming protein called *perforin*.[9-11] These pore-forming proteins assemble in the target cell membrane to form transmembrane channels. The killing reaction is mediated by the *major histocompatibility complex (MHC)*. Cytotoxic T-cells recognize foreign antigen in association with MHC.[12,13]

This chapter explores what membranes are, how they are organized, and how cell injury affects the integrity of membranes. It will not discuss the mechanisms by which such destruction may occur.

THE LIPID BILAYER

X-ray diffraction studies of myelin, an assembly of surface membranes around the nerve fibers, have clearly shown that membranes consist of bilayers of lipids. These bilayers can split across the middle when membranes are freeze fractured, a technique used in freeze fracture electron microscopy of the ultrastructure of membranes. Phospholipids dissolved in an aqueous environment have the property of self-assembly. Assembly in sheets of lipid bilayers forming internal and external membranes is one form of self-assembly. Another form of assembly in an aqueous environment is micelle formation. Along with the property of self-assembly comes the property of fluidity and of self-sealing of all compartments if injury occurs. Studies by Kornberg and McConnel have suggested that lipid bilayers form a two-dimensional fluid.[14] Fifty percent of the mass of membranes consists of lipids. The plasma membrane includes three species of amphipathic lipids: phospholipids, cholesterol, and glycolipids. The term *amphipathic* expresses the ability of these lipids to have both hydrophilic and hydrophobic domains.[2]

Phospholipids

Spin label studies have demonstrated that phospholipids rarely migrate from one layer to the other within the two halves of the bilayer, a process called *flip-flop movement*. But lateral movement (that is, neighbors trading places) is common at about 10^7 times/second. The diffusion coefficient is 10^{-8} cm^2/second. Each of the phospholipid molecules is mobile at body temperature, showing flexion and rotation.[2] The plasma membrane is asymmetrical. Phosphotidylcholine and sphingomyelin is found on the outer surface, and phosphotidylserine and phosphoethanolamine are located on the inner surfaces of red blood cells and, probably, of other eukaryotic cells as well. This corresponds to the asymmetric distribution of proteins. In plasma membranes fluidity is controlled by the temperature-dependent property of phospholipids, which is described by the term *phase transition*. It is also controlled by the intermixing with cholesterol. (See Table 1-2.)

Cholesterol

The ratio of cholesterol to phospholipid in the plasma membrane of eukaryotic cells is 1 : 1. Cholesterol controls fluidity and adds mechanical stability to membranes, possibly due to the rigid planes of the steroid ring structure positioned close to the polar head group.[2] The cholesterol molecules are closely interspaced with phospholipid molecules in the plasma membrane. This is possible since esterified cholesterol also forms amphipathic molecules similar in dimension and size to phospholipids. The addition of cholesterol molecules eliminates the phase transition process in plasma membranes. It is thought that this interaction of

TABLE 1-2

Contents of Lipids of Different Internal and External Membranes

						Lipids	
Membranes	Chol	P-Tidyl ethanol	P-Tidyl serine	P-Tidyl choline	Sphingo myelin	Glyco lipids	Others
Liver plasma membrane	17	7	4	24	19	7	22
Red blood cells	23	18	7	17	19	7	13
Mitochondria	3	35	2	39	0	?	21
Endoplasmic reticulum	6	17	5	40	5	?	27

*Modified from Alberts et al., 1983.[2]

relatively immobile cholesterol characterizes the stability of membranes since increased stability is noted upon the addition of cholesterol. Cholesterol prevents interaction of hydrocarbon chains. It also causes the cryostabilization of membranes. An elaborate system controls the insertion and turnover of cholesterol in the membrane, interacting with endoplasmic reticulum and cytosolic enzymes. Removal of cholesterol from plasma membranes involves the high density lipoproteins (HDL).

Glycolipids

Glycolipids are found on the surface of the plasma membrane of eukaryotic cells, where they make up 5% of the surface molecules of the outer layer.[15] The amino alcohol sphingosine is a prominent species in mammalian membranes. The role of glycolipids is presently uncertain. Distinct distributions — for example, galactocerebroside in myelin and ganglioside in the plasma membrane of neurons — suggest that glycolipids have a distinct role in shaping membranes and in membrane labeling. In particular the role of galactocerebroside in myelin sheets is suspected to have such function in directing the enveloping and stalking of myelin in nerve cells. For other membranes, special orientation functions are suspected, but not precisely known. Glycolipids are supposed to affect cellular interaction and cellular differentiation, and to mark oncogenetic trends.[16,17]

MEMBRANE PROTEINS

Lipids of the membranes have been defined as the solvent for proteins.[2] Singer and Nicolson identified lipid–protein relationships and proposed the fluid mosaic model of the structure of membranes.[18] It has become clear that certain proteins require specific phospholipid head groups, such as choline, serine, and ethanolamine. It is thought that some forms of cell injury disturb this specific protein phospholipid interaction.

Characteristics of Membrane Proteins

Lipids determine the shape of membranes, but proteins determine their function, such as receptors, enzymes, cell-to-cell connection, and transporter. Proteins account for 50% of the mass of the plasma membranes and add many distinct functions. Many proteins of the membranes are amphipathic, like lipids. Their hydrophilic components interact with the hydrophilic region of the lipid bilayer; the hydrophobic components, with the hydrophobic region.[3,18] It is also thought that covalently bound fatty acids anchor the protein into the lipid bilayer. Proteins can be associated with the lipid bilayer of the membranes in five ways: (1) as a single helix; (2) as multiple helixes that span the lipid bilayer of the plasma membrane (single- or multiple-pass transmembrane proteins); (3) as polypeptide chains that may be attached to only one surface of the membrane; (4) as a polysaccharide that may form the link between phospholipid and protein; and (5) as proteins that may be attached to other transmembrane proteins.[2,19] One particular type of membrane-associated protein found in red blood cells is *spectrin*. Spectrin is located at the cytoplasmic site of red blood cells. Similar structural proteins with related functions are seen in the plasma membrane of other cells. Spectrin is a long fibrous protein functioning as the struts of the membrane dome. The assembled spectrin complex consists of two polypeptide chains coiled around each other. Spectrin filaments form a cross-linked system, or meshwork. At certain points, a so-called anchoring protein (ankyrin) links spectrin to other membrane proteins, such as band 3 transport protein. Band 3 transport protein is an anion channel protein.[2] Spectrin is also linked to actin by a junctional complex. Spectrin has a structural function related to deformability of red blood cells and stability of the cell shape. Spectrin is more a component of the cytoskeleton than the membrane.[20]

Unrestricted and Restricted Lateral Mobility

Lateral mobility describes movement within the lipid layers or layers of the membrane.

Protein Clusters and Capping

Lipids and many proteins have unrestricted lateral mobility. This process can be seen in the clustering of membrane proteins that interact with antibodies or lectins. A consequence of this process is the phenomenon of "capping," in which clusters of membrane proteins are found at one pole of a cell and are frequently removed by endocytosis.

Restricted Lateral Mobility

Some membrane proteins are known to be restricted in their lateral mobility. For example, the transport enzymes of certain epithelial cells are restricted to either the apex, the lateral border, or the base of asymmetrically organized epithelial cells. Alteration of localized transport proteins may be the reason for selected swelling and bleb formation at the apex of renal tubular epithelial cells, and at the base of gastrointestinal epithelial cells.[21] Another example of such restricted mobility is found in the cell junctional proteins. Such restriction may be due, in part, to the interaction with membrane-associated portions of the cytoskeleton.

MEMBRANE CARBOHYDRATES

All eukaryotic cells have carbohydrates on their cell surface as glycoproteins. The surface carbohydrates stain with ruthenium red and are visible, in electron micrographs, mainly on the noncytosolic cell surface. They consist of intrinsic glycoproteins, glygolipids, and adsorbed glycoproteins and proteoglycans, the sum of which is called the *glycocalyx*. Seven monosaccharides are found to be associated with membranes, such as galactose, mannose, fucose, galactosamine, glucosamine, glucose, and sialic acid. The role of these complex carbohydrate assemblies is unknown. A structural role has been proposed for the orientation of membrane-associated proteins as well as a role in cell-to-cell recognition.

MEMBRANE TRANSPORT OF SMALL MOLECULES

Transport Mechanisms and Transport Proteins

Model lipid phospholipid bilayers are impermeable to large uncharged polar molecules such as glucose and sucrose. They are also impermeable to ions such as protons, sodium ions, potassium, calcium, chloride, and magnesium. The same membrane models are permeable to hydrophobic molecules such as oxygen, nitrogen, and benzene. Small uncharged polar molecules are also able to pass such molecules as water, urea, glycerol, and carbon dioxide. In biological membranes, transport of impermeable solutes such as amino acids, glucose, and electrolytes nucleotides may occur with the help of transport proteins. The transport proteins receive their designation from their functional characteristics. Transport proteins that simply bring solute from one side to the other are called *uniport systems*. Transport that depends on the simultaneous transport of another solute is called a *symport system*, and systems that transport solutes in two different directions across the membrane are called *antiport systems*.

Many transport systems allow transposition of solutes across the membrane by passive transport. In passive transport systems, the difference in concentration is the sole determinant of the direction of the translocation (see Table 1-3). If a solute carries a net charge, then both concentration gradient and total electrical gradient across the membrane determine the direction of transport. Proteins that allow passive transport through aqueous channels are called *channel proteins*. Others, called *carrier proteins*, bind a substance and bring it across the membrane by a process called *facilitated diffusion*[2]. Some carrier proteins drive transport across the membranes against an electrical or concentration gradient, under expenditure of energy. The proteins are called *pumps* and the process is termed *active transport*.[21] In injury affecting the bioenergetic mechanisms, this active transport by pumps is most vulnerable and is likely the initiator of a series of degradative steps leading to cell death.

Antiports in the plasma membrane regulate *intracellular pH (pH_i)*. Mammalian cells contain a Na^+-H^+ exchange carrier. This carrier couples the efflux of H^+ to the efflux of Na^+ and maintains the cellular pH_i at around 7.1–7.2. An alternative route of regulation is the $Cl^--HCO_3^-$ exchanger. The $Cl^--HCO_3^-$ exchanger is activated if the pH_i rises, increasing the extrusion of HCO_3^-. In cell injury, control over protons concentration is lost.

The Na^+/K^+ ATPase and Cellular Volume Regulation

The sodium–potassium pump function is one of the most essential activities of the plasma membrane, since it maintains the membrane potential of the plasma membrane. Two transport proteins are responsible for this function: The Na^+/K^+ pump, which actively pumps sodium out of the cell and potassium into the cell, and the K^+ leak channel. The ubiquitous Na^+/K^+ pump is an ATPase, hydrolyzing ATP to ADP and phosphate. It functions optimally in the presence of both Na^+ and K^+. It is essential to the events in ischemic and toxic cell injury that the Na^+/K^+–ATPase helps to control cell volume by controlling the solute concentration inside cells. This control regulates the osmotic functions that make the cell swell or shrink. Inhibition of this ATPase by toxic substances or by ATP deficiency causes cells to loose volume control. Cell swelling, the universal sign of severe and advanced cell injury, is a consequence of this alteration.[22,23]

Excitable Membranes

The voltage difference across the plasma membrane depends on the membrane po-

TABLE 1-3

Ion Concentration inside and outside a Mammalian Parenchymal Cell

Electrolyte	Intracellular concentration in mM	Extracellular concentration in mM
Sodium	5–15	145
Potassium	140	5
Magnesium	30	1–2
Calcium	1–2	2.5–5.0
Chloride	4	110

*Modified from Alberts et al., 1983.[2]

tential. The membrane potential is determined by the distribution of electrical charge. This charge is carried back and forth across excitable membranes by small electrolytes: Na^+, K^+, Cl^-, and Ca^{2+}. The ions traverse the membrane through selective ion channels. The opening and closing of these channels determines the distribution of electrolytes, hence the name *gated ion channels*. Two classes of gated ion channels have been described: (1) the voltage gated channel and (2) the ligand gated channel. The voltage gated channel plays a role in propagating the action potential along nerve axons and Purkinje fibers. Ligand gated channels translate transmembrane chemical signals into electrical signals.[24] The flow of ions through a membrane channel is driven by the electrochemical gradient. The strength of the electrochemical gradient is composed of the *voltage* gradient and the *concentration* gradient of a given electrolyte. The voltage gradient across the cell membrane is also called *membrane potential.* The *membrane capacitance* determines the charge required to create a given voltage difference (cell membranes typically have a capacitance of 1 $\mu F/cm^2$). The ion current is proportional to the driving force multiplied by the membrane *conductance*. The conductance for a given ion is proportional to the number of ion-specific channels open at a given time.[2]

Cell injury affects the membrane potential in two specific ways. Initially, the disturbance of cell metabolism collapses the energy-dependent membrane functions of electrolytes and charge separation. Later, the disturbance of the membrane organization interferes with effective formation of gradients.

Calcium Metabolism

Eukaryotic cells contain a low intracellular calcium ion concentration of about 10^{-7} mol, while the extracellular calcium ion concentration is 10^{-3} mol. The gradient between the interior of cells is maintained by calcium pumps. These pumps remove calcium from the cell under consumption of energy. Upon selected signals, the cytosol calcium concentration may change, expressing the transmission to an intracellular signal (the messenger function of calcium). In nerves and other excitable tissues, such as the conduction system of the heart, some of the transmembrane protein channels for electrolytes are gated and open only transiently.

Asymmetrically Distributed Ion Channels

Most cells have polarity, and there is evidence that the ion channels are unequally distributed on the cell surface as well.[25] This could cause transcellular ion currents and play a role in distinctive cellular functions and orientation, such as transporting epithelial cells of the gastrointestinal tract and renal tubular cells.[15]

EFFECTS OF CELL INJURY

Isolated plasma membrane injury is rare. Many forms of injury primarily involve intracellular organelles. This effect deprives cells of distinctive organelle functions such as biosynthesis or bioenergetic functions. Examples of primary plasma membrane injury are the injury induced by the membrane attack complex of complement mentioned previously and the osmotic lysis of cell shown by the osmotic lysis of red blood cells.[26] The consequences of the resulting disturbance of membrane functions follow.

A Loss of the Barrier Function between Inside and Outside

Loss of the barrier function results in alteration of the electrochemical gradients that are created by carrier proteins coupled to an energy source, called *pumps*. The gradients are maintained by the barrier of the phospholipid bilayer. In cell injury, cell functions that depend on these gradients are interrupted. Injured cells swell, for example, indicating altered *volume control* as described earlier. The underlying mechanism is as follows: In uninjured cells, the activity of pumps counteracts the high osmotic pressure inside cells. The internal osmotic pressure is determined by macromolecules and metabolites. The insides of cells contain macromolecules that attract, based on their charge, a great number of counter ions. These ions enhance the low osmotic activity of macromolecules notably. Metabolites and small intermediary metabolic products also exert osmotic activity, again augmented by their counter ions, which are attracted by the charge of these small molecules.[2]

Loss of Selective Extrusion

Selective *extrusion* of certain ions such as calcium or sodium plays a special role in the regulation of normal cell metabolism. It is maintained by membrane-bound ATPase with specific affinity toward certain controlled ions. A loss of membrane function will disrupt this control and allow for a rise in cytosolic calcium and similar electrolytes.

Osmotic Lysis of Red Blood Cells

Osmotic lysis is a simple model of plasma membrane injury. It is seen when red blood cells are incubated in hypotonic salt solution. If red blood cells are placed into a hypotonic salt solution, the volume control of such red cells disintegrates. Water rushes into the interior of these cells. The red blood cells first swell and become round, and then burst. The cells release the cytoplasmic proteins, such as hemoglobin, and remain membrane particles or ghosts. These ghosts can reseal again.[2] Recently, Jennings et al.

have postulated that the ischemic cell death also represents such osmotic lysis of parenchymal cells, such as cardiac myocytes.[27,28]

When tightly controlled cations or anions rise in the intracellular space, this increase will initiate uncontrolled catalytic actions. The targets of such lysis are membranes, membrane-associated complexes, and cytosol. The outcomes of such lysis are altered membranes and organelle functions, and an altered cytoskeleton. Examples of such actions are the calcium-activated phospholipase, the calcium-related disassembly of actin and microtubles, and the phosphate-related dissociation of creatine kinase from mitochondria.

Depressed Synthesis

The synthetic functions of enzymes are reduced and depressed when cations that are essential for synthetic enzymes leave the cell space. Such cations are potassium and magnesium; their loss causes a lasting depression of ATP synthesis, protein synthesis, and cell maintenance.

Leakage of Large Molecules

With excessive disruption of membranes, large molecules finally can penetrate the cell space from the extracellular compartment or escape from cells into the interstitium, lymphatics, and bloodstream. Such events may indicate irreversible injury. They are used clinically to diagnose cell necrosis in myocardial infarct (MB–CPK) and experimentally to identify cell necrosis using lactate dehydrogenase. Testing for entry of labeled albumin into cells is another method of using the development of large membrane leaks as indicators of cell necrosis.

Intracellular Membranes

Loss of membrane-dependent synthetic activity is seen in membrane complexes that are driven by energized gradients. This effect is mainly noticed in intracellular membranes of organelles such as the endoplasmic reticulum and mitochondria, and affects protein synthesis and energy metabolism. Mitochondria loose the ability to form proton gradients at an ischemic time interval when ATP synthesis declines. It is our proposal that this indicates severe damage to the inner membrane of mitochondria and to the ATP synthesis function, although the details of this are still not clear.

Inactivation and dissociation of ribosomes from endoplasmic reticulum may be related to an alteration of the membrane-linked protein synthesis machinery. In normal cells, the binding of ribosomes to the endoplasmic reticulum is initiated by so-called signal peptides that direct ribosomes to the receptor proteins. These receptor proteins serve as transport proteins for newly generated polypeptide chains. The process is a dynamic one, with a steady state of interaction and a highly sensitive change of the electrolyte balance, particularly a change in sodium concentration in the cytosol.

Active Transport of Solute

One mechanism of glucose uptake occurs through the symport with sodium uptake. An alternative transmembrane transport system occurs through the carrier protein — mediated facilitated passive diffusion. There is no good evidence, however, that substrate entry into cells is interrupted during early and decisive phases of ischemic cell injury. The inhibition of the metabolic machinery inside of cells appears to overpower the potential for internal starvation.

Loss of Cell-to-Cell Connections

Specialized cell junctions play an important role in the formation and cohesion of tissues. Severe degrees of cell injury can disrupt cell-to-cell connection, which may play a role in the delayed recovery of organ functions. The following cell junctions are recognized:

1. *Occluding, or tight, junctions.* Tight junctions connect epithelial cells and separate two spaces from each other; for example, the intestinal lumen from the interstitium. Freeze fracture study has shown that encircling sealing strands are present in the apical portion of epithelial sheets. Separation of occluding junctions causes leakage between tissue spaces, such as endothelium-lined blood spaces and the interstitium, leading to leakage of serum proteins and basal edema.

2. *Anchoring junctions (actin or intermediate filament dependent).* Anchoring filaments connect cells to each other or connect cells to intercellular proteins. The structural transmembrane complex interacts with the cytoskeleton of adjacent cells. *Attachment proteins* connect the junctional complex to specific elements of the cytoskeleton (actin, intermediate filaments). *Transmembrane linker glycoproteins* bridge the membrane and connect several attachment proteins with the extracellular matrix or transmembrane linker proteins of other cells. In this way, anchoring junctions provide coherence and resistance to displacement in mechanically stressed tissues. The anchoring junctions fall into two groups: the adherence junctions and the desmosomes. The separation of anchoring junctions allows cells to be removed from the tissue organization.

3. *Communicating junctions, such as gap junctions, and chemical synapsi.* Experiments reported by Loewenstein, Furshpan and Potter, Gilula et al., and Hooper and Subak-Sharpe demonstrated that cell-to-cell communication through gap junctions allows even small fluorescent dye molecules to pass from cell to cell.[29-31] The size of the pores has been determined to be such that 1,500-dalton polypeptides can transfer from one cell to another.[32] Gap junctions are constructed from transmembrane proteins, the so-called connexons. Six of

these connexons align themselves to form a low-resistance pathway between cells for ions and small molecules. The permeability of gap junctions is regulated. Gap junctions can open and close similarly to gated ion channels. It has been demonstrated that sick or dying cells become disconnected at the gap junctions. The gap junctions become impermeable in response to injury, preventing the injury from spreading to other, noninjured cells by closure of the channel.[32]

The Role of Cell Surface Receptors

The receptor proteins are transmembrane proteins that, upon contact with a signaling molecule, transmit the message and activate cellular mechanisms. Three classes of cell surface receptors are known: channel linked, G-protein linked, and catalytic. Channel linked receptor proteins support neurotransmission as transmitter gated ion channels. Catalytic receptors are linked to enzymes, mainly tyrosine-specific protein kinase. The G-protein linked receptor allows the activation of two pathways following the binding of the signal molecule: the cAMP pathway and the Ca^{2+} pathway. It is thought that cell injury induces specific, mediator-related dysfunction.

Cytoskeletal Network and Plasma Membrane

Mammalian cell membranes are strongly supported by the cytoskeleton. For example, actin forms a dense network, called the *cell cortex*, just beneath the cell membrane surface. The cell cortex adds mechanical strength to the cell surface. Two types of protein attachments serve as links between cytoskeleton fibers and the plasma membrane: the cap ends of actin, which allow attachment of single actin fibers to the membrane; and spectrin, which forms at-

tachment sites within the plasma membrane. In cell injury, frequent calcium-induced alterations of the actin and microtubular network will also modify this mechanical support of cell membranes. The formation of large blebs on the cell surface is the end result of such membrane injury. The network of actin is associated with structural and catabolic proteins. Filamin serves to cross link actin filaments. Gelsolin-induced solution, limiting cross linking, is activated by calcium.

Other Forms of Membrane Injury

Radiation Injury

Cellular nucleic acids are the obvious, well-described targets of radiation injury, and cause long-term injuries and genetic defects. Recent studies of radiation injury under conditions of limited free radical detoxification have suggested that ionizing radiation can also contribute to acute membrane injury. Oxygen free radicals can be generated through radiolysis of water and can subsequently cause membrane alteration through lipid peroxidation.[33]

Poisons and Heavy Metals

Poisons that affect membrane proteins can inhibit selective membrane functions. Such injury has been described by Trump et al. in the proximal tubule of the kidney where, as a consequence of binding of membrane protein sulfhydryl groups, massive calcium influx occurred, along with subsequent massive loading of injured epithelial cell mitochondria by calcium.[22] Tetrodoxin and saxitoxin, among others, are poisons selective for the sodium channel of the plasma membrane,[34] which affects one of the most basic functions of membranes: the generation of bioelectricity.[35]

Ionophores

Ionophores are small hydrophobic molecules that dissolve in lipid bilayers, either as mobile carriers or as channel formers, and increase the permeability of lipid bilayers to certain preferred solutes. They discharge the transmembrane concentration gradients that inactivate membrane functions are dependent on. Examples of these ionophores are: valinomycin, a mobile carrier for K^+; A23187, another mobile carrier for calcium; and Gramicidin, a channel-forming ionophore for monovalent cations.

MEMBRANE TRANSPORT OF MACROMOLECULES: EXOCYTOSIS AND ENDOCYTOSIS

Most cells secrete and ingest macromolecules. These processes are called *exocytosis* and *endocytosis*, respectively. The process by which macromolecules are taken up or ejected by plasma membranes is different from that of the uptake and transport of small solutes. For the purpose of uptake, cells form vesicles and keep their content sequestered from the interior of the cells by the vesicle wall. These vesicles are moved inside the cell and fuse only with certain membrane structures, such as primary lysosomes.

Endocytosis

Most animal cells are continuously endocytosing segments of their plasma membrane. This requires that localized regions of the plasma membrane invaginate and form pinocytotic or phagocytotic vesicles that travel into the cell interior, where they fuse with primary lysosomes to form secondary lysosomes.[36] Depending on the size and diameter of the particles, either large phagocytic or small pinocytotic vacuoles form. The process of membrane fusion involves the joining of different bilayers between vesicles.[37,38] Some of the endocytotic vesicles form from smooth segments of the plasma membrane, but many of the vesicles develop from specialized regions of the plasma membrane, forming depressions that are coated on the cytoplasmic side. These pits form coated vesicles. The coating protein is clathrin, which, together with a small polypeptide, develops a basketlike mesh of hexagons and pentagons. The coated vesicles are thought to interact with the cytoskeleton. Coated pits and vesicles provide a specialized pathway for receptor-mediated endocytosis of specific macromolecules,[2,39-42] although some macromolecules can penetrate cell membranes directly.[43] Specialized phagocytic cells digest particles that bind to specialized receptors and to their surface.[2] There is a continuous shuttle of membrane-coated vesicles between cell compartments and organelles.

Exocytosis

Similarly, intracellular transfer of substances, between the endoplasmic reticulum and the Golgi apparatus occurs by the same orderly process. The subsequent secretion requires fusion of the vesicle with the plasma membrane. Some of these vesicles contain packaged, newly synthesized proteins and can serve as storage vesicles, holding their content and releasing it through the plasma membrane in a triggered, calcium-mediated response. Since some of the storage vesicles contain highly catalytic enzymes, it was once thought that membrane lysis of these vesicles under circumstances of ischemia or toxic injury would induce a highly destructive rampage by releasing these uncontrolled lytic enzymes. Detailed research by Hawkins and Trump, however, has shown that such rupture likely occurs only under special circumstances and is not a common pathway to cell death.[27] Membrane components are constantly recycled. After exocytic fusion of membranes, the rest of the surface membranes are removed from the plasma membrane.

MEMBRANE SYNTHESIS AND REPAIR

The endoplasmic reticulum is a membrane factory that synthesizes membrane proteins[2]; phospholipid and cholesterol are also synthesized on the membrane of the endoplasmic reticulum. The synthesis of one of the major membrane phospholipids, phosphatidyl choline, is conducted by enzymes that face the cytosolic side of the endoplasmic reticulum, where all substrates are present. It appears that all major phospholipids—phosphotidyl choline (PC), phosphotidyl ethanol annine (PE), phosphotidyl serine (PS), and phosphotidyl inositol (PI)—are formed in a similar fashion. During the synthesis process, some of the phospholipids are brought to the other site of the lipid bilayers. This flip-flop movement, which does not occur in other membranes, is thought to be the work of a *phospholipid translocation*. The phospholipid translocation is head group specific. The endoplasmic reticulum contains a translocation specific for phosphotidyl choline transporting, or flipping, PC to the luminal site of the endoplasmic reticulum. This translocation is responsible for the asymmetric distribution of phospholipids in membranes.[44,45] Phospholipid exchange proteins transport phospholipids from the endoplasmic reticulum to mitochondria and peroxisomes,[46,47] while the plasma membrane, the Golgi apparatus, and lysosomes communicate with the endoplasmic reticulum by precursors of transport vesicles. In endoplasmic reticulum–mitochondrial phospholipid transfers, water-soluble carrier proteins (phospholipid exchange proteins) can transfer phospholipid molecules between membranes. The glycolipid is inserted on the luminal side of the endoplasmic reticulum. The endoplasmic reticulum contains enzymes that are dissolved in the bilayer of the endoplasmic reticulum bilayer.

The lipid synthesis occurs as follows: From substrate dissolved in the cytosol, fatty acid CoA is linked with glycerol phosphate by an acyl transferase to form phosphatidic acid. Phosphatidic acid is hydrophobic and reamins in the membrane. In step two, a diglyceride is formed from the phosphatidic acid substrate, and in step three the phosphoend group is added. Alterations in the membrane biosynthesis are more likely in sublethal and chronic cell injury. However, since lipid metabolism is highly dynamic, defective maintenance may also be a factor in the ability of cells to recover from injury. More details of this thought are offered in the following chapters.

CONCLUSION

Cell membrane injury is a term that describes the alteration of complex biological structures: plasma membranes and intracellular membranes. With alterations of these membranes, cell maintenance, cell structure, and the numerous membrane-associated functions are affected. Most likely, there exists a rank order, or hierarchy, of distorted functions that allows a window of recovery. The most severe form of injury to plasma membranes may occur if the separating function between the exterior and interior is permanently (for the life of the cell) interrupted. Immune injury by complement or by natural killer cells seem to use such mechanisms to inflict certain death to target cells. Other forms of injury may affect viability through catabolic mechanisms that cause irreversible destruction of the cell membranes.

REFERENCES

1. Ostro MJ. *Liposomes.* New York: Marcell Dekker; 1983.
2. Alberts B, Bray D, Lewis J, Raff M, Roberts K, Watson JD. *The Molecular Biology of the Cell.* 1st ed. New York: Garland Publishing; 1983.
3. Bretscher MS, Raff MC. Mammalian plasma membranes. *Nature.* 1975;258:43–49.
4. Finnean JB, Coleman R, Michelle RH. *Membranes and Their Cellular Function* 2nd ed. London: Blackwell. 1978.

5. Harrison R, Lunt GG. *Biological Membranes.* Glasgow: Blackie, 1980.

6. Harold FM. *The Vital Force: A Study of Bioenergetics.* New York: WH Freeman Co; 1986.

7. Müller-Eberhard HJ. The membrane attack complex of complement, *Annu Rev Immunol.* 1986; 4:503–528.

8. Reid KBM, Porter RR. The proteolytic activation systems of complement. *Annu Rev Biochem.* 1988; 57:321–348.

9. Herberman RB, Reynolds CW, Ortaldo J. Mechanisms of cytotoxicity by natural killer (NK) cells, *Annu Rev Immunol.* 1986;4:651–680.

10. Möller G, ed. Molecular mechanisms of T-cell mediated lysis. *Immunol Rev.* 1988;103:1–211.

11. Young JD, Sachs DH. Multiple mechanisms of lymphocyte mediated killing. *Immunol Today.* 1988;9:140–145.

12. Widera G. Molecular biology of the histocompatibility complex. *Science.* 1986;233:437–443.

13. Bach FH, Sachs DH. Transplantation immunology. *N Engl J Med.* 1987;317:489–492.

14. Kornberg RD, McConnell HM. Lateral diffusion of phospholipids in vesicle membranes. *Proc Natl Acad Sci USA.* 1975;68;2564–2568.

15. Wagner R, Gabbert H, Hohn P. The mechanism of epithelial shedding after damage to the small intestine mucosa: a light and electron microscopic investigation. *Virchow Arch (Cell Pathol).* 1979; 30:25.

16. Fishman PH, Brady RO. Biosynthesis and function of gangliosides. *Science.* 1976;194:906–915.

17. Hakomori S. Glycosphingolipids in cellular interaction, differentiation, and oncogenesis. *Annu Rev Biochem.* 1981;50:733–764.

18. Singer SJ, Nicolson GL. The fluid mosaic model of the structure of cell membranes. *Science* 1972; 175:720–731.

19. Eisenberg D. Three-dimensional structure of membrane and surface proteins. *Annu Rev Biochem.* 1984;53:595–623.

20. Byers T, Branton D. Visualization of the protein associations in the erythrocyte membrane skeleton. *Proc Natl Acad Sci USA.* 1985;82:6153–6157.

21. Semenza G, Kinne R. Membrane transport driven by ion gradients, *Ann NY Acad Sci.* 1985;456:1.

22. Trump BF, Croker BPJ, Mergner WJ. The role of energy metabolism, ion and water shifts in the pathogenesis of ischemic cell injury. In: Richter GW, Scarpelli DG, eds. *Cell Membranes: Biological and Pathological Aspects.* Baltimore: Williams & Wilkins; 1971:84.

23. Jennings RB. Cell volume regulation. *Acta Med Scand.* 1976;587:83–93.

24. Stevens CF. The neuron. *Sci Am.* 1979;241:54.

25. Jaffe LF. Control of the development by steady ionic currents. *Fed. Proc.* 1981;40:125–127.

26. Sweadner KJ, Goldin SM. Active transport of sodium and potassium ion: mechanism, function and regulation. *N Engl J Med.* 1980;302;777–783.

27. Hawkins HK, Ericsson JLE, Biberfeld P, Trump BF. Lysosome and phagosome stability in lethal cell injury, *Am J Pathol.* 1972;68:253–288.

28. Hawkins HK, Lowe JE, Hill ML, Jennings RB. Loss of cell volume and electrolyte control in slices of ischemic myocardium. *Am J Pathol.* 1977; 86:44a.

29. Loewenstein WR. The cell-to-cell channel of gap junctions. *Cell.* 1987;48:725–726.

30. Furshpan EJ, Potter DD. Low-resistance junctions between cells in embryos and tissue culture. *Curr Top Dev Biol.* 1968;3:95–127.

31. Gilula NB, Reeves OR, Steinbach A. Metabolic coupling, ion coupling and cell contacts. *Nature.* 1972;235:262–265.

32. Rose B, Loewenstein WR. Permeability of cell junctions depends on local cytoplasmic calcium activity. *Nature.* 1972;254:250–252.

33. Rubin E, Farber JL. Environmental and nutritional pathology. In: Rubin E, Farber JL, eds. *Pathology.* Philadelphia: JB Lippincott; 1988.

34. Kao CY, Levinson SR. Tetrodotoxin, saxitoxin and the molecular biology of the sodium channel, *Ann NY Acad Sci.* 1986;479:1.

35. Fuhrman FA. Tetrodotoxin, tarichatoxin and chiriquitoxin: historical perspectives, in tetrodotoxin, saxitoxin and the molecular biology of the sodium channel. *Ann NY Acad Sci.* 1986;479:1.

36. Silverstein SC, Steinman RM, Cohn ZA. Endocytosis. *Annu Rev Biochem.* 1977;46:669–772.

37. Palade G. Intracellular aspects of the process of protein synthesis. *Science.* 1975;189:347–358.

38. Simionescu N, Simionescu M, Palade G. Permeability of muscle capillaries to small heme-peptides: evidence for the existence of patent transendothelial channels. *J Cell Biol.* 1975;64: 586–607.

39. Roth TF, Porter KR. Yolk protein uptake in oocyte of the mosquito, *Aedes aegypti. J Cell Biol.* 1964;20:313–332.

40. Goldstein JL, Anderson RGW, Brown MS. Coated pits, coated vesicles and receptor mediated endocytosis. *Nature.* 1979;279:679–685.

41. Pearse BMF, Bretscher MS. Membrane recycling by coated vesicles. *Annu Rev Biochem.* 1981; 50:85–101.

42. Ungewickell E, Branton D. Assembly units of clathrin coats. *Nature.* 1981;289:420–422.

43. Smith MW, Wehman N, Mergner WJ. Studies on representative isolation of mitochondria from ischemic myocardium. *Fed Proc.* 1981;40:750.

44. Rothman JE, Lenard J. Membrane asymmetry. *Science.* 1977;195:743–753.

45. Bishop WR, Bell RM. Assembly of phospholipids into cellular membranes: biosynthesis, transmembrane movements and intracellular translocation. *Annu Rev Cell Biol.* 1988;4:579–610.

46. Yaffe MP, Kennedy EP. Intracellular phospholipid movement and the role of phospholipid transfer proteins in animal cells. *Biochemistry.* 1983; 22:1497–1507.

47. Dawidowicz EA. Lipid exchange: transmembrane movement, spontaneous movement, and protein-mediated transfer of lipids and cholesterol. *Curr Topics Memb Transp.* 1987;29:175–202.

2

Cellular Ion Regulation

Mohammed M. Sayeed

INTRODUCTION

A Historical Perspective

The first indication that living cells regulate their internal ions came from the studies of Liebig in 1847.[1] Liebig observed that muscle tissue contains greater amounts of potassium and lesser amounts of sodium salts than does the blood from which the muscle was built. Liebig did not know of the existence of entities called *ions* in blood or muscle, nor did he know of any definitive physiological role of potassium or sodium salts. The importance of inorganic salts in the maintenance of appropriate osmotic pressures in tissues was recognized by physiologists in the 1850s through 1870s.[2-5] However, a specific physiological role of an inorganic salt was not shown until in the 1880s Ringer demonstrated that calcium salt is required to sustain the beating of an isolated frog heart.[6] The role of inorganic ions (rather than their salts) as agents essential for physiological activities became evident in 1887, when Arrhenius showed that strong electrolytes in solutions spontaneously dissociate into ions and that each ion has a characteristic property independent of its counter ion.[7] Thus, the contributions of ions as initiators or regulators of physiological activities became evident in the late nineteenth century.

That the ions themselves are "regulated" such that they are more concentrated in cells than in interstitial fluids or *vice versa* was evident by 1908. Potassium, magnesium, and phosphate were shown to be the major ions of the muscle cells; sodium, calcium, and chloride in the interstitial fluids.[8-10]

In the first three decades of the twentieth century, several investigators hypothesized and demonstrated a semipermeable membrane surrounding the cell that would restrict potassium and phosphate ions within the cell and sodium and chloride outside the cell.[11,12] Subsequent studies supported the role of the outer cell membrane (the plasma membrane) in the distribution of ions between the intra- and intercellular compartment.[13] Research in the 1940s and 1950s greatly enhanced our understanding of the regulation of Na^+ and K^+ concentrations within cells and of the control-elements within the plasma membrane that are responsible for the transmembrane distributions of Na^+ and K^+.[14-19] The ion control elements discovered were membrane "pumps" capable of supporting uphill ion transport and membrane ion channels that permit downhill flow of ions. Their specific purpose is to maintain electrical potentials and to modulate these potentials to produce regenerative, self-propagating action potentials.

Today the membrane ion pumps and channels are no longer just so-called functional entities within the membranes; they are real molecular entities that can be isolated from membranes so that their physical and chemical properties can be characterized. Although originally discovered in the excitable cells, where their main function is membrane potential modulation, the ionic channels are found in all cells and serve to mediate nonelectrical processes such as secretion and absorption of substances and cell volume regulation in the so called nonexcitable cells. The regulation of transmembrane distribution of calcium has been extensively investigated since 1960.[20-22] There is evidence now that the overall cellular regulation of calcium involves movements of this ion not only across the plasma membrane but also across organelle membranes (e.g., mitochondrial and endoplasmic/sarcoplasmic reticular membranes).

Ion Regulation and Cellular Function in the Resting versus Activated Cell

In their resting state, cells generally maintain a steady transmembrane distribution of ions. In cells activated by various external signals (e.g., electrical, mechanical, thermal, or chemical), there occur transient and/or sustained changes in intracellular and intraorganellular concentrations of ions. Physiologically, cells revert to their resting state after external signals are withdrawn. Thus, cells are endowed with the ability to regulate transmembrane distribution of ions not only in the resting state but also during their activation and recovery.

An important development in physiology in the last two decades has been the recognition of a coupling between cellular ion regulation and cellular functions such as contraction, secretion, cell-to-cell communication, growth, and intermediary metabolism.[21,22] This development has revealed the utility of the elaborate membrane mechanisms that exist in cells for the purpose of regulating various ions. During cell activation, the transmission of external signals to cellular responses often occurs as a result of a coupling between cellular ion regulation and cellular functions. In such instances, the intracellular ionic transients serve as messengers of external signals and interact with the response elements within the cells. The role of cellular ion regulation in cell function is exemplified by the involvement of cellular calcium regulation in the activation of Ca^{2+}-dependent cellular processes. The wealth of information available on the role of calcium in cellular functions and their regulation is not just due to an extensive research in this area but also due to an extensive involvement of calcium in physiological processes.

Figure 2-1 is a generalized diagram of the various regulatory elements that determine the set-points of the transmembrane ion concentration gradient or the intracellular concentration of the major ions (Na^+, K^+, H^+, Ca^{2+}, Cl^-). The membrane ion paths — namely, ion channels, pumps, and other transporters — serve as controllers (control elements) of the ion set-point. Dynamic fluctuations in one control element are compensated for by adjustments by other controller(s). Thus, activities of controllers are autoregulated within a physiological range. In the resting state of the cell, the ion set-point is important for the maintenance of basal cellular activities such as metabolism, protein synthesis, and cell volume regulation. If two ions share the same membrane path, a change in the intra- or extracellular concentration of one ion affects the set-point of the other ion. This would affect the basal cellular activities associated with the resting state. Such ion changes in the cellular or extracellular compartment serve as modifiers of the controllers. Figure 2-1 also shows the sequence of events in ion regulation that occur during cell activation. External signals, which activate cells, often function as modulators of the ion set-point, and modulated set-points elicit cellular responses such as excitation, contraction, secretion, and metabolism. Ion set-point modulations are mediated by intracellular messengers (second messengers), including AMP, GMP, and inositol triphosphate derived from membrane metabolism of phosphatidyl inositol. Also, influx of extracellular Ca^{2+} subsequent to cell activation by external messenger effects change in intracellular Ca^{2+} concentration, allowing Ca^{2+} to serve as a coupling factor between

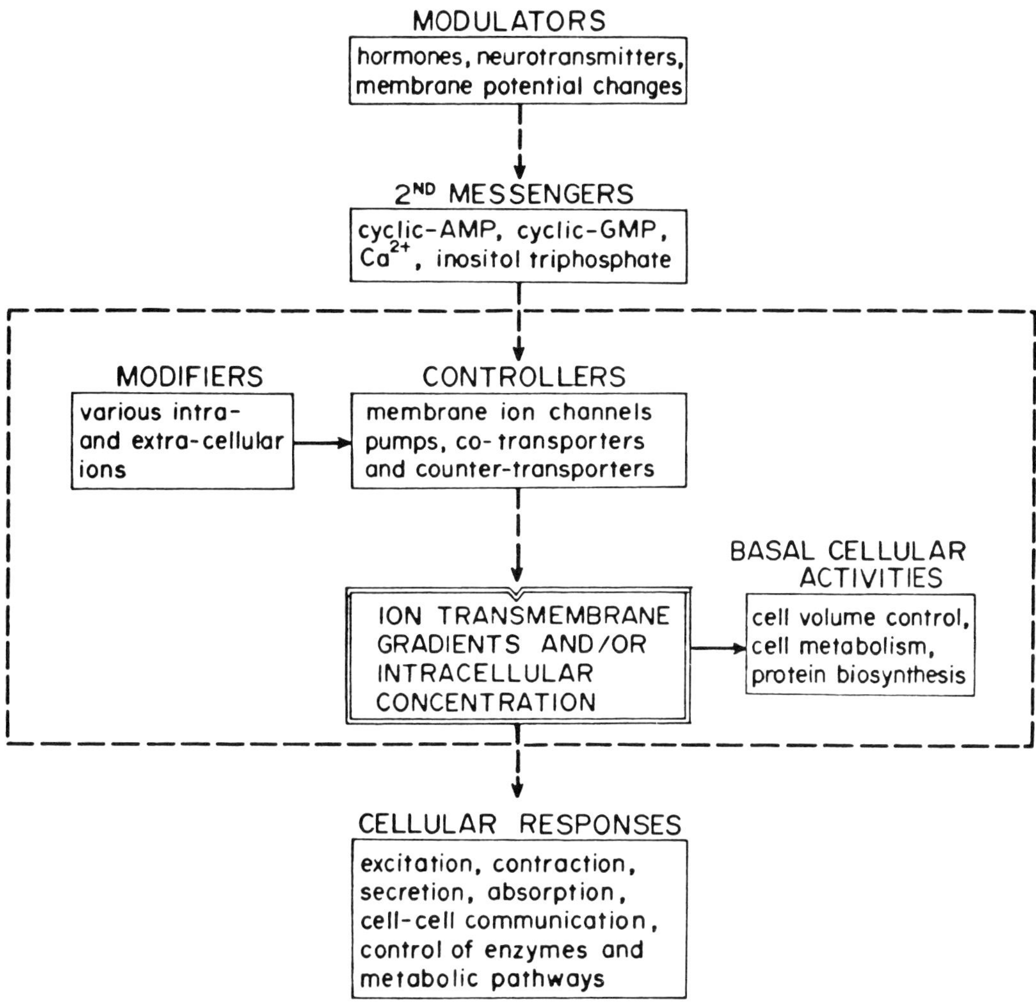

Figure 2 1. Flow diagram showing cellular and extracellular elements that provide for *control, modification,* and *modulation* of cellular ions and ion-dependent cellular functions. The broken line encloses the ion regulatory events and associated cellular activities in the resting state of cells.

an external messenger and a cellular response. These relationships between ion regulatory elements show that disturbances in ion controllers, such as might occur under pathophysiological conditions, can adversely affect both basal and external messenger–induced cellular functions.

Review Objectives

The foregoing survey of the development of the concept of cellular ion regulation and its role in the regulation of cellular activities makes it apparent that alterations in ion regulatory mechanisms under pathophysiological conditions would impair ion-dependent cellular functions and their modulation. Many investigators have indeed shown such alterations in a variety of pathophysiological states. These investigations also imply that cellular ionic distributions contribute to the cell injury in pathologic states. The purpose of this review is to describe and discuss these investigations. In addition, a discussion of potential consequences of changes in the various membrane control elements is also provided. The review is organized to present relevant details of the physiology of membrane regulation of major ionic systems (e.g., Na^+, K^+, and Ca^{2+}) prior to discussions of alterations in the ion regulatory processes. Ionic alterations in the liver and skeletal muscle are emphasized, as cellular ion pathophysiology in other organs is dealt with in other chapters of this book. The review is primarily concerned with ionic mechanisms in the pathophysiological state of hemorrhagic, endotoxic, and septic shock.

PHYSIOLOGY OF ION REGULATION

Ion Driving Forces

The fundamental driving forces for passive ion fluxes across membranes are transmembrane concentration and electrical potential differences. Figure 2-2 shows steady state intra- and extracellular concentrations of various ions (Na^+, K^+, H^+, and Ca^{2+}, and Cl^-), and the resting transmembrane potential difference in an idealized mammalian

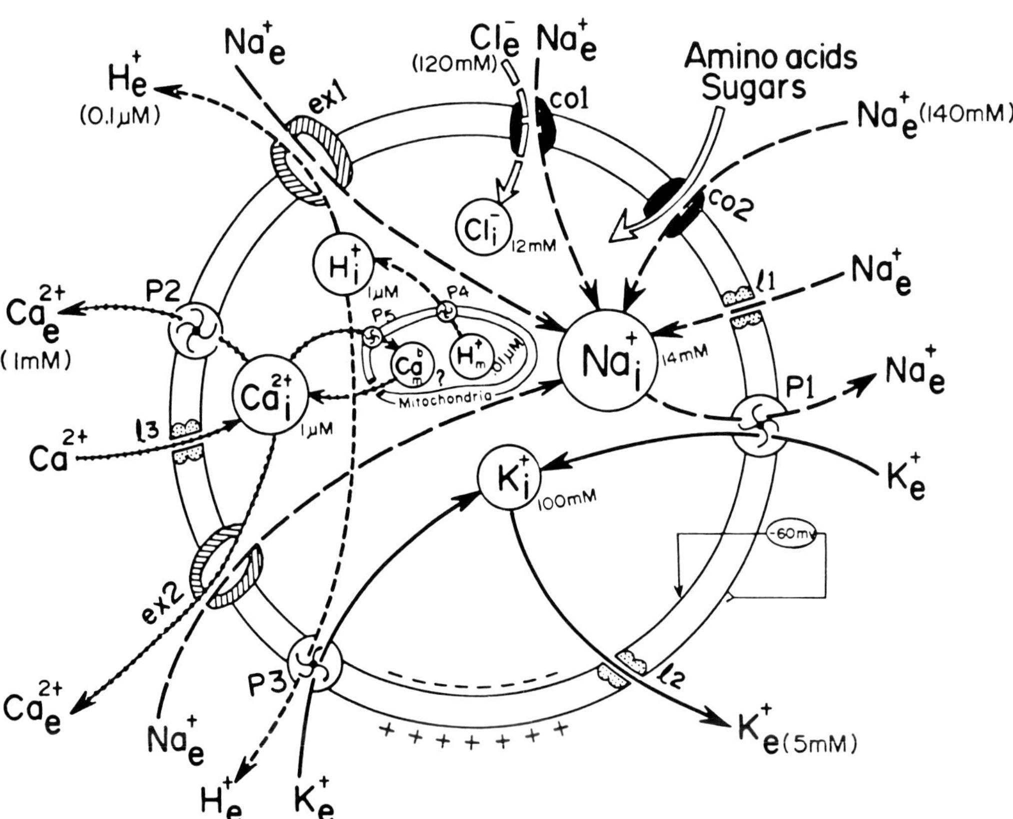

Figure 2-2. Schematic diagram of membrane ion paths in an idealized cell system. The diagram emphasizes the relationships between networks of ion paths and the interactions between ions as they traverse cellular membranes. "Spin-wheels" represent primary ion pumps; ionic paths lined by closed brackets are co-transporters; cross-hatched brackets represent ion counter-transport systems; dotted brackets depict leak or regulated channels for ions. The intraorganellar, cytosolic, and extracellular concentrations of ions and the transmembrane potential, though not typical for any one natural cell system, are intermediate between the excitable and the nonexcitable tissues.

cell system. The transmembrane potential value of -60 mV is intermediate between that found in nerve and muscle cells (-80 mV), and the epithelial and smooth muscle cells or the hepatocytes (-40 mV). In most mammalian cells, the resting membrane potential (E_m) is primarily a consequence of the tendency of certain ions to passively move across the membrane according to their electrochemical gradients. This tendency of ions is dependent both on the magnitude of ion electrochemical gradients and permeability of the membrane to these ions. The steady state maintenance of E_m requires that the net passive movement of ions that contribute to E_m be matched by active ion movements in the opposite direction.

Whether the membrane permeable ions, Na^+, K^+, Cl^- are distributed across the plasma membrane due to purely passive movements or due to a combination of active and passive movements can be ascertained from estimations of the free energy (ΔG) of ion electrochemical gradients. The free energy of ion electrochemical gradient, which indicates passive ion movements, is equal to the sum of ΔG due to the transmembrane concentration gradient (ΔG_c) plus that due to electrical potential gradient (ΔG_e) of a given ion. If $\Delta G_c + \Delta G_e$ is equal to zero or close to it, that ion is distributed passively across the membrane. According to the Nernst equation,

$$(1) \qquad E = \frac{1}{z} \times 60 \log \frac{[\text{ion}]_i}{[\text{ion}]_e}$$

where E = electrical potential (in millivolts) due to transmembrane distribution of an ion; for a cation, the value is to be multiplied by a minus sign

$[\text{ion}]$ = ion concentration on the inside (i) or outside (e) of the membrane

z = ion valence

a tenfold concentration difference of a monovalent ion across a membrane is associated with a 60-mV membrane potential difference. The 60-mV potential and tenfold concentration difference yield ΔG_e ($= .096 \times z \times E$) and ΔG_c ($= 5.7 \times \log \times [\text{ion}]_i/[\text{ion}]_e$), values that are both equal to 5.7 kJ/mol.

For Na^+ ion, the concentration and E_m values given in Figure 2-2 indicate that the concentration difference (10-fold) and E_m (-60 mV) would both passively drive the ion inward with the sum of ΔG_e plus ΔG_c: $(-5.7 \text{ kJ/mol}) + (-5.7 \text{ kJ/mol}) = -11.4$ kJ/mol. Thus, for steady state of Na^+, an active outward movement of ion would be required. For K^+, in which the 20-fold concentration difference would passively drive the ion outward with a decrease in free energy, $\Delta G_c = 2 \times (5.7) = 11.4$ kJ/mol; the E_m (-60 mV) would support an inward K^+ movement with a decrease in $\Delta G_e = -5.7$ kJ/mol. The net K^+ passive flux would be outward, with $\Delta G_c + \Delta G_e = (11.4) + (-5.7) = 5.7$ kJ/mol. An active inward movement of K^+ would be needed to balance the net passive efflux due to the resultant electrical and concentration difference. In the case of Cl^-, a passive influx would be predicted due to the tenfold inwardly directed concentration difference and an equivalent passive efflux due to E_m of -60 mV. Since $\Delta G_e + \Delta G_c = (-5.7) + (5.7) = 0$, Cl^- would be expected to distribute itself due to passive forces, without the need of active movements. Thus, in the case of the cell schematized in Figure 2-2, Na^+ and K^+, but not Cl^-, distribution require active ion transport. The equivalency of passive Cl^- influx and passive Cl^- efflux means an absence of a net passive Cl^- movement to contribute to the resting membrane potential, E_m. Apparently, net passive Na^+ and K^+ fluxes could contribute to E_m. The extent of contributions of Na^+ and K^+ to E_m depends on membrane permeability to these cations. If the E_m can be accounted for solely by the Na^+ and K^+ concentration gradients and membrane permeability to these ions, it is referred to as the Na^+ and K^+ diffusion potential (E_d) and can be calculated by the Goldman-Hodgkin-Katz (GHK) equation:

$$(2) \quad E_d = 60 \log \frac{[K^+]_e + (P_{Na^+}/P_{K^+})\,[Na^+]_e}{[K^+]_i + (P_{Na^+}/P_{K^+})\,[Na^+]_i}$$

where

P_{Na^+}/P_{K^+} = the ratio of permeability (sodium to potassium)

In some cell systems (skeletal muscle, epithelial, smooth muscle cells, and hepatocytes), E_m is more negative than the Na^+/K^+ diffusion potential (E_d). In such cases, an electrogenic ion pump may contribute to E_m. The Na^+/K^+ pump can contribute to negative potential if the stoichiometry of active Na^+/K^+ movement is 3 Na^+:2 K^+. Thus, the resting membrane potential can be due to both passive diffusions and active movements of ions.

Membrane Ion Paths

Figure 2-2 also depicts various known vectorial membrane ion paths, namely, channels, pumps, and co- and counter transporters. Not all of the membrane paths shown in Figure 2-2 are found in any one given cell type. For a given ion, the membrane ionic paths, together with the ion activities (concentrations) in the intra- and intercellular compartments, can be taken to represent the cellular network of that ion. The ion network concept prescribes that the cellular regulation of ion is a result of a network action rather than ionic movements through individual pathways along a single ion electrochemical gradient. As schematized in Figure 2-2, more than one ion sometimes share the same path. This indicates obligatory interactions between different ion networks. Thus, the overall regulation of a given ion depends not only on factors controlling the cellular network of that ion but also on interactions with network(s) of one or more other ions. The concept of an ion network is comparable to ionic relationships, and is emphasized by the term *ionic net*, which was introduced by Rasmussen.[23]

Leak Channels

Ionic movements across the membrane lipid bilayer dictate that the hydrophilic ions be kept from coming into contact with the hydrophobic environment in the interior of the membrane. In the case of active ion transport, energy-dependent conformation changes in an amphipathic intramembrane protein accomplish ion pumping. Passive ion movements must take place through membrane pores lined with amphipathic proteins such that the hydrophilic protein surface faces the lumen and the hydrophobic surface is apposed against the intramembrane lipids. The membrane pores together with the liner proteins are referred to as *ion channels* and constitute the physical basis of membrane ion permeability. In the resting state of cells, ions tend to move downhill through membrane channels and thus generate the membrane potential difference. Such channels are referred to as *leak channels*. In Figure 2-2, l_1, l_2, l_3 represent leak channels for Na^+, K^+, and Ca^{2+}, respectively. The ion specificity of such channels is not definitive. To be sure, they permit different ions to move at different velocities. The leak channels offer much greater resistance to Na^+ than to K^+, and therefore confer greater membrane permeability to K^+ than to Na^+. The higher permeability to K^+ could be due to the smaller size of hydrated K^+. The leak channels for all ions presumably remain open during the resting and activated states of cells. The characteristics of the leak channels can be determined by the isotopic flux (passive) studies or by studies of current–voltage relationships. In the latter approach, effects of ion-specific channels have to be eliminated to measure the behavior of leak channels.

Ion-specific Channels

Downhill movements of ion through ion-specific channels occur during activation of cells by external signals. The ion-specific channels are presumably closed during the resting state and are opened through an action with external signals. Almost all known ionic channels show the "gating" phenomenon.[24] *Gating* refers to transitions in the protein liner of channels to one or more open or closed conformations that are elicited by external signals. The external signals may be a change in voltage across the membrane or the binding of a ligand to the membrane. In short, unlike the leak channel, ion-specific channels are regulated. To date, Na^+-, K^+,- Cl^-, and Ca^{2+}-specific channels have been identified in excitable cells such as nerve and muscle, and in epithelial cell membranes.[25,26] The evidence for the regulated ion channels comes from recordings of ionic currents through such channels by means of so-called noise analysis or the patch clamp technique. Ion channels have also been identified via selective pharmacological triggering or blocking of the channel currents.[27] Also, channel proteins have been isolated from natural membranes and reincorporated into artificial membranes for the purpose of their characterization.[28] The ionic channels play a fundamental role in the generation of action potential in excitable cells. They also play a significant role in the control of secretion in endocrine and exocrine glands, and in the control of absorbtion of electrolytes in epithelia.[29-31]

In tight epithelia (e.g., colon, distal kidney tubles, and collecting ducts), the absorption of Na^+ seems to depend on a Na^+ channel in the apical membrane. The channel supports an energy-independent electrogenic uptake of Na^+ into the mucosal cells. This epithelial Na^+ channel is distinguishable from a Na^+ leak channel by its characteristic regulation via intra- or extracellular Na^+, aldosterone, and by its specific inhibition by the pyrazine diuretic amiloride.[32-34] Though dependent on both influx into mucosal cells through the channel and efflux from the cell via the serosal Na^+ pump, the overall transepithelial transport of Na^+ is determined primarily by the influx through the channel. Thus, the Na^+ channel is the rate-limiting step in the transepithelial Na^+ transport and in the fluid reabsorption associated with it.

A distinctive feature of the Na^+ channel is generation of a current to depolarize the apical membrane of the epithelial cell. Thus, the apical membrane exhibits a Na^+ diffusion potential rather than the Na^+ and K^+ diffusion potential exhibited across the serosal or basal epithelial membrane. On average, the apical membrane exhibits a potential that is 10–30 mV less negative (cell interior negative with respect to the lumen) than that across the basal membrane. As a result of the different potentials across the two membranes, there exists a potential difference of about 10–30 mV (lumen negative with respect to serosal) across the epithelial layer of cells. This potential, referred to as *transepithelial* or *transmural potential difference*, is maintained due to the Na^+ channel across the apical cell membrane and due to ion impermeability of tight junctions between adjacent epithelial cells. The latter qualify the epithelia as "tight" as opposed to "leaky." The transepithelial potential is an indicator of an optimal activity of the Na^+ channel and of a well-maintained transepithelial Na^+ and fluid reabsorption. There is an indication that, like aldosterone, ADH also promotes the Na^+ channel activity to support water reabsorption across the tight epithelia.[35] The transmural potential also plays a role in excretion of K^+ in the distal kidney tubules and colon.

Primary Active Transport Pumps

Membrane ion pumps serve to maintain intracellular ion homeostasis and to restore it after ionic disturbances due to cell activation or reversible cell injury. They support ion movement against ion electrochemical gradients and thus are essential in the maintenance of steady state ionic activities within the cells. The primary active transport pump is directly dependent on cellular metabolic energy, namely, high energy phosphates. Five such pumps are depicted as P_1-P_5 in Figure 2-2. P_1-P_5, respectively, represent the Na^+/K^+ pump, the Ca^{2+} pump, the H^+/K^+ pump, the mitochondrial proton pump, and the active Ca^{2+} uptake mechanism. Of these, the Na^+/K^+, Ca^{2+}, H^+/K^+, and proton pumps are transport ATPases. Each of these ATPases is an integral membrane protein spanning the entire thickness of the plasma membrane.

Na^+/K^+-ATPase. The Na^+/K^+-ATPase is ubiquitous in animal tissues and is the principal mechanism for maintaining high intracellular K^+ and low intracellular Na^+. The enzyme consists of a catalytic subunit (alpha -unit, M.W. 100,000 daltons) and a glycoprotein subunit (beta-unit, M.W. 50,000 daltons) of unknown function.[36] A proteolipid component (M.W. 12,000 daltons) is also present but is not essential for the transport activity. A diagramatic representation of the Na^+/K^+-ATPase is shown in Figure 2-3. The molecular mechanism and the control of the Na^+/K^+-ATPase has been investigated extensively over the last three decades.

The pumping of Na^+ out of cells and of K^+ into them is probably the result of conformational changes in the enzyme in response to binding and hydrolysis of ATP and to binding of cations and ligands. Na^+ binding to enzyme at its intracellular surface facilitates ATP binding to the enzyme and the generation of phosphorylated form of the enzyme intermediate (Figure 2-4, E~P-ADP). Thus, the transport enzyme is activated. In a subsequent step, the energy-rich phosphate (~P) bond between the enzyme and ATP is reduced to an energy-poor (-P) bond with a simultaneous release of ADP; this leads to another change in enzyme molecular configuration and to release of enzyme-bound Na^+ on the extracellular side of the membrane. The energy-poor enzyme-phosphate (Figure 2-4, E-P) conformation leads to binding of K^+ at an externally situated enzyme site. The K^+-bound enzyme-phosphate (K^+-E-P) is subsequently dephosphorylated, which causes release of enzyme-bound K^+ to the cell interior. Thus,

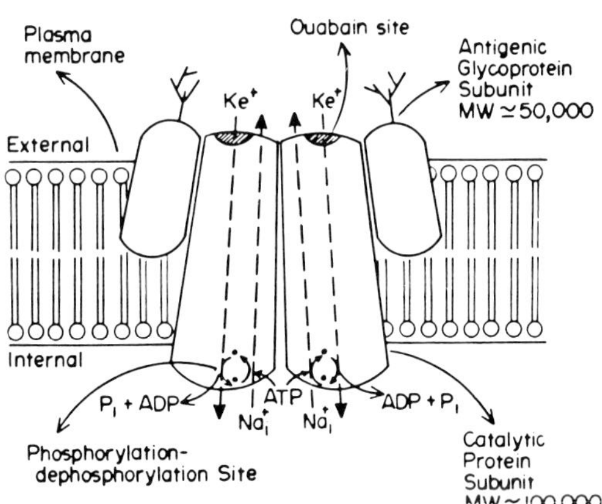

Figure 2-3. Schematic representation of structural and functional components of the cell membrane-bound transport enzyme $(Na^+ + K^+)$-adenosine triphosphatase. K^+_e, extracellular K^+; Na^+_i, intracellular Na^+; P_i, inorganic phosphate. ATP and ADP have their usual meaning. (From Sayeed, MM. Membrane Na^+-K^+ transport and ancillary phenomena in circulatory shock. In: Cowley RA, Trump BF, eds. *Pathophysiology of Shock, Anoxia, and Ischemia.* Baltimore: Williams & Wilkins; 1982.)

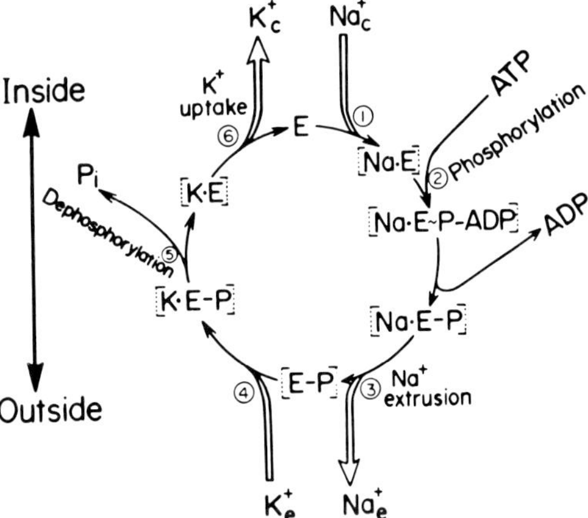

Figure 2-4. Proposed cycle of Na^+/K^+ reactions involving the hydrolysis of ATP and the translocation of NA^+ and K^+ across the plasma membrane. Reaction 1 = Na^+ binding to the enzyme (E) to form the complex, Na-E; Reaction 2 = phosphorylation of the Na-E complex followed by ATP hydrolysis; Reaction 3 = Na^+ extrusion; Reaction 4 and 5 = K^+-dependent dephosphorylation of enzyme; Reaction 6 = K^+ uptake into cells and release of free enzyme. The asymmetric nature of reactions occuring inside versus outside the plasma membrane is apparent.

transitions between various conformations of the enzyme control the ion pumping mechanism.[37] The stochiometry of the transport process is such that for every two moles of K^+ pumped in, three mole of Na^+ are extruded at the expense of one mole of ATP. The loss of an extra mole of Na^+ over K^+ accumulated hyperpolarizes the membrane by a potential difference of from -2 to -6 mV.[38] For this reason, the Na^+/K^+ pump is referred to as an *electrogenic process.*

The loss of tissue potassium and its buildup in plasma occurs after tissue ischemia.[39] This is generally attributed to a failure of the Na^+/K^+ pump subsequent to ATP depletion in the tissue. However, the loss of K^+ and the gain of Na^+ by tissue may also occur without a critical ATP depletion during cell injury due to a disturbance in the structural and/or functional integrity of the transport enzyme within the membrane.

The cardiac glycoside ouabain binds to the portion of the enzyme present on the external cell surface and thereby inhibits the Na^+/K^+ pump process.[40] Modulation of enzymes by intra- and extracellular ligands permits cells to control Na^+/K^+ transport by this enzyme on a short-term basis. Increases in intracellular Na^+ or extracellular K^+ increase the transport activity. On a long-term basis, triiodothyronine and aldosterone increase the enzyme activity in mammals.[41] Recent studies have shown that natriuretic factors may bind at the ouabain binding site and thereby inhibit the transport function.[42] A failure of the Na^+/K^+ pump would affect intracellular Na^+ and K^+ concentrations, and thus the cation-dependent function in all types of cells. Its failure in kidney tubule cells and intestinal mucosal cells leads to impaired reabsorption of Na^+ and water into the body.

Ca^{2+} – ATPase. There is a steep electrochemical gradient for the entry of calcium into the cell. The inward concentration gradient is 10,000-fold and should support calcium influx with a decrease in free energy of -22.8 kJ/mole. E_m would also drive the influx of Ca^{2+} with a $\Delta G_e = -11.4$ kJ/mol. The net effect would be Ca^{2+} influx with $\Delta G_c + \Delta G_e = (-22.8) - (-11.4) = -34.2$ kJ/mol. This driving force is far greater than that impinging on Na^+. However, the influx of Ca^{2+} is from 100 to 1,000 times smaller than the leak of Na^+ into cells. The low membrane permeability to Ca^{2+} is the cause of the low level of calcium influx into cells via calcium leak channels.[20] Consequently, a low level of efflux of calcium from cells via active transport is required for the steady state transmembrane Ca^{2+} concentration gradient.

A primary active calcium transport via Ca^{2+} – ATPase has been characterized in the plasma membrane in a variety of mammalian cells (Figure 2-2, P_2).[36] The low-level requirements of active calcium transport are correlated with a low density of Ca^{2+} – ATPase copies per unit surface area of membrane and by a low consumption of cellular metabolic energy by the calcium transport process. Presumably, less than 1% of the total cellular energy consumed is adequate for the active efflux of calcium from cells.[20,22]

Relatively little is known about the structure of the plasma membrane Ca^{2+} – ATPase. More is known about the structural and functional characteristics of a membrane-bound Ca^{2+} – ATPase in the sarcoplasmic reticulum of muscle cells. Sarcoplasmic reticulum Ca^{2+} – ATPase pumps Ca^{2+} from the cytosol into the sarcoplasmic reticulum lumen. Normally there is a tight coupling between the number of Ca^{2+} pumped into sarcoplasmic reticulum and ATP hydrolyzed (2 Ca^{2+}/ATP).[36]

Like the Na^+/K^+ – ATPase, sarcoplasmic reticulum Ca^{2+} – ATPase contains a 100,000-dalton catalytic subunit that has the phosphorylation site.[36] The Ca^{2+} – ATPase, however, lacks the 50,000-dalton glycoprotein subunit that is an integral part of the Na^+/K^+ – ATPase. A 12,000-dalton proteolipid component of Ca^{2+} – ATPase plays an ionophoric (ion-carrying) role. The proposed sequence of Ca^{2+}–ATPase reactions that occur during Ca^{2+} translocation are somewhat similar to those proposed for the Na^+/K^+–ATPase action. The proposed sequential reactions are: (1) binding of ATP and Ca^{2+}, and phosphorylation of the enzyme; (2) conformational change with the turning of Ca^{2+} binding site to the opposite site and release of Ca^{2+}; and (3) dephosphorylation of the phosphoenzyme.

Mitochondrial Active Ca^{2+} Transport. A mechanism similar to the uptake of Ca^{2+} by sarcoplasmic reticulum is also present in the mitochondrial inner membrane (Figure 2-2, P_4). Mitochondria actively take up Ca^{2+} with phosphate in a $1:1$ ratio.[43] The mitochondrial Ca^{2+} uptake is driven by substrate oxidation or ATP hydrolysis. During mitochondrial Ca^{2+} uptake *in vivo*, the energy of electron transport is shunted away from ATP synthesis and channeled into the uphill Ca^{2+} transport.

Mitochondrial ATPase. An enzyme similar to the Na^+/K^+– and Ca^{2+}–ATPase is located in the inner mitochondrial membrane (Figure 2-2, P_5). However, it normally works in a reverse direction. Instead of coupling ATP hydrolysis to ion pumping, the mitochondral ATPase couples an existing downhill flow of ions (H^+ influx into mitochondrial matrix) to ATP synthesis. This enzyme is also referred to as *ATP-synthase.*[44] The H^+ gradient across the inner mitochondrial membrane is initially set up by the electron transport system.

H^+/K^+–ATPase. This transport enzyme is localized in the gastric mucosal cell membranes of a tubulovesicular organ and is believed to be involved in the active transport of H^+ out of cells (Figure 2-2, P_3).[36] The H^+/K^+–ATPase transports H^+ out of, and K^+ into, cells in a $1:1$ ratio. Four H^+ are exchanged for four K^+ per ATP hydrolyzed. Thus, this ATPase operates as an electroneutral pump. The enzyme contains a 100,000-dalton subunit that is phosphorylated and dephosphorylated, presumably to accomplish the counter-transport of H^+ and K^+. So far this transport enzyme has not been demonstrated in any mammalian cell system other than the gastric mucosal cells.

Co-Transporters

The active transport of an ion unaccompanied by a counter ion or without an exchange with another ion of the same charge contributes to the membrane potential. The membrane potential and the concentration gradient of the ion, taken together, constitute the so-called electrochemical gradient of that ion. Since the electrochemical gradient stores energy derived from ATP, its dissipation (like ATP hydrolysis) can be coupled to other energy-requiring transport processes within the same membrane system. ATP and ion electrochemical gradients are similarly exploited by the cell.

The Na^+ concentration gradient and membrane potential, the two components of the Na^+ electrochemical gradient, are frequently the source of energy for the uphill transport of other ions and metabolites. Since ATP establishes the Na^+ electrochemical gradient to begin with, the Na^+–gradient linked uphill transport is secondarily dependent on ATP. Therefore, this uphill transport is sometimes referred to as *secondary active transport.* Besides the input of free energy, the transport process is dependent on a specific membrane protein that presumably binds to Na^+ and the other substrate, and translocates them in a manner somewhat similar to the ion-specific ATPases. The Na^+–gradient linked transporter proteins have been isolated and partially characterized.[45,46]

The Na^+-linked transport system falls into two categories. In one category, the downhill Na^+ movement supports an uphill movement of an anion or a neutral metabolite in the same direction as the Na^+ movement. This is the so-called mediated co-transport system. In the case of an accompanying anion, there is no change in membrane potential and the co-transport is thus electroneutral. The co-transport of a neutral substance along with Na^+ leads to a change in membrane potential and is therefore electrogenic. Two such co-transport systems are shown in Figure 2-2, label co1 and co2. Co1 represents a system responsible for a simultaneous inward movement of Na^+ and Cl^-. Such a co-transport system exists in the apical or luminal membrane of mucosal cells in the so-called leaky epithelia (e.g., small intestine, gall bladder, and proximal kidney tubules).[47,48] In the leaky epithelia, the NaCl co-transport is important in the transmural (transepithelial) absorption of isotonic NaCl solution. Once taken up into mucosal cells, Na^+ is pumped out of cells at the basal or serosal border via an ouabain-sensitive Na^+ pump. Cl^- may be transported out of mucosal cell via an electrophoretic transport into intercellular space. The accompanying Na^+ and Cl^- movements probably drive water movement from the intestinal or kidney tubular lumen to the serosal spaces.

In Figure 2-2, co2 represents the Na^+-linked co-transport of amino acids or hexoses. This transport system is found in skeletal muscle and in the apical membranes of small intestinal and proximal kidney tubule epithelial cells.[49-51] This co-transporter functions to move substrates into cells for cellular metabolic needs and to transport nutrients across the epithelial layer into circulation.

Ion Counter-Transporters

In the second category of Na^+-linked transport process, the free energy of the Na^+ gradient is replaced by the uphill transport of a cation in the opposite direction. Such an exchange of ion mediated by a membrane protein is referred to as *ion counter transport*. Na^+-linked uphill transports of H^+ and Ca^{2+} in the direction opposite to

that of Na^+ movement are labeled as exchange 1 (ex 1) and ex 2 in Figure 2-2. The Na^+-Ca^{2+} exchange plays an important role in maintaining intracellular $[Ca^{2+}]$ in cardiac nerve and smooth muscle cells.[52] The Na^+-Ca^{2+} exchange mechanism is probably not present in liver or adipose tissue.[53,54] Ca^{2+} efflux via Na^+-Ca^{2+} exchange decreases as extracellular $[Na^+]$ is decreased. The transporter exchanges one internal Ca^{2+} for two Na^+, in which case the ionic exchange is electroneutral. The stoichiometry of exchange depends on the intracellular Na^+. With increasing intracellular Na^+, the exchange is slowed down and there is an increase in both intracellular Na^+ and Ca^{2+}. At high intracellular Na^+ concentrations, there may be a reversal of exchange; that is, an exchange of internal Na^+ (an uphill process) for external Ca^{2+}. The reversal would be expected to lead to increased Ca^{2+} influx into cells. In cardiac cells, inhibition of the Na^+ pump via ouabain leads to decreased Ca^{2+} influx, both of which would raise intracellular Ca^{2+}, exerting a positive inotropic effect. The Na^+-Ca^{2+} exchange mechanism may be the target of the beta-adrenergic agonists in smooth muscle cells. The beta-adrenoreceptor stimulation via isoproterenol activates the Na^+/K^+ pump and enhances the Na^+ electrochemical gradient. The activation of this pump presumably promotes Ca^{2+} efflux and decreases intracellular Ca^{2+}, leading to an attenuation of the contractile force development in smooth muscle. Thus, the beta-adrenoreceptor relaxation of smooth muscle could be effected through the Na^+-Ca^{2+} exchange mechanism.[55]

The ex 2 system (Figure 2-2) serves to actively extrude H^+ out of cells. Ordinarily $[H^+]$ is less than 10-fold higher inside the cell than outside. The efflux of H^+ due to concentration gradient would be less than influx due to a membrane potential of -60 mV, which is a driving force equal to a 10-fold concentration gradient. This would cause a net passive influx of H^+. The Na^+-

H^+ exchange mechanism provides for extrusion of H^+ to maintain intracellular pH. Such an exchange mechanism is known to be operative in skeletal muscle.[56,57] An inhibition of the Na^+ pump and subsequent loss of the Na^+ electrochemical gradient can cause increases in intracellular $[H^+]$—a nonmetabolic intracellular acidosis that can deleterously affect many a cellular process. On the other hand, if a primary alteration in the membrane is an alteration in membrane permeability to Na^+ relative to K^+ (i.e., an increase in P_{Na+}/P_{K+}) leading to membrane depolarization (rather than membrane depolarization due to Na^+ pump failure), then the enhancement in H^+ efflux would increase influx of Na^+ via Na^+–H^+ exchange and lead to an increase in intracellular Na^+.

Ion Networks: Intra- and Inter-Network Relationships

Having discussed the ion driving forces and the various intramembrane ion control elements, it is instructive to reconsider individual ion networks and the relationships between them from the standpoint of their regulation and functions.

Na^+ (its intracellular concentration and transmembrane concentration gradient) is controlled by the flow of the ion through various membrane paths. Na^+ enters cells through leak channels, Na^+-specific channels, and co-transporters and/or counter-transporters. In all cases, the influx of ion is driven by its electrochemical gradient. These various Na^+ influx paths serve different physiological roles. Na^+ influx through the leak channel(s) leads to Na^+ uptake and related osmotic fluid flux into cells, affecting cell volume. The leak channel conductance to Na^+ contributes to the development of the resting membrane potential. Influx through the ion-specific channel, which has the highest conductance to Na^+ of any influx path, generates current and depolarizes the membrane. Of utmost importance in the

generation of action potential in excitable cells, Na^+ channels also serve a role in Na^+ transport and fluid absorption in nonexcitable epithelial cells. Downhill Na^+ flux during co-transport or counter-transport contribute to intracellular homeostasis of other ions, such as Ca^{2+} and H^+. In spite of the multiple paths for influx of Na^+, its efflux is universally mediated by the ouabain-sensitive Na^+/K^+–ATPase. Perhaps an ouabain-insensitive ethacrynic acid–sensitive NaCl pump also helps extrude Na^+ out of cells in the epithelia.[58] Nevertheless, Na^+/K^+–ATPase is the major mechanism to counter influxes through various paths and plays a crucial role in controlling all of the various Na^+ influxes and the various influx path–associated functions. Transcellular Na^+ and fluid uptake, cell volume control, and maintenance of membrane potential and ionic gradients are all ultimately dependent on the operation of the Na^+ pump. Whereas an alteration in one of the Na^+ influx channels affects a specific cellular function, altered Na^+ pumps would affect cellular functions associated with all influx paths.

One unique metabolic contribution of the Na^+/K^+–ATPase is its utilization of ATP and generation of ADP, which is the most important stimulator of oxidative phosphorylation.[59] Other functional roles of Na^+/K^+ pumps are due to obligatory uphill transport of K^+ and maintenance of high intracellular $[K^+]$. The maintenance of both low intracellular Na^+ and high K^+ are presumably required for optimal enzyme activities. High intracellular $[K^+]$ is required for gluconeogenic, protein biosynthetic, and oxidative phosphorylative reactions.[60–62] Increased intracellular Na^+ could adversely affect oxidative phosphorylation.[62]

Ca^{2+} enters the resting cells mainly through a leak channel. In cells activated by membrane voltage changes (e.g., nerve and muscle cells), calcium enters through voltage-activated Ca^{2+} channels.[38] The actual amount of calcium entering the cell via ion-specific channels is minute compared with total intracellular calcium. Also, calcium enters cells via receptor-activated Ca^{2+} channels. Such receptor-activated channels are presumably found in smooth, skeletal, and cardiac muscle, pancreatic exocrine and endocrine cells, and possibly in liver and kidney cells.[21] The evidence for the presence of receptor-activated or voltage-activated calcium channels in different cell systems is obtained by selectively "blocking" the Ca^{2+} channels with agents such as verapamil, nifedipine, or diltiazem and/or "opening" channels by specific ligands such as catecholamines, vasopressin, or acetylcholine. The restoration of intracellular calcium after influx due to leak or Ca^{2+}-specific channels is apparently under the control of plasma membrane as well as organelle uphill Ca^{2+} transport systems. Thus, the overall regulation of intracelluar $[Ca^{2+}]$ and the Ca^{2+}-dependent cellular functions (contraction, secretion, cell metabolism) are dependent not only on uphill efflux paths but also on controlled influx paths. Moreover, as emphasized earlier, the Na^+ gradient plays an important role in the maintenance of intracellular $[Ca^{2+}]$.

A link between ion networks is also indicated in the Ca^{2+} regulation of a K^+ channel. K^+ enters cells through the Na^+/K^+ pumps and leaves via leak- and K^+-specific channels. One such K^+ channel is the so-called Ca^{2+}-activated K^+ channel.[25] In addition to being activated by a voltage change in excitable cells, the channel is sensitive to influx of Ca^{2+} in a variety of other cell systems.[63,64] Whereas Ca^{2+}-activated K^+ channels regulate excitability in the excitable cells, it plays a significant role in the control of secretion by endocrine and exocrine glands, and of mitogenesis in lymphocytes.[30,65] It is reasonable to speculate that the functional role of K^+ channels in nonexcitable cells could be to hyperpolarize the membrane and thereby influence the electrochemical gradients of other electrolytes and, therefore, the glandular secretions of fluids containing those electrolytes.

Regulation of Intracellular Ca²⁺: General Considerations

Among the various cellular ion systems, the Ca^{2+} system seems to occupy a unique position. This uniqucncss is due both to an elaborate cellular mechanism for its own regulation and to a multiplicity of cellular functions that are regulated by it. Whereas a long-term regulation of intracellular Ca^{2+} is determined by plasma membrane influx and efflux paths, acute cellular regulation is dependent on influx and efflux paths in both the plasma membrane and in organelle membranes such as mitochondrial inner membrane and sarcoplasmic or endoplasmic reticular membranes. The role of intracellular calcium chelator calmodulin, albeit minor in the overall acute regulation of intracellular Ca^{2+}, is important in the regulation of cellular functions. The external signal–mediated release of Ca^{2+} from sarcoplasmic or endoplasmic reticulum to give rise to a transient increase in intracellular $[Ca^{2+}]$ serves as an intracellular trigger for cellular responses such as contraction in muscle cells, and secretion in endocrine and exocrine cells.[21] The external signal that mediates Ca^{2+} release from sarcoplasmic reticulum in cardiac muscle is the slow inward calcium current via the Ca^{2+} channel.[38] In hepatocytes, alpha-adrenoreceptor agonists (catecholamines, phenylephrine), vasopressin, and angiotension II cause the release of Ca^{2+} from the endoplasmic reticulum to effect an activation of metabolic pathways such as glycogenolysis and gluconeogenesis.[66-68] The role of mitochondrial inner membrane in the uptake and release of Ca^{2+} seems to be crucial in the regulation of the intracellular Ca^{2+} level both in resting and in signal-activated states of cells.[69,70] Thus, mitochondria mainly guard against an inordinate rise in cytosolic Ca^{2+}, protecting cells against the imminent toxic effects of high cytosolic Ca^{2+}. Increases in cytosolic Ca^{2+} concentration beyond the physiological ranges in the resting or activated state of cells presumably activate lysosomal phospholipases and proteases, and thus a cellular autolytic process.[71,72]

PATHOPHYSIOLOGY OF ION REGULATION

Cellular Ion Dysregulation and Its Etiologies in Cell Injury of Sepsis and Shock: General Considerations

In view of the ion dependency of the various basal and signal-induced cellular functions, it is reasonable to hypothesize that a disturbance in cellular function in cell injury involves ion dysregulation. Ion regulatory elements probably continue to operate in tissues in an adaptive manner in the early stages following a pathogenic insult to tissues. The adaptive ion regulation may maintain cellular function at a near-normal level and prevent the progress of the pathogenic effects to cell injury. Alternatively, the pathogenic process may exacerbate adaptive ion changes and lead to cellular dysfunction and injury. In some cases, cell injury itself may be reversible and may be followed by a recovery of cellular function. Under these circumstances, cellular ion regulation should also recover. Although factors other than cellular ion regulation may contribute to cellular dysfunction and injury, a gross disruption of membrane ion control could presumably play a primary role in the irreversibility of the injury. Thus, a pattern of both an adaptive change and a gross impairment of ion control mechanisms may emerge in the course of development of cell injury.

The changes in ion regulation in cell injury could be due to a variety of etiologies. To date, there are studies to indicate that an injury to cell membranes and their functions could be due to: (1) anoxia or acidosis in tissues; (2) a chemical, thermal, or radiation trauma to tissue; (3) an infective invasion of the host organism by bacteria or viruses; or (4) immunologic reactions in the host. Although derangements in the cardiovascular system of mammals are generally the causes of tissue hypoperfusion and anoxia in vital organs, tissue hypoperfusion also follows many of the inflammatory and immunologic signals (e.g., eicosanoids, free radicals, lysosomal enzymes, and activated complement components) that are encountered in bacterial infections and immunologic reactions. The immediate causes of membrane transport dysfunction with either a primary cardiovascular or immunologic/infective injury can be postulated to be (1) anoxic depletion of high-energy phosphates; (2) inappropriate release of neuroendocrine ligands and their actions at the membrane level; (3) bacterial products (e.g., endotoxin and exotoxins); (4) membranolytic inflammatory mediators (e.g., eicosanoids, free radicals, lysosomal enzymes, and activated complement components).

In humans, the sepsis and the ensuing state of shock occur more often with gram-negative bacteremias than with other infective conditions. For this reason, there has been an intense investigation of the effects of gram-negative sepsis and the circulatory shock produced by the gram-negative bacterial product, endotoxin.[73] The pathophysiology of the cardiovascular collapse that is seen in the late stages of human sepsis seems to be simulated in animal models of hemorrhagic/hypovolemic shock.[74] Thus, experimental models of gram-negative endotoxic and hypovolemic shock simulate various facets of human sepsis and shock. The cell membrane transport changes in endotoxic and hypovolemic shock may well represent critical alterations that contribute to irreversible cell injury in vital organs during septic shock. The investigation of the altered cellular ion regulation may elucidate those steps in the pathogenesis of cell injury in shock that, if corrected, may convert the irreversible injury into a reversible phenomenon. Selective pharmacological interventions to block membrane Ca^{2+} channels have indeed been indicated as therapy against ischemic cell injury in the myocardium, which was previously considered an irreversible process.[75]

Na$^+$/K$^+$ Transport in Shock and Shocklike States

Effect of Shocklike States

Studies by Fuhrman and Crimson in the 1950s measured the Na$^+$ and K$^+$ content of muscles that were subjected to ischemia by means of a tourniquet.[76] After release of the tourniquet, there was an increase in Na$^+$ and a decrease in K$^+$ in the muscles. Fuhrman raised the possibility that failure of the metabolically dependent Na$^+$/K$^+$ pump explained Na$^+$ and K$^+$ changes in the injured muscles.[77] This appears to be the first implication of active Na$^+$/K$^+$ transport failure in muscle due to an ischemic injury. Later on, other investigators showed the effect of anoxia and cold on skeletal muscle electrolyte distribution.[78,79] They also found an increase in Na$^+$ and a decrease in K$^+$. Similar changes in Na$^+$ and K$^+$ were also noted in the liver after anoxia.[80] Petterson, et al. quantitated Na$^+$ and K$^+$, and the Na$^+$/K$^+$–ATPase activity in dog kidneys subjected to ischemia.[81] In these experiments, there was evidence of both active Na$^+$ and K$^+$ transport and the activity of the transport enzyme, Na$^+$/K$^+$–ATPase. These studies demonstrated that the decrease in activity of the ATPase was due to the loss of the ouabain-sensitive enzyme activity.

Studies by Trump et al. examined the effects of metabolic inhibition, increased permeability to Na$^+$, and anoxia on Na$^+$ transport and volume regulation in epithelial cells (tight epithelia of amphibian urinary bladder and flounder kidney tubules).[82–84] In these cells, mucosal Na$^+$ influx occurs via the energy-independent Na$^+$-specific channel, and serosal efflux is due to the energy-dependent Na$^+$/K$^+$–ATPase. In flounder kidney, the epithelial cell flooding with Na$^+$ and cell swelling occured more rapidly with protein inhibitors such as PCMBS (p-chloromercurobenzosulfonate) than with depressed ATP levels, due to metabolic inhibition via potassium cyanide. Since PCMBS acts on membrane peripheral proteins, it presumably augmented the Na$^+$ influx by increasing muscle permeability to Na$^+$ such that the influx exceeded the serosal efflux by Na$^+$/K$^+$–ATPase. Therefore, this study suggests that flooding of cells with Na$^+$ and cell swelling can occur in epithelial cells due to altered membrane permeability without alteration in cellular ATP and serosal Na$^+$ pump activity. In the amphibian urinary bladder, Trump and co-workers found decreased transepithelial Na$^+$ flux with anoxia. Presumably, anoxia had not significantly affected the serosal Na$^+$ pump, as ATP content was only slightly altered at the time of decreased transepithelial Na$^+$ flux. The investigators hypothesized that anoxia decreased transepithelial flux by increasing mucosal membrane resistance to Na$^+$ influx and that decreased mucosal influx afforded protection against cell swelling. However, the finding of increased membrane resistance was based on transepithelial short circuit current measurement rather than mucosal Na$^+$ influx; the anoxic injury could plausibly injure basilar uphill Na$^+$ efflux and cause the decrease in short circuit current with normal or increased mucosal Na$^+$ influx. In any event, these experiments are possibly the only attempts to elucidate membrane ion channel modifications that might occur with cell injury.

Effect of Hemorrhagic Shock

Studies in the Skeletal Muscle. Haljamae showed that ionic alterations were present in skeletal muscle and tissue fluid of dogs subjected to hemorrhagic shock (45 mm Hg hypotension for 135 minutes).[85] He employed an isolated muscle fiber preparation for the determination of *in situ* changes in cellular Na$^+$ and K$^+$, and for the measurement of K$^+$ transport.[86,87] Changes in interstitial fluid electrolytes during shock were measured in interstital fluid samples obtained by means of a micropipet. There was a 26% decrease in muscle K^+. Muscle interstitial fluid K^+ increased from 4.5 mmol to 10 mmol after shock, and muscle cell Na^+ increased while tissue fluid Na^+ decreased. *In vitro* accumulation of K^+ by skeletal muscle fiber was considered active transport. However, since in skeletal muscle influx of K^+ due to passive forces may be nearly equal to its passive efflux, the K^+ uptake may not represent K^+ movement against an electrochemical gradient.

Campion et al. measured resting membrane potential in the skeletal muscles of animals subjected to hemorrhagic shock.[88] The membrane potential was found to decrease from -90 mV in controls to about -60 mV after hemorrhage in rats and dogs. The investigators assumed a passive transmembrane distribution of Cl^- and, using the measured value of membrane potential, calculated the intracellular concentration of Cl^-. The calculated intracellular Cl^- was higher after shock than in controls. This led the investigators to conclude that Cl^- entered skeletal muscle cells along with Na^+ during shock. Whether or not active Na^+ pumping was affected could not be ascertained as the observed membrane potential change could be due to changes in membrane ion permeability alone. Trunkey et al. measured membrane potential in the skeletal muscle of the monkey during hemorrhagic shock and the effect of fluid resuscitation of animals for several hours to several days after shock.[88] The transmembrane potential decreased from about -90 mV prior to shock to -60 mV after 3 hours of shock. The intracellular Cl^- and the muscle extracellular space were calculated by use of the Nernst equation. Intracellular Na^+ and K^+ were estimated from measured muscle Na^+ and K^+ contents, and calculated extracellular space value. These estimations showed an increase in intracellular Na^+ and a decrease in K^+. Trunkey et al. suggested that the change in intracellular Na^+ and K^+ was either due to an altered Na^+/K^+ pump or to membrane ion permeability in shock.[89]

Trump et al. investigated Na^+, Cl^-, and K^+ distribution in liver and skeletal muscle by means of x-ray microanalysis.[90] In freeze-dried thin sections of liver and muscle from rats subjected to a shock protocol of acute blood loss, the quantities of cellular Na^+ and Cl^- were found to be higher, and K^+ lower, than in controls. These changes were found as early as 15 minutes after the acute loss of blood. These findings indicated an early loss of Na^+/K^+ transport in muscle and liver after hemorrhage.

Studies in the Liver. Sayeed and co-workers studied the effects of hemorrhagic shock in the liver by evaluating (1) intracellular $[Na^+]$, $[K^+]$, and $[Cl^-]$[91]; (2) net active and passive movements of Na^+ and K^+ *in vitro*,[92]; (3) unidirectional radioisotopic Na^+ fluxes, *in vitro*[93]; (4) hepatic resting membrane potential *in vivo*[91]; and (5) Na^+/K^+-ATPase activity.[94] The hemorrhagic shock model consisted of bleeding rats to a mean arterial pressure of 40 mm Hg. The animals were referred to as *early shock animals* when a decrease in blood pressure to 40 mm Hg was first observed. *Intermediate shock* was designated by blood pressure maintained at 40 mm Hg by a gradual reinfusion of from 25% to 30% of shed blood, which required approximately 1 hour. *Late shock* referred to the stage of shock when from 60% to 70% of the shed blood was returned, to maintain pressure at 40 mm Hg; this usually took approximately 2 hours from the initiation of bleeding. Animals in early or intermediate shock were killed before or after rapid infusion of the shed blood (or the remaining shed blood) plus from 3 ml to 4 ml of Ringer's lactate solution.

Intracellular ion concentrations were analyzed in livers excised immediately after the animals were killed. The hepatic extracellular space was measured by means of insulin. The intracellular ion concentration was the difference between the total tissue and the extracellular concentration of the ion. Changes in hepatic intracellular ion concentrations in shocked animals are shown in Figure 2-5. No difference was found in

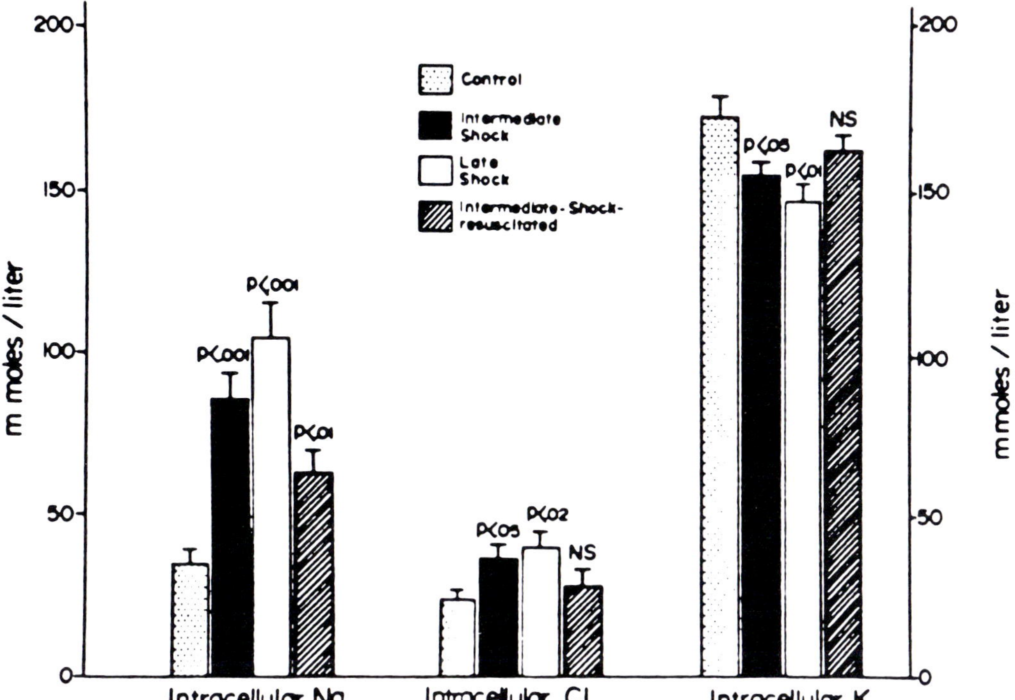

Figure 2-5. Hepatic intracellular ion concentrations in control and hemorrhagic shock animals. (From Sayeed et al., 1981.)[91]

plasma Na^+ or Cl^- concentrations of control and shocked animals; plasma $[K^+]$ was higher in the late shock animals than in control and intermediate shock groups. The hepatic intracellular $[Na^+]$ of all shocked animals was higher than that of controls. The accompanying decrease in intracellular K^+ and the increase in intracellular Cl^- could be due to a disturbance in coupled Na^+/K^+ movement plus an impairment of active movement of NaCl. Russo et al. demonstrated that the outward transport Na^+ is due both to an ouabain-sensitive coupled Na^+/K^+ movement and to an ouabain-insensitive active co-transport of Na^+ and Cl^-.[95] Resuscitation of rats after intermediate shock resulted in a complete recovery of cell $[K^+]$ and $[Cl^-]$, but not $[Na^+]$. This suggests a recovery of Na^+-K^+ transport but not the NaCl extrusion.

The net transmembrane movements of Na^+ and K^+ were assessed *in vitro* during incubations of thin liver slices (from 0.3 mm to 0.5 mm) in oxygenated Krebs-Ringer media. Net passive movements were measured by incubating liver slices either at 0.5°C or 37°C. Figure 2-6 shows the effect of late shock on net active movements of Na^+ and K^+. In both control and shocked rat liver slices, Na^+ increased and K^+ decreased with chilling. Upon rewarming, there was a reversal of changes in Na^+ and K^+ in the controls. Na^+ continued to increase, and K^+ was unaltered in the shocked rats. Figures 2-7 and 2-8 show net active Na^+ and K^+ movements in livers of early and intermediate shock animals. The effect of treatment or resuscitation of the animals is also seen in Figures 2-7 and 2-8. Although Na^+/K^+ active transport was inhibited in early shock, it could be restored to control level with resuscitation. The recovery with resuscitation was limited in the intermediate shock animals. Therefore, it appears that the

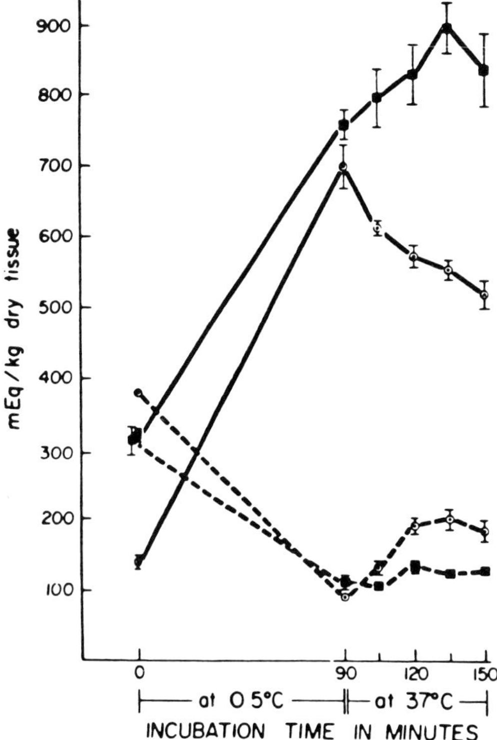

Figure 2-6. Na(———) and K (— — —) content of rat liver slices from control ( ) and shocked ( ) animals at end of chilling at 0.5°C and during rewarming at 37°C. Cation contents given for time 0 were determined in freshly prepared slices. Means ± SE of n samples (two samples per animal) are shown. Control, $n = 6$; shock, $n = 10$. SEM values less than 5 are not shown. (From Sayeed et al. 1973.)[92]

Na^+/K^+ transport impairment becomes irreversible at the intermediate shock state.

The unidirectional Na^+ efflux was measured in liver slices that were initially incubated in a Krebs-Ringer medium containing a trace quantity of ^{24}Na. Radioisotopic labeled slices were subsequently washed in media containing nonradioactive Krebs-Ringer solutions. The washout of ^{24}Na into the medium measured the efflux activity. Figures 2-9 and 2-10 show ^{24}Na efflux in liver slices of control rats and of rats in intermediate shock. The initial rapid loss of ^{24}Na represented a loss of isotope from the extracellular space. This was followed by a slower loss of ^{24}Na, representing efflux from the intracellular space. The rate of efflux, expressed in terms of log cpm/0.5 hour in liver slices from shocked rats, was about one-half of that found in controls. Whereas ouabain inhibited the cellular efflux of ^{24}Na to about 50% in controls, no significant inhibition of ^{24}Na efflux by ouabain in liver slices from rats in intermediate shock was shown. These data show an inhibition of ouabain-sensitive Na^+/K^+ pump in shock.

Another aspect of altered Na^+ transport in livers of shocked rats was discovered during unidirectional flux measurements. We found that whereas ouabain-sensitive Na^+ transport was not affected in extracellular $[Na^+]$ in liver slices of control animals, shock liver slices exhibited a dependency of Na^+ efflux on extracellular $[Na^+]$. This suggested an emergence of a Na^+-for-Na^+ ex-

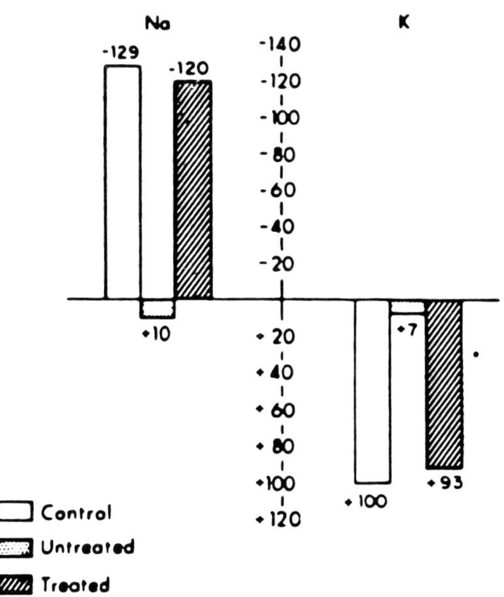

Figure 2-7. The transport capability of liver slices from control, untreated, and treated animals in early shock is expressed as the mEq kg dry wt of Na^+ extruded (−) and K^+ reincorporated (+) after incubation at 37°C for 60 min preceded by chilling to 0.5°C for 90 min. In untreated shock, Na was +10 and K was only +7, indicating lack of active transport. (From Baue AE, Wurth MA, Chaudry IH, and Sayeed MM., "Impairement of cell membrane transport during shock and after treatment." Impairment of cell membrane transport during shock and after treatment. *Ann Surg.* 1973;178:412.)

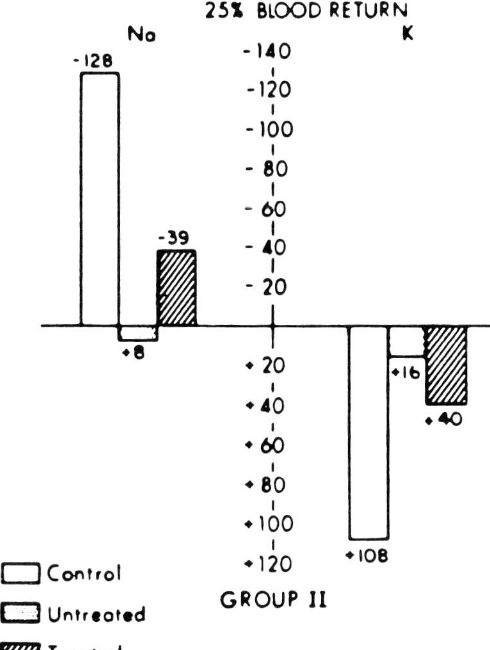

Figure 2-8. Plot of net ion (Na$^+$ and K$^+$) change in liver slices after rewarming (60 minutes at 37°C) of previously chilled (90 minutes at 0.5°C) slices. Net extrusion of Na (minus values) and net uptake of K (plus values) were seen in control and in treated intermediate hemorrhagic shock rats. Small uptake of Na$^+$ and of K$^+$ occurred in untreated intermediate hemorrhagic shock rats. All values have units of mmol/kg dry tissue weight. (From Baue AE et al. *Ann Surg.* 1973;178:412.)

change process in the shocked animal livers. The shift of the Na$^+$ pump from the Na$^+$/K$^+$ exchange to Na$^+$/Na$^+$ exchange mode might be indicative of a possible loss of the electrogenicity of the Na$^+$ pump.

The hepatic resting membrane potentials (E_m) were measured in the livers of control and shocked rats by means of intracellular glass microelectrodes. Table 2-1 shows the values of hepatic E_m in control and shocked animals. In the livers of intermediate shock animals, there was a depolarization of the membrane by approximately 10 mV. A further depolarization was noted in late shock. Rats, after resuscitation from intermediate shock, showed a partial recovery of E_m. The change in E_m with shock indicates an *in vivo* alteration in membrane Na$^+$/K$^+$ transport

or altered membrane permeability to Na$^+$ and K$^+$. Both the intracellular and the extracellular (i.e., plasma) concentrations of Na$^+$, K$^+$, and Cl$^-$ and the membrane potentials were measured *in situ* in the livers of various control and shock animals.

With an electroneutral coupling (1 : 1) of Na$^+$ to K$^+$ transport, the hepatic membrane potential would be determined by Na$^+$ and K$^+$ concentration gradients and membrane permeabilities to Na$^+$ and K$^+$. If the Na$^+$/K$^+$ pump (whether electroneutral or electrogenic) stops, then the ion gradients dissipate, which is followed by a change in the resting membrane potential; that is, depolarization. In this case, there may not be any change in

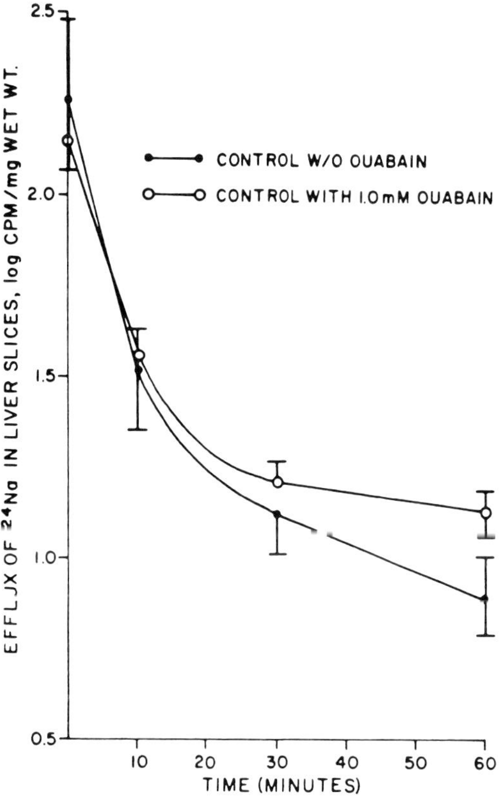

Figure 2-9. The effect of ouabain on Na$^+$ efflux from liver slices of control rats. Efflux was measured during washout of slices previously loaded with ^{24}Na. The initial 10-minute washout represents release of ^{24}Na from the liver slice extracellular compartment. (From Sayeed MM. In: Crowley RA, Trump BF, eds. *Pathophysiology of Shock, Anoxia, and Ischemia.* Baltimore: Williams & Wilkins; 1982.)

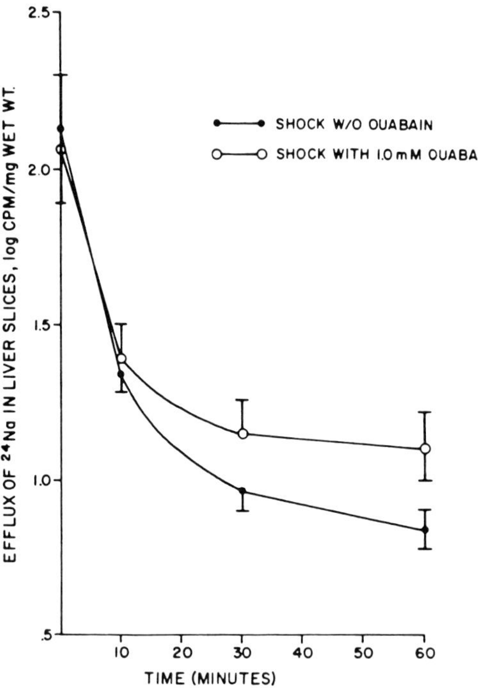

Figure 2-10. The effect of ouabain (10^{-3} on Na$^+$ efflux from liver slices of hemorrhagic shock rats. Efflux was measured during washout of slices previously loaded with ^{24}Na. The initial 10-minute washout represents release of ^{24}Na from the liver slice extracellular compartment. (From Sayeed MM. Membrane Na$^+$ − K$^+$ transport and ancillary phenomena in circulatory shock. In: Crowley RA, Trump BF, eds. *Pathophysiology of Shock, Anoxia, and Ischemia*. Baltimore: Williams & Wilkins; 1982.)

permeability of the membrane to Na$^+$ and K$^+$. On the other hand, if the membrane permeability changes first, the membrane potential would be affected without any change in the Na$^+$/K$^+$ pump. Thus, a finding of altered resting membrane potential alone does not indicate whether the change was due to the pump or the permeabilities. A direct analysis of both the pump and the membrane permeability characteristics are fundamental to assessment of the integrity of the membrane in its handling of Na$^+$ and K$^+$ transport.

In the case of an electrogenic Na$^+$/K$^+$ pump, passive diffusional forces and the Na$^+$ pump contribute to the resting membrane potential. Mullins and Noda have developed an equation that relates the resting membrane potential to diffusional forces as well as the contribution of electrogenic Na$^+$ extrusion (unaccompanied by K$^+$) to the potential.[96] This equation is similar to the GHK equation (Equation 2) except that it takes into account the ratio of coupling between Na$^+$ and K$^+$ transport. The Mullins and Noda (MN) equation is:

$$(3)\ E_m = 60 \log \frac{\gamma\, [\text{K}^+]_e + (P_{\text{Na}^+}/P_{\text{K}^+})\, [\text{Na}^+]_e}{\gamma\, [\text{K}^+]_i + (P_{\text{Na}^+}/P_{\text{K}^+})\, [\text{Na}^+]_i}$$

TABLE 2-1

Transmembrane Potential (E_m) and Cl$^-$ Equilibrium Potential (E_{cl}) in the Rat Liver[a]

	E_m (mV)	E_{Cl^-} (mV)
Control	−40.4 ± 0.4 (90) $N = 21$	−39.5 ± 2.1 $N = 8$
Intermediate shock	−30.9 ± 2.0 (36)* $N = 8$	−32.1 ± 1.2 $N = 6$
Late shock	−18.8 ± 0.4 (64)* $N = 11$	−26.6 ± 1.5** $N = 7$
Intermediate shock — resuscitated	−36.3 ± 1.0 (38)* $N = 8$	−33.6 ± 1.8 $N = 8$

*($p < 0.05$), control versus shock.

**($p < 0.05$), from E_m in late shock.

[a]All values are means ± SE; N = number of animals; values in parentheses are number of cell penetrations; four to five penetrations/animal. Means and SE were calculated with values from all penetrations.

From Sayeed et al., 1981.[91]

where

$$\gamma = (\text{active Na}^+ \text{ efflux})/(\text{active K}^+ \text{ influx})$$

Both the GHK and MN equations were used to calculate P_{Na^+}/P_{K^+}. In control livers, the coupling ratio of Na^+/K^+ transport was $3:2$; therefore, P_{Na^+}/P_{K^+} was assessed by using the MN equation (Equation 3). However, the P_{Na^+}/P_{K^+} in livers of intermediate shock animals before and after resuscitation could be calculated only by using the GHK equation (Equation 2) because the coupling ratio was $1:1$ in livers of these animals. The calculated value of P_{Na^+}/P_{K^+} in control and intermediate shocked animals was 0.3. This indicated that gross changes in cell permeability to Na^+ relative to K^+ did not occur, at least not in an intermediate stage of shock.

These data support the concept that hemorrhagic shock primarily alters an electrogenic component of the Na^+/K^+ pump that precedes membrane depolarization and shifts in intracellular ion concentrations. However, the extent of depolarization caused only by cessation of the electrogenic pump cannot be determined. A failure of an electrogenic Na^+ pumping in the liver would lead to more rapid depolarization of the membrane potential and ion concentration gradients, and of the cellular processes dependent on these gradients, than that caused by a derangement of an electroneutral Na^+ pump. The *in vivo* failure of the hepatic Na^+/K^+ pump in shock is supported by *in vitro* measurements of active Na^+ and K^+ movements, and by unidirectional ouabain-sensitive Na^+ effluxes in liver slices from animals in shock.

An altered pumping of Na^+ and K^+ would be associated with an alteration in the activity of Na^+/K^+–ATPase. In the studies of Sayeed et al., Na^+/K^+–ATPase activity increased in the liver in shock when Na^+/K^+ transport was shown to be decreased (Figure 2-11).[92-94] It is plausible that the Na^+-for-Na^+ exchange process, which was found in the shock livers but not in control livers,

may be related to the increased NA^+/K^+–ATPase. Another possible explanation for the increased Na^+/K^+–ATPase in the face of decreased Na^+ pumping would be an "uncoupling" of the ATP-hydrolyzing activity of the enzyme from the actual translocation on ions. This could be due to an increase in the Na^+-sensitive phosphorylation of the Na^+/K^+–ATPase, at the expense of ATP, on the internal surface of the plasma membrane and without the subsequent dephosphorylation of the enzyme or the uphill translocations of Na^+ and K^+.

Effect of Endotoxin and Septic Shock

In experimental studies, both endotoxin and the parent live bacteria have been widely used (via intravenous or intraperitoneal injections to animals) to simulate the state of sepsis and the ensuing septic shock in human patients. Endotoxin is the high molecular weight lipopolysaccharide–protein complex found in the cell wall of gram-negative bacteria of genuses such as *Escherichia* and *Salmonella*. As indicated earlier, endotoxin produces hypovolemia and hypoperfusion in vital organs, as is found in hemorrhagic shock. However, the circulatory disturbance in endotoxic shock is preceded by a hematologic and metabolic dysfunction, as well as a neuroendocrine and inflammatory response that is typical of human sepsis and septic shock.[73]

Studies in the Skeletal Muscle and Cultured Cells. Flear et al. carried out some of the earlier studies of the endotoxin-mediated effects on skeletal muscle Na^+/K^+ distribution and transport.[97,98] Serum obtained from animals injected with gram-negative bacterial endotoxin was added to skeletal muscle of control animals. In these experiments, Flear and co-workers found an increase in Na^+ permeability in skeletal muscles exposed to shock animal plasma. The serum of septic animals may contain a wide variety of substances released during the endotoxin-mediated inflammatory and

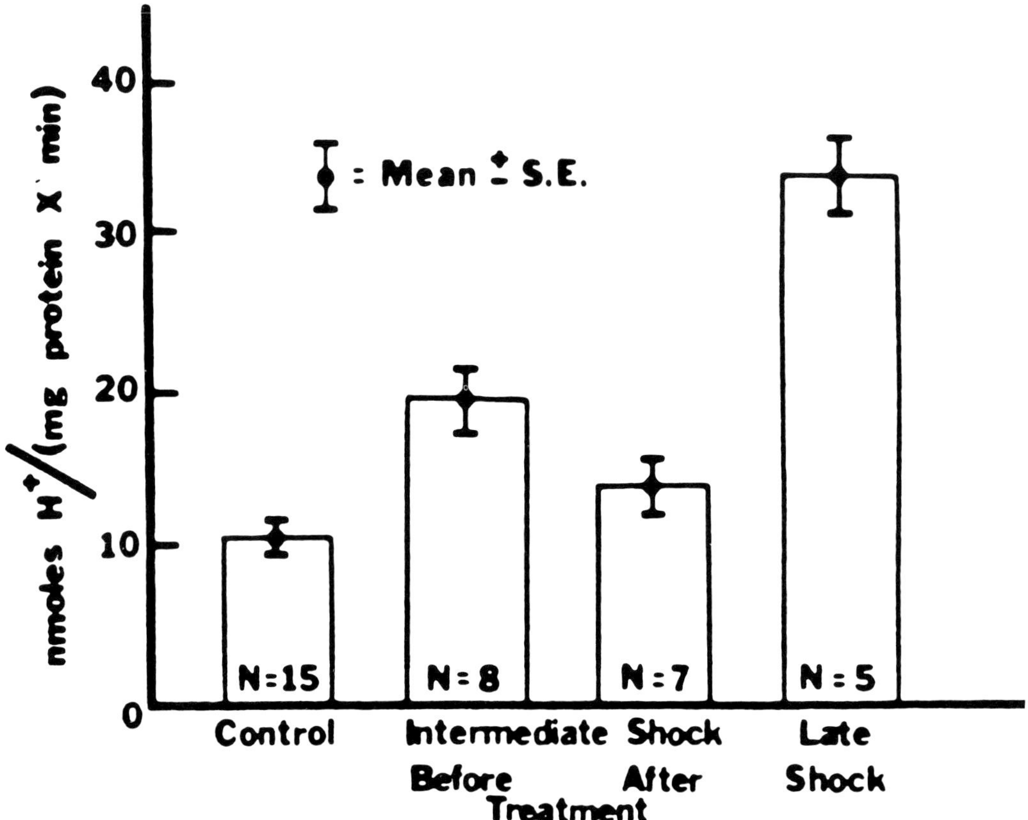

Figure 2-11. Na^+/K^+-ATPase activity in livers of rats in intermediate and late shock. N, number of animals. The values in intermediate shock and late shock were significantly higher ($p < 0.001$) than control values. The activity after treatment in intermediate shock was significantly lower ($p < 0.05$) than the activity before treatment. (From Sayeed et al., 1974.)[94]

neuroendocrine response. While these studies do not indicate which serum substance or the type of serum substance that affects membrane ion permeability, they suggest that membrane dysfunction may be effected as a result of factors other than anoxia and tissue energy depletion.[97-98]

Silver has reported direct effects of gram-negative bacterial endotoxin on a variety of cells in culture (horse lung fibroblasts, mouse brain neuroblasts, and neuroblastoma cells). These cells generally leaked K^+ when exposed to endotoxin.[99] The leakage of the K^+ could be due to: (1) a cessation of active Na^+/K^+ transport subsequent to depletion of cellular energy, (2) a derangement of membrane-bound Na^+/K^+ ATP without

the critical decrease in cellular energy, or (3) a gross increase in membrane ion permeability. Derangements in membrane Na^+/K^+ transport and/or selective membrane ion permeability are also indicated by measurements of skeletal muscle resting membrane potentials. Gibson et al. reported skeletal muscle resting potential measurements in dogs injected with endotoxin.[100] They found a decrease in potential from a control value of approximately −88 mV to −55 mV during endotoxic shock. Trunkey et al. measured skeletal muscle resting membrane potential in baboons injected with live *Escherichia coli*.[101] These investigations found a decrease in potential from −91 mV in controls to −67 mV in bacteremic baboons. The decrease in potential was correlated with an increase in intracellular Na^+ and Cl^-, and a decrease in K^+. The intracellular electrolytes were calculated from direct measurement of potential differences and blood electrolytes. These calculations are based on the assumptions of passive Cl^- distribution in skeletal muscle. The assessments of resting membrane potentials and estimates of transmembrane Na^+ and K^+ distributions support alterations in Na^+/K^+ transport mechanism but do not indicate whether the Na^+ pump or the membrane Na^+ or K^+ leaks are altered. Illner and Shires also reported a depression of skeletal muscle resting membrane potential by between 10 mV and 15 mV, from a control value of about -90 mV, in rabbits injected with *Escherichia coli*.[102] In a majority of bacteremic rabbits, membrane potential depression occured prior to any significant level of hypotension. Furthermore, high energy phosphates were maintained at a normal level in the skeletal muscle, even in the terminal stage of septic shock. These studies indicate that membrane potential alterations that are related to membrane Na^+/K^+ transport occur prior to a cellular energy depletion and a state of low blood flow. Thus, membrane potential shift may be a cause rather than an effect of tissue injury, paralleling low flow and metabolic energy depletion

Studies in the Liver. Sayeed has studied both the active and passive hepatic Na^+ and K^+ movements in endotoxic shock.[103] Rats were given intravenous injections of *Salmonella enteritidis* endotoxin, which produced shock symptoms in the animals 5 hours later. The active net *Na^+ and K^+* movements were measured in liver slices that were initially chilled (to 0.5 °C) to suppress active ion movement and then rewarmed (to 37 °C) to quantitate the active net extrusion of Na^+ and reaccumulation of K^+. This technique is the same as employed in the study of the effect of hemorrhagic shock on hepatic Na^+/K^+ transport.[92] Active Na^+ ex-

trusion and K^+ reaccumulation by liver slices were inhibited in endotoxin shock. Thus, the failure of Na^+/K^+ active movement in endotoxin is very similar to that found in hemorrhagic shock.

Passive Na^+ and K^+ movements were assessed by measuring changes in tissue Na^+ and K^+ content in liver slices during incubation in oxygenated KRB medium at 0.5 °C. The increase in liver slice Na^+ with time results from a flux of Na^+ from a large size Na^+ reservoir (incubation medium) into an approximately 200-fold smaller cellular reservoir of Na^+. The initial flux would therefore be independent of medium Na^+ concentration and be calculable from the linear portion of the tissue Na^+ uptake curve. The initial net uptake represents a primarily passive unidirectional Na^+ influx. The increase in liver slice Na^+ content (in units of mmol/kg dry weight) appeared linear during from 5 to 25 minutes of incubation (Figure 2-12). The slopes of linear re-

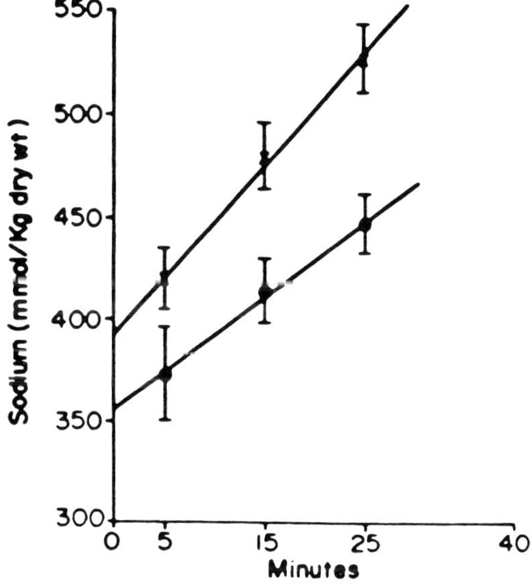

Figure 2-12. Liver slice Na^+ at varying times of incubation. Closed circles are values (mean ± SE) from liver slices of control rats ($n = 13$), and multiplication signs are values of slices from endotoxic rats ($n = 9$). Equations of regression lines are $y = 3.7x + 355$ for controls and $y = 5.4x + 392$ for the endotoxic group. Regression coefficient (r) was 0.99 and was significant at $p < .01$ for both groups. (From Sayeed, 1984.)[103]

gression lines drawn through Na$^+$ contents at 5-, 15-, and 25-minute incubations were the rate of Na$^+$ entry into cells. The rate of Na$^+$ entry was corrected to per unit cell surface area to approximate the passive influx of Na$^+$ (J_i^{Na}) in units of moles $\times$ cm^{-2} $\times$ sec^{-1}. To estimate the cell surface area, we assumed liver cells to be spherical, as has been done by other investigators.[104] From the regression lines shown in Figure 2-12, Na$^+$ uptake rate was calculated to be 3.8 (mmol $\times$ kg^{-1} dry weight $\times$ minute^{-1}) in controls and 5.3 in the shock group. The observed linearity of the rate of Na$^+$ uptake during the from 5- to 25-minutes of incubation of liver slices of control or shock group indicated a constant J_i^{Na} in either group for the entire incubation period. The estimated mean, ($\pm$SE) J_i^{Na}, was 5.4 $\pm$ .12 (pmol $\times$ second^{-1} $\times$ cm^{-2}) in controls and 7.9 $\pm$ .38 in the shock group. The sodium flux in the shock group was significantly higher ($P <$ .05) than controls.

The decrease in tissue K$^+$ content at 0.5°C estimated the passive efflux of K$^+$ out of liver cells. Since K$^+$ was released from a small size cellular compartment into a larger medium compartment, intracellular K$^+$ concentration is expected to change substantially and to effect an exponential decline in K$^+$ release. To determine the exponential release of K$^+$, a base 10 logarithm of K$^+$ content was plotted against incubation time. An initial monoexponential K$^+$ release was evident from 5 to 25 minutes (Figure 2-13). Rate constants (λ) were calculated from the initial monoexponential K$^+$ release curves. K$^+$ effluxes ($J_e^{K^+}$) (in units of mol $\times$ cm^{-2} $\times$ sec^{-1}) at 5, 15, and 25 minutes of incubation were approximated using the following equation:

$$(4) \quad J_e^{K^+} = \lambda \times [K^+] \times \frac{1}{S.A.}$$

where

λ = rate constant of K$^+$ release, minute^{-1}

[K$^+$] = intracellular K$^+$ content, mmol $\times$ kg^{-1} dry weight

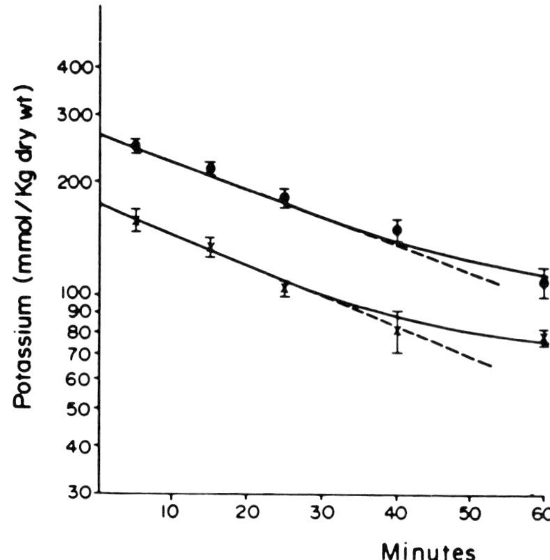

Figure 2-13. Semilogarithmic plots of liver slices K$^+$ versus time of incubation at 0.5°C. Values are means $\pm$ SE. Closed circles are values from control rats ($n = 13$), and multiplication signs are endotoxic group values ($n = 9$). (From Sayeed, 1984.)[103]

S.A. = surface area of liver cells, cm^{-2} $\times$ kg^{-1} dry weight

Rate constants, computed from linear portion of the semilogarithmic plots (Figure 2-13), were nearly the same for the control and the shock livers (.016/minute in control versus .018/minute in shock).

Changes in tissue chloride content were measured, along with the extracellular (insulin) space in slices. From these and the tissue K$^+$ content measurements, intracellular contents (mmol/kg dry weight) and intracellular concentrations (mmol/L) of K$^+$ and chloride in liver slices at from 5 to 25 minutes at 0.5°C were estimated (Table 2-2). The changes in intracellular K$^+$ content in liver slices and the estimated rate constants of K$^+$ release from liver slices permitted an approximation of $J_e^{K^+}$ at 5, 15, and 25 minutes during 0.5°C incubation slices. The average $J_e^{K^+}$, in units of pmol $\times$ second^{-1} $\times$ cm^{-2}, were as shown in Table 2-3.

TABLE 2-2
Intracellular Electrolytes in Rat Liver Slices at 0.5°C[a]

	n	Minutes at 0.5°C	Content (mmol/kg dry weight)		Concentration (mmol/L)	
			K⁺	CL⁻	K⁺	CL⁻
Control	8	5	242 ± 5	101 ± 7	116 ± 3	48 ± 4
		15	200 ± 4	135 ± 7	88 ± 5	59 ± 5
		25	177 ± 5	170 ± 6	75 ± 4	76 ± 5
Endotoxic shock	9	5	154 ± 7*	149 ± 6*	75 ± 6*	73 ± 4*
		15	128 ± 3*	186 ± 8*	55 ± 7**	80 ± 4**
		25	96 ± 6*	211 ± 9**	40 ± 3*	86 ± 6***

*$p < 0.001$

**$p < 0.01$

***$p < 0.005$ (Not significantly different.)

[a]Values are means $\pm$ SE; n = number of rats. Control and endotoxic shock rats as described in Table 2-1. From Sayeed, 1984.[103]

There appeared to be a decrease in $J_e^{K^+}$ with the progress of cold (0.5°C) incubation in both the control and shock groups. Furthermore, $J_e^{K^+}$ in the shock group was lower than in controls at all three incubation times.

To assess whether or not the altered passive ion fluxes were due to altered membrane ion conductance, permeabilities were evaluated to Na⁺ (P_{Na^+}) and K⁺ (P_{K^+}). After assuming independent migrations of cations through the membrane, the permeability coefficient for Na⁺ and K⁺ was calculated by means of the Goldman-Hodgkin-Katz flux equation (GHKf) as follows[105]:

$$(5) \quad P_{cation} + \frac{J_{cation}}{[cation]} \times \frac{RT}{E_m F} \times (1 - e^{-E_m F/RT})$$

where
R, T, and F have their usual meaning

Since chloride ions are passively distributed across the hepatic cell membrane and do not contribute to membrane currents, the transmembrane distribution of chloride, calculated from intracellular chloride concentration (Table 2-2) and the chloride concentration in the incubation media, were used to estimate the resting membrane potential (E_m) by the use of the Nernst equation (Equation 1).[106] Such estimated E_m values (in mV) were as shown in Table 2-4.

TABLE 2-3
Average $J_e^{K^+}$ (in pmol $\times$ second⁻¹ $\times$ cm⁻²)

Time at 0.5°C (minutes)	Control	Shock
5	6.05	4.37
15	4.64	3.33
25	4.19	2.39

TABLE 2-4
Estimated E_m Values (in mV)

Time at 0.5°C (minutes)	Control	Shock
5	−23.1	−13.4
15	−18.3	−11.2
25	−12.3	−9.4

These estimates indicate a decrease in membrane potential between the control and shock groups at each incubation time and within the control or shock group with increasing incubation time. Using values of estimated E_m, of Na^+ and K^+ fluxes, and of extracellular Na^+ and intracellular K^+ concentrations, P_{N+} and P_{K+} (in units of $10^{-8} \times$ cm/second) was calculated (see Table 2-5).

A cation permeability coefficient arrived at by using the GHKf equation is valid only if the E_m is equal to the Na^+/K^+ diffusion potential. In control rat livers, E_m is comprised of both a diffusion potential and an electrogenic Na^+ pump potential.[107] If hepatic E_m is measured in the presence of ouabain, it is reduced to a Na^+/K^+ diffusion potential. The use of the reduced E_m (with ouabain) in the GHK equation yields the ratio of P_{Na^+} to P_{K^+} to be 0.3.[107] Since our estimates of P_{Na^+}/P_{K^+} in control chilled liver slices were also around 0.3, it is believed that chilling affected the electrogenic pump but not the selective membrane permeability of liver to K^+ relative to Na^+ in control slices. The apparent increase of the permeability ratio to 0.56 in liver slices of shocked rats indicates a disruption of membrane differential permeability to Na^+ and K^+. Furthermore, the increase in P_{Na^+}/P_{K^+} *with shock was related to increased* P_{Na^+} but not to a change in P_{K^+}.

The observed increase in passive Na^+ influx in endotoxic liver can therefore be explained as due to decreased membrane resistance to Na^+. The observed decrease in K^+ efflux was likely due to decreased intracellular K^+ concentrations resulting from failure of active transport. The overall finding from assessments of net active and passive movements is that there was both an increased membrane Na^+ permeability and an impaired active Na^+/K^+ transport in livers of endotoxic rats. Each of these alterations can independently contribute to hepatic membrane depolarization in the state of endotoxic shock.

In experimental studies, inhibition of hepatic active transport was found along with the metabolic deteriorations (hypoglycemia and hyperlactic acidemia).[103] Whether the systemic metabolic changes precede the hepatic membrane transport alteration or vice versa could not be definitively determined. However, the partial inhibition of active K^+ transport before the onset of hypoglycemia (Table 2-6) at 2½ hours after endotoxin, 10 mg/kg, suggests that the hepatic ion transport derangement precedes metabolic changes and progresses into the shock phase. Furthermore, there is evidence to support that altered hepatic cation transport and membrane permeability may be causally related to altered hepatic glucose production. Studies in control rat livers *in vitro* have shown that glucose production varies linearly with hepatic membrane potential; hyperpolarization was associated with increased glucose production, and depolarization with decreased glucose production.[108] Moreover, hepatic membrane hyperpolarization and increased glucose production occur after applying to the liver (*in vitro*) either an alpha- or a beta-adrenergic agonist.[109,110] The beta-adrenergic hyperpolarization was mediated via: (1) an ouabain-

TABLE 2-5
P_{Na^+} and P_{K+} (in $10^{-8} \times$ cm/second)

Time at 0.5°C (minutes)	Control			Shock		
	P_{K+}	P_{Na^+}	P_{Na^+}/P_{K+}	P_{K+}	P_{Na^+}	P_{Na^+}/P_{K+}
5	8.6	2.7	0.30	7.8	4.5	0.57
15	8.0	2.7	0.33	7.7	4.3	0.55
20	7.4	3.1	0.40	7.4	4.1	0.56

TABLE 2-6

Net Hepatic Active K^+ Transport *in vitro* and Blood Glucose Levels in Endotoxin-Injected Rats[a]

	Blood Glucose, mg/100 ml	Net K^+ Transport, mmol $\times$ kg dry weight^{-1} $\times$ hours^{-1}
Control	95 ± 6 (6)	97 ± 10 (14)
2½ hours post-endotoxin 10 mg/kg	112 ± 11*** (12)	57 ± 5** (8)
4½ hours post-endotoxin 5 mg/kg	113 ± 8*** (5)	88 ± 8*** (12)
10 mg/kg (shock group)	38 ± 6* (20)	12 ± 4* (13)

*$p < 0.001$

**$p < 0.01$

***$p < 0.05$ (Not significantly different.)

[a]Values are means ± SE; number in parentheses represents number of rats. Control rats received saline and were killed after 4½ hours. Net K^+ transport was the difference in K^+ contents of chilled (90 minutes at 0.5°C) and rewarmed (37°C for 60 minutes preceded by 90 minutes at 0.5°CC) liver slices for 1 hour after chilling.

From Sayeed, 1984.[103]

sensitive Na^+/K^+, ATPase-dependent process (presumably an activation of electrogenic Na^+/K^+ pumping),[111] or (2) a change in membrane ion permeability (P_{K+} or P_{Na+}/P_{K+}).[112,113] These studies imply that hepatic membrane depolarization, increased ratio of P_{Na+}/P_{K+}, and impaired active Na^+/K^+ transport, as was found in endotoxin shock, would not only impair basal but would attenuate agonist-stimulated hepatic glucose production.

Relationships between Hepatic Cellular Energy and Na^+/K^+ Transport in Shock

An important concern regarding the shock-induced alteration in active Na^+/K^+ transport that needs clarification is whether or not this is due primarily to decreased cellular ATP synthesis. The loss of active transport function after ischemia or anoxia in kidney and skeletal muscle cells has been shown to be due to ATP depletion in these cells. However, this may not be the case in hepatic Na^+/K^+ transport in hemorrhagic shock. The energy requirements of hepatic Na^+/K^+ transport are relatively low. One-half of maximal Na^+/K^+ transport activity persisted when ATP synthesis was reduced from aerobic levels to just above the anaerobic level. This may indicate that Na^+/K^+ transport has a higher affinity for cellular ATP than for other energy-dependent cellular processes. Experimental data concerning hepatic mitochondrial energy metabolism in hemorrhagic shock also support the concept that suppression of aerobic ATP production may not contribute primarily to altered active Na^+/K^+ transport.[114] Sayeed and Baue found that liver mitochondrial metabolism was depressed in the intermediate hemorrhagic shock stages but was restored to control levels after resuscitation of animals at this stage. Both respiratory control and ADP-stimulated respiration by isolated mitochondria were restored to near control levels. A similar resuscitation of animals failed to restore the Na^+/K^+ pump activity.[92] This supports the view that active Na^+/K^+ transport could not be restored even after the recovery of the mitochondrial mechanisms responsible for the ATP synthesis. Measurements of tissue adenine nucleotide content of hemorrhagic shock liver showed a decrease *in* hepatic ATP from 1.80 ± 0.03 μmol/g (mean ± SE) in controls to 0.35 ± 0.04 μmol/g in intermediate shock.[94] The decreased level of ATP in the liver in intermediate shock was above the K_m value of hepatic Na^+/K^+ transport for ATP, which was reported to be 0.20 μmol/g.[115] Although we do not have direct evidence of Na^+/K^+ transport failure prior to ATP changes in shock, this suggests that membrane active Na^+/K^+ transport is affected in the absence of significant alterations in hepatic ATP. There is also a possibility that ATP alterations occur secondarily to an impairment of active Na^+/K^+ transport. A change in the cytoplasmic cationic milieu, owing to failure of transport, can adversely affect mitochondrial respiration, as increased intracellular Na^+ has a detrimental effect on the respiratory control of rat liver mitochondria.[62]

The endotoxin-related decrease in net active Na^+/K^+ transport could be due to a depletion of hepatic cellular energy due to ischemia that might prevail during shock and be characterized by hyperlacticacidemia, hypoglycemia, and the visible "ischemic" lesions in the bowel. However, the ATP content of endotoxic liver tissue, initially chilled and rewarmed under optimal incubation conditions, was equal to 80% of similarly incubated control liver tissue.[103] Previous studies had shown that active Na^+/K^+ transport by rat liver slices continues at one-half the maximal rate even when ATP decreases to between 25% and 30% of the control level.[115] These considerations suggest that the hepatic oxidative and energy synthetic capabilities were reversibly affected by the grade of ischemia seen in shock, and that active cation transport was affected irreversibly.

Plausible Mechanisms of Altered Na^+/K^+ Transport in Shock

A tenable hypothesis to account for the altered hepatic Na^+/K^+ transport in hemorrhagic shock is that of an alteration in the structure and/or function of the membrane-bound enzyme, Na^+/K^+–ATPase. The mechanism by which hemorrhagic shock may induce alteration in the Na^+/K^+–ATPase remains restricted to hypothetical consideration. It may involve hemorrhagic shock-induced acidosis and lysosomal enzyme activation.[116] The lysosomal enzymes (phosphoolipases and proteases) are capable of disrupting the structure and function of membranes, and thus the membrane components such as the Na^+/K^+–ATPase.

Disruptions of membrane structure and function may also be responsible for increased Na^+ permeability in shock.

In endotoxic/septic shock, we can speculate that altered hepatic Na^+/K^+ transport and membrane permeability in endotoxic shock were due to a direct effect of endotoxin molecule or to breakdown products of the endotoxin molecule, such as lipid A, on the hepatic plasma membrane. Either the intact endotoxin molecule or its lipophilic breakdown products may "intercalate" themselves into the membrane phospholipid bilayer,[117] and thus indirectly alter the conformation of membrane protein complexes such as the Na^+/K^+–ATPase and/or those lining the ionic channels. At the present time, however, there is a controversy as to whether there is a direct endotoxin effect of target tissue.[118] Finally, endotoxin may interact with serum components and/or with the host defense system and generate mediators that interfere with membrane structure–function.[119]

Cellular Ca^{2+} Regulation: Effect of Endotoxin and Endotoxic Shock

Studies in Nonhepatic Tissues

Several studies have implicated endotoxin-mediated alterations in transmembrane Ca^{2+} movements and cellular Ca^{2+} homeostasis. Most of these studies have determined the effect of endotoxin on the transmembrane fluxes of Ca^{2+} in isolated tissues and cells. For example, Connor et al. showed an endotoxin-mediated increase in inward calcium flux (inward calcium current) in excitable cells (smooth muscle, squid giant synapse, and frog atrial fiber).[120] Hess et al. measured active Ca^{2+} uptake by isolated cardiac sarcoplasmic reticulum and the sarcoplasmic reticulum Ca^{2+}–ATPase.[121,122] Sarcoplasmic reticulum Ca^{2+} uptake and the Ca^{2+}–ATPase were measured in sarcoplasmic reticulum exposed to endotoxin either *in vivo* or *in vitro*. They found that endotoxin *in vitro* effected a depression of both Ca²⁺ uptake and the AT-Pase. In contrast the *in vivo* effect of endotoxin was a depression in Ca²⁺ but not in the ATPase activity. These studies support the concept that endotoxin molecules can exert direct toxic effects on membranes, and that these effects may or may not manifest in endotoxicosis or endotoxic shock *in vivo*. Nelson and Spitzer determined ⁴⁵Ca exchange by using isolated adipocytes. When adipocytes from endotoxic rats were studied, they exhibited an increased level of ⁴⁵Ca uptake compared to adipocytes from control rats.[123] Increased uptake of calcium was due to both an uptake into the intracellular compartment and an increased binding to external cell membrane sites. Endotoxin *in vitro* failed to affect ⁴⁵Ca exchange in isolated adipocytes. These studies suggest that there is an increase in intracellular calcium in adipocytes during endotoxicosis.

Studies in the Liver

In his laboratory, Sayeed assessed hepatic cellular calcium movements in liver slices of control and endotoxic rats.[124] Intravenous injections of *Salmonella* endotoxin into rats produced a shocklike state at between 5 and 6 hours post injection. The shock state was characterized by hypoglycemia, lactic acidemia, and hemorrhagic lesion in the bowels.

Basal Cellular Calcium Efflux. The cellular efflux of Ca²⁺ was assessed in liver slices that were initially equilibrated with ⁴⁵Ca at 37°C. The equilibration period of 60 minutes was carefully selected to allow ⁴⁵Ca labeling of (1) the rapidly exchanging extracellular Ca²⁺ binding sites on the plasma membrane; and (2) the slowly exchanging intracellular pools of Ca²⁺, namely, mitochondrial, endoplasmic reticulum, and the cytosol. The nuclear and other intracellular pools are presumably not significantly equilibrated during a 60-minute labeling period. The ⁴⁵Ca-equilibrated liver slices were sequentially washed through a series of flasks containing nonradioactive media. The re-

lease of ⁴⁵Ca into the washout media quantitated the Ca²⁺ efflux from hepatocytes. The energy dependency of cellular Ca²⁺ efflux was ascertained by measuring efflux in the presence of the metabolic inhibitor iodoacetate.

Figure 2-14 shows the effects of temperature and metabolic inhibition on ⁴⁵Ca efflux in control rat liver slices. The efflux activity was expressed as the rate of appearance of ⁴⁵Ca [cpm/(minutes × g tissue)] into washout media. In iodoacetate-free media, ⁴⁵Ca efflux declined progressively with the continuation of washout at 0.5°C. The increase in medium temperature to 37°C after the initial 20 minutes at 0.5°C resulted in a nearly 10-fold increase in ⁴⁵Ca efflux. Also, effluxes during the 20- to 70-minute washouts were significantly ($p < .05$) higher at 37°C than at 0.5°C. ⁴⁵Ca effluxes at 37°C appeared to be substantially lower in the

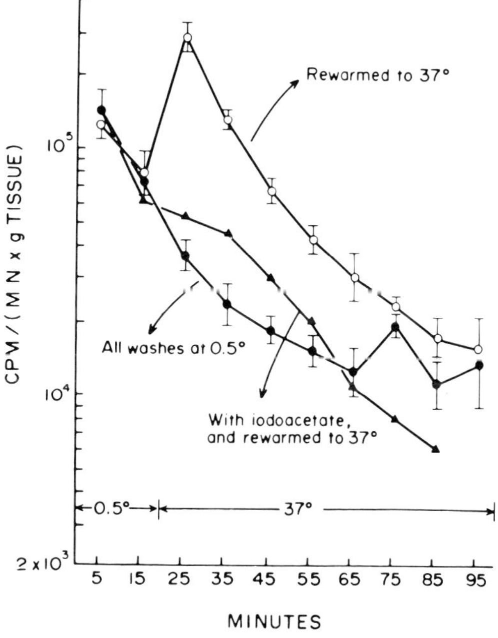

Figure 2-14. Calcium efflux from control rat liver slices. Open circles show control effluxes (mean ± SE, $n = 11$); closed circles show effluxes ($n = 11$); all of which were measured at 0.5°C. Closed triangles show means of effluxes ($n = 3$) at 0.5°C and 37°C in the presence of 0.1/mmol iodoacetate. (From Sayeed, 1986.)[124]

presence of iodoacetate (10^{-4} M) than in its absence during the from 20- to 60-minute washouts.

The low rate of isotope exchange at 0.5°C during the between 0- and 20-minute washouts would prolong the exchange time of the rapidly exchanging extracellular pool beyond the expected exchange time for this pool at 37°C, that is, 10 minutes.[125] Thus at 0.5°C, efflux beyond 10 minutes is due not only to release of ^{45}Ca from the slow intracellular calcium compartment, but also from the rapid extracellular pool. As was found in subsequent measurements (see Figure 2-18, p. 60), the $t_{1/2}$ of the rapid extracellular pool was approximately doubled ($t_{1/2}$ was 5.3 at 0.5°C as compared to $t_{1/2} =$ 2.7 at 37°C),[125] indicating that the release of ^{45}Ca from the extracellular pool at 0.5°C would be nearly complete by about 20 minutes. Therefore, effluxes after 20 minutes would mostly be due to transmembrane flux out of the intracellular compartment. The effects of temperature elevation from 0.5°C to 37°C were evaluated using estimates of Q_{10}. Whereas iodoacetate Q_{10} ranges from 1.13 to 1.24, in the absence of iodoacetate the range was 1.81 to 1.94. The latter values are substantially higher than those that are typical for the diffusion of inorganic electrolytes (i.e., 1.24).[126] Values estimated in the presence of iodoacetate are comparable to those for the inorganic ion self-diffusion in the aqueous phase. The observed changes in Q_{10} suggest that metabolic inhibition via iodoacetate reduces efflux of calcium to an energy-dependent diffusive process. Thus, the basal efflux of calcium through the plasma membrane of liver cells would appear to require cellular metabolic energy. Previous studies have supported the presence of a plasma membrane active Ca^{2+} transport system in the liver.[127,128]

Figure 2-15 shows ^{45}Ca effluxes by liver slices from the endotoxic rats. As described before, slices were washed initially at 0.5°C and then at 37°C. Effluxes in endotoxic rat liver slices were not different from those measured in control rat liver slices.

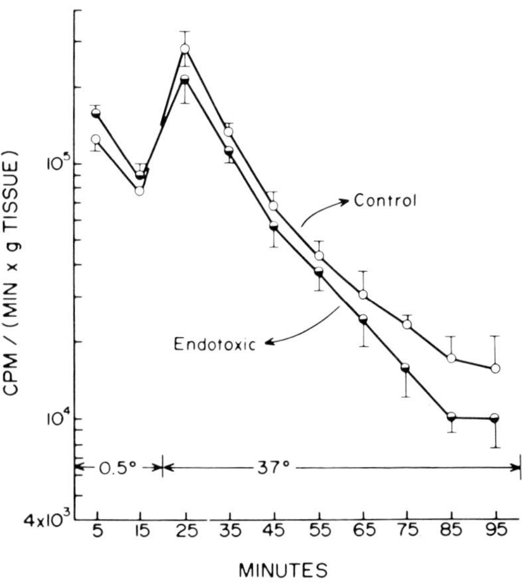

Figure 2-15. Calcium efflux from rat liver slices of control ($n = 11$) and endotoxic rats ($n = 7$). Mean ($\pm$ SE) values of effluxes were plotted at the midpoints of 10-minute intervals. (From Sayeed, 1986.)[124]

Calcium Uptake by Endoplasmic Reticulum. The uptake of ^{45}Ca by isolated hepatic endoplasmic reticulum was measured as another variable of hepatocyte Ca^{2+} regulation. Isolated endoplasmic reticulum were incubated with ^{45}Ca in the presence of 5 mmol ATP, and calcium uptake into endoplasmic reticulum vesicles was followed as a function of time. It was ascertained that endoplasmic reticulum Ca^{2+} uptake was ATP dependent. Figure 2-16 shows measurement of ATP-dependent calcium uptake by hepatic endoplasmic reticulum from control and endotoxic animals. There was no demonstrable change in calcium uptake by endotoxic hepatic endoplasmic reticulum.

Effect of Norepinephrine on Cellular Efflux. Figure 2-17 shows the effect of 1 μmol norepinephrine (NE) on the efflux of calcium. Prior to addition of NE, ^{45}Ca efflux by control rat liver slices was not different from that by liver slices of endotoxic rats. After NE, however, significant differences were observed between the control and the endotoxic group. Compared to its value mea-

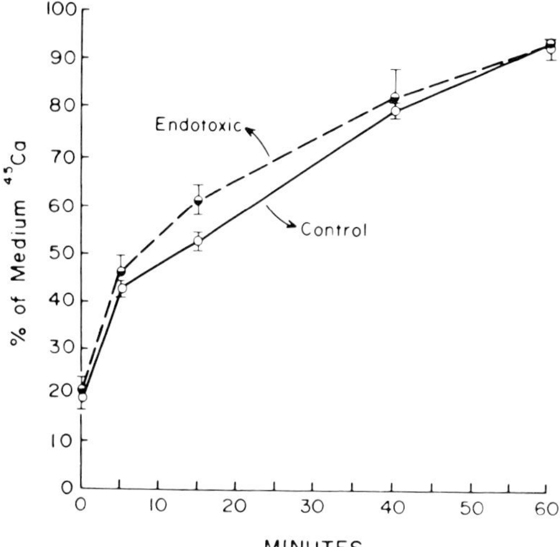

Figure 2-16. Calcium uptake by hepatic endoplasmic reticulum from control (open circles, $n = 6$) and endotoxic rats (semiclosed circles, $n = 6$). Values are mean $\pm$ SE. (From Sayeed, 1986.)[124]

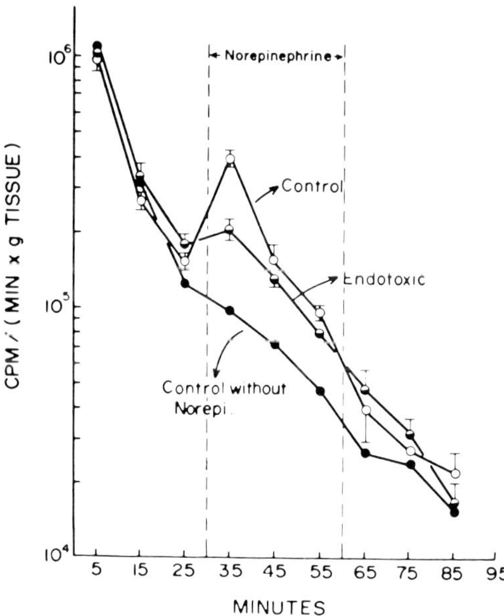

Figure 2-17. Effect of norepinephrine on calcium efflux (mean $\pm$ SE) in liver slices of control (open circles, $n = 7$) and endotoxic rats (semiclosed circles, $n = 5$). Closed circles are effluxes in control rat liver slices without norepinephrine, values are the mean of measurements in three rats. (From Sayeed, 1986.)[124]

sured just before NE, control liver slice ^{45}Ca efflux after NE was 105% higher; it was approximately 300% higher than efflux during the same washout period in the absence of NE. In the endotoxic group, ^{45}Ca efflux measured after NE was not significantly different from that measured prior to NE; it was, however, about 100% higher than control liver slice efflux measured in the absence of NE during the same washout period.

Although during the 20- to 40-minute washouts NE caused a substantially greater increase in ^{45}Ca efflux in controls than that seen in endotoxic rat liver slices, effluxes in the two groups were comparable during the from 40- to 50-minute and the from 50- to 60-minute washout periods. These results indicate that in control rat liver slices NE first elicited a high magnitude yet transient net calcium efflux. This response apparently lasted several minutes and was followed by effluxes that were higher than those measured in the absence of NE but smaller in magnitude than the initial response. The net efflux of calcium initially seen could be due to either mobilization of calcium from the slowly exchanging intracellular pools with subsequent stimulation of active calcium efflux across the plasma membrane or to NE directly stimulating the membrane calcium transport system. The latter explanation would predict an NE-mediated decrease in cytosolic calcium, which is contrary to recent evidence. Norepinephrine mobilization of calcium may occur from intracellular subcompartments such as endoplasmic reticula, mitochondria, or the inner plasma membrane. In endotoxic rat liver slices, there was clearly an attenuation in the NE-induced high magnitude net efflux without any significant effect on the subsequent effluxes in the presence of NE.

Intracellular Exchangeable Calcium. The semilogarithmic plots of ^{45}Ca remaining in the tissue (disintegrations per minute/g tissue) as a function of washout time are shown in Figure 2-18. The time-related de-

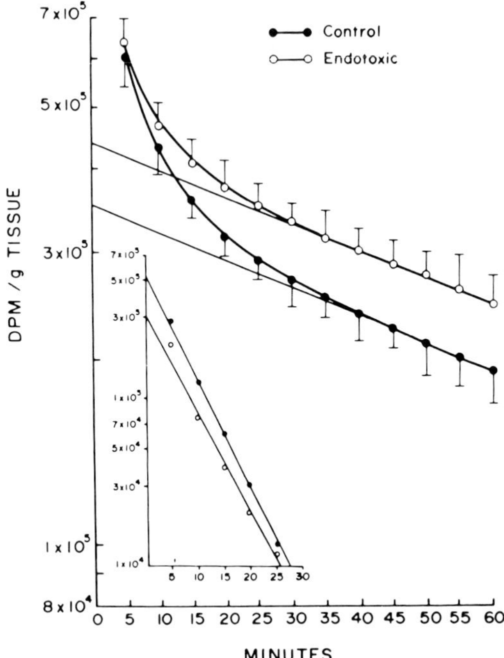

Figure 2-18. Liver slice ^{45}Ca content (mean ± SE; control, $n = 15$; endotoxic, $n = 9$) during washout of the radioisotope. Inset: Calculated values (means) of loss of ^{45}Ca from the "rapid" extracellular calcium pool. (From Sayeed, 1986.)[124]

cline in radioactivity of liver slices of control and endotoxic rats was resolved into two monoexponential functions corresponding to ^{45}Ca release from the rapid extracellular exchange and the slow intracellular exchange compartment. Tissue efflux changes due to simultaneous washout of ^{45}Ca from both the slow and the rapid exchange pools are represented by the following equation:

$$(6) \qquad ^{45}Ca_{TT,t} = (^{45}Ca_{r,t0} \times e^{-t\lambda r}) + (^{45}Ca_{s,t0} \times e^{-t\lambda r})$$

where

$^{45}Ca_{TT,t}$ = total tissue (TT) radiocalcium (dpm/g tissue) at a given washout time (t)

$^{45}Ca_{r,t0}$ = radiocalcium in the rapid exchange (r) compartment at the initiation of the washout (= zero-washout time, t_0)

$^{45}Ca_{s,t0}$ = radiocalcium in the slow exchange (s) compartment at t_0

λ_r and λ_s = rate coefficients (% minutes^{-1}) for effluxes at 0.5°C out of rapid (r) and slow (s) exchange compartments

The monoexponential nature of ^{45}Ca efflux from the intracellular pool was ascertained by means of a "curve peeling" analysis on data from liver slices of individual control and endotoxic animals. Figure 2-18 shows that visually drawn lines gave reasonable fits to slow and rapid effluxes in the control and endotoxic groups.

The effect of low temperature on the washout of ^{45}Ca from the rapid extracellular calcium pool would be to simply decrease the isotope exchange rate and prolonging the half-life ($t_{1/2}$) of the efflux process. This is due to the fact that there is no involvement of membrane transport or cellular metabolism in the washout of ^{45}Ca from the extracellular membrane-bound pool. On the other hand, the low temperature (0.5°C) would affect efflux from the intracellular compartment not only due to a decrease in the isotope exchange rate *per se* but also due to suppression of active intracellular (uptake by endoplasmic reticulum and mitochondria) and transplasma membrane calcium movements. The suppression of active plasma membrane transport would play a minor role in the net entrance of calcium into the cytoplasmic compartment rather than the release of calcium from endoplasmic reticula and mitochondria. The hepatic endoplasmic reticulum and inner mitochondrial membrane fluxes of calcium are from 10 to 15 times greater than the hepatic plasma membrane flux.[21,125] Thus at 0.5°C, there would be a net calcium accumulation into the cytoplasmic compartment with no real net calcium movements across the plasma membrane. Under these conditions, ^{45}Ca efflux out of cells, along the passive calcium efflux pathway, would essentially be due to isotope exchange but not to net calcium movement out of cells. And since this isotope exchange between medium and the slow intracellular compartment was a monoexponential function, it permitted us to make valid assessments of the size of the total intracellular exchangeable calcium pool. The intracellullar content of exchangeable calcium was quantitated by extrapolating the slow monoexponential curve to zero-washout time. The intracellular pool, thus estimated, consisted of mainly the endoplasmic reticulum, the mitochondrial, the inner plasma membrane–bound pool, and the cytosolic-free Ca^{2+}. The intracellular calcium pool size in liver slices of control rats (mean $\pm$ SE $= 413 \pm 17$ μmol/kg tissue) was comparable to previous measurements[125] but was significantly different ($P <$ .05) from that in endotoxic rats (553 ± 23).

Significance of Endotoxic Alterations in Hepatocyte Calcium Regulation. Although the basal uphill calcium efflux was unaltered in liver slices of endotoxic rats, NE stimulation of efflux was grossly reduced in livers of these animals. A number of investigators have shown that alpha-adrenergic (e.g., NE) and certain other agonists (vasopressin and angiotensin II) act on liver cells to enhance calcium efflux.[66,67,129,130] Recent studies have demonstrated the net calcium efflux to occur along with an elevation of cytosolic-free Ca^{2+} concentration.[132] The increase in cytosolic-free calcium, which occurs transiently, initiates the hepatocyte's metabolic response, namely, glucose production. However, a sustained metabolic response, with continuous agonist occupation of hepatocyte plasma membrane receptors, is believed to require an influx of extracellular Ca^{2+} as well. In any event, the agonist increased cytosolic-free Ca^{2+} serves as the activator calcium for the metabolic response.

In hepatocytes, the activator calcium could be derived from intracellular pools such as mitochondria or endoplasmic reticulum, or from the calcium bound to inner plasma membrane. Although several studies have indicated that liver mitochondria released the activator calcium,[133] more recent studies have shown endoplasmic reticulum to be its major source.[67,129,130] The agonist-mediated elevation of cytosolic-free calcium and net efflux of calcium would lead to depletion of the activator calcium pool. Moreover, once depleted of calcium, the intracellular activator calcium pool is not recharged with calcium unless agonist binding is terminated. Furthermore, the metabolic response cannot occur with subsequent applications of the agonist until there is resequestration of calcium within the intracellular activator calcium pool. These findings imply that an excessive and/or repetitive stimulation of liver cells by NE would lead to depletion or inadequate repletion of the hepatic intracellular activator calcium pool. Presumably, both the depletion and inadequate repletion would occur primarily due to excessive and/or repetitive release of calcium out of the activator pool without an alteration in active calcium uptake into the pool. The attenuation of NE-induced calcium efflux from liver slices of endotoxic rats may well be due to depletion or inadequate repletion of the activator calcium pool. Such an alteration seems probable given the high blood catecholamine levels that are found in the endotoxic rats.[134] There may be a washout of some of the catecholamine, which presumably binds to endotoxic animal liver cells *in vivo*, during the preparation and incubation of liver slices in agonist-free Krebs-Ringer media. However, more catecholamine is likely to remain bound to endotoxic animal tissues than to tissues of control animals.

The assessments of the size of the intracellular calcium pool indicated that it was significantly greater in the livers of endotoxic animals than in controls. Apparently, the increase in the exchangeable intracellular pool size with endotoxic shock could result from a net shift of extracellular Ca^{2+} *in vivo*, either from the extracellular fluid or from the extracellular membrane-bound pool of calcium. In control animal livers, although there is an influx of extracellular calcium into cells following their activation by the agonists, most of this calcium is acutely taken up by mitochondria and then eventually extruded out of mitochondria and out of cells during a period of recovery after the termination of stimulation by the agonist.[22] It is conceivable that the increase in the size of the intracellular calcium pools in livers of endotoxic rats was due to an inadequate recovery of cells during their excessive/repetitive stimulation by catecholamines. Rasmussen and Barrett have estimated that the sum of the extracellular calcium that enters the cell and the calcium released from intracellular sources (e.g., endoplasmic reticulum) during a maximal cell activation is far in excess of the increase in cytosolic-free $[Ca^{2+}]$ (approximately $100: < 1$).[22] However, since mitochondria serve as a sink for this calcium, they acutely prevent an intracellular calcium "overload" during cell activation. This suggests that the endotoxic shock-related increase in the cytoplasmic exchangeable calcium mainly represents an accumulation of calcium into mitochondria.

Whereas the consequence of the depletion or inadequate repletion of activator calcium pools would be an attenuation of the hepatic metabolic response, namely, glucose production, the mitochondrial accumulation of calcium may potentially lead to a compromise of their ability to store calcium in the face of an intracellular calcium overload. There is ample evidence of a deleterious and lethal effect of uncompensated overload of calcium in cells when the presumed mitochondrial capacity for calcium storage is exceeded.[135] Previous studies have also shown occurrences of calcium overload in endotoxemia, and in other models of cell injury.[136,137]

ACKNOWLEDGMENT

I thank Julia Guszcza for typing this manuscript. At the time of writing, author's research was supported by National Institutes of Health Grants GM-32288 and HL-31163.

REFERENCES

1. Liebig J. Über die bestandtheile der flüssigkeiten des fleisches, *Ann d Chem u Pharm*. 1847;62:257.
2. Von Mohl H. *Grundzuge der Anatomie und Physiologie der vegatabilischen zelle*. Braunschweig: 1851.
3. Nägeli K, Cramer C. *Pflanzenphysiologische Untersuchungen*. Zürich: 1855.
4. Fick AE. Über Diffusion. *Poggendorff's Annalen*. 1855;94:59.
5. Pfeffer W. *Osmotische Untersuchungen*. Leipzig: Engelmann; 1877.
6. Ringer S. A further contribution regarding the influence of different constituents of the blood on the contraction of the heart. *J Physiol*. 1983;4:29.
7. Arrhenius SA. Über die dissociation der in wasser gelösten stoffe. *Z Phys Chem*. 1887;1:631.
8. Katz J. Die mineralischen bestandteile des muskelfleisches. *Arch f d Physiol*. 1896;63:1.
9. Urano F, Nachtrag zu: Neue versuche über die salze des muskels. *Ztschr f Biol*. 1908;51:483.
10. Urano F. Neue versuche über die salze des muskels. *Ztschr f Biol*. 1908;50:212.
11. Berstein J. Untersuchungen zur thermodynamik der biolelektrischen ströme. *Pfluger's Arch*. 1902;92:521.
12. Hoeber R. *Physikalische Chemie der Zelle und der Gewebe*. Leipzig: Engelmann; 1926.
13. Kotyk A, Janacek K. *Membrane Transport*. New York: Plenum Press; 1977.
14. Steinbach HB. Sodium and potassium in frog muscle. *J Biol Chem*. 1940;133:695.
15. Boyle PJ, Conway EJ. Potassium accumulation in muscle and associated changes. *J Physiol*. 1941;100:1.
16. Dean RB. Theories of electrolyte equilibrium in muscle. *Biol Symp*. 1941;3:331.
17. Hodgkin AL, Huxley AF. Resting and action potentials in single nerve fiber. *J Physiol*. 1945;104:176.
18. Hodgkin AL, Keynes RD. Active transport of cation in giant axons from Sepia and Loligo. *J Physiol*. 1955;128:28.
19. Hodgkin AL, Huxley AF. Currents carried by sodium and potassium through the membrane of the giant axon of Loligo. *J Physiol*. 1952;116:449.
20. Borle AB. Control, modulation and regulation of cell calcium. *Rev Physiol Biochem Pharmacol*. 1981;190:13.
21. Rasmussen H. *Calcium and cAMP as Synarchic Messengers*. New York; John Wiley; 1981.

22. Rasmussen H, Barrett PQ. Calcium messenger system: an integrated view. *Physiol Rev.* 1984; 64:938.

23. Rasmussen H, Jenson P, Lake W, Goodman DBP. Calcium ion as a second messenger. *Clin Endocrinol.* 1976;5(suppl):11s.

24. Miller C. Integral membrane channels: studies in model membranes. *Physiol Rev.* 1983;63:1209.

25. Hille B. *Ionic Channels of Excitable Membranes.* Sunderland, Mass.:Sinauer, Sunderland; 1984: chaps 3–6.

26. Van Driessche W, Zeiske W. Ionic channels in epithelial membranes. *Physiol Rev.* 1985;65:833.

27. Hille B. Ionic selectivity of sodium and potassium channels of nerve membranes. In: Eisenmann G, ed. *Membranes, Lipid Bilayers and Biological Membranes: Dynamic Properties.* New York: Marcel Dekker; 1975:255.

28. Sariban-Sohraby S, Latorre R, Burg MB, Olans L, Benos D. Amiloride-sensitive epithelial sodium channels reconstituted into planar lipid bilayer membranes. *Nature.* 1984;308:80.

29. Maruyama Y, Peterson OH. Single-channel currents in isolated patches of plasma membrane from basal surfaces of pancreatic acini. *Nature.* 1982;299:159.

30. Lebrun P, Malaise WJ, Herschultz A. Activation, but not inhibition, by glucose of calcium-dependent potassium permeability in the rat pancreatic beta-cell. *Biochem Biophys Acta.* 1983;731:145.

31. Lindemann B, Van Driessche W. The mechanism of sodium uptake through sodium selective channels in the epithlium of frog skin. In: Hoffman JF, ed. *Membrane Transport Processes.* New York: raven; 1978:155.

32. Turnheim R, Frizzell RA, Schultz SG. Interaction between cell sodium and amiloride-sensitive sodium entry step in rabbit colon. *J Membr Biol.* 1978;39:233.

33. Frizzell RA, Schultz SG. Effect of Aldosterone on ion transport by rabbit colon, *in vitro. J Membr Biol.* 1978;39:1.

34. Turnheim K, Thompson SM, Schultz SG. Current-voltage relations of apical sodium transport in rabbit descending colon: effect of varying luminal sodium concentrations. *Pflugen Arch.* 1983;322:R13.

35. Li JHY, Palmer LG, Edelman IS, Lindemann B. The role of sodium-channel density in the natriferic response of the toad urinary bladder to antidiuretic hormone. *J Membr Biol.* 1982;64:77.

36. Stekhoven FS, Bonting SL. Transport adenosine triphosphateses: properties and functions. *Physiol Rev.* 1981;61:1.

37. Hoffman JF, Forbush B. Structure, mechanism and function of the sodium/potassium pump. In: *Current Topics in Membranes and Transport.* Vol. 19. New York: Academic Press; 1983.

38. Sperelakis N. Electrical properties of cells at rest and maintenance of the ion distributions. In: Sperelakis N, ed. *Physiology and Pathophysiology of the Heart.* Boston: Marinus Nijhoff; 1984: chap 4.

39. Opie LH, Naylor W, Geyers W. *Pumps, Channels and Currents in the Heart,* Opie LH, ed. London: Grune & Stratton; 1984: chap 4.

40. Hegywary C, Jorgensen PL. Conformational changes of renal sodium plus potassium ion-transport adenosine triphosphatase labeled with fluorescein. *J Biol Chem.* 1981;256:6296.

41. Lo CS, Lo TN. Effect of triiodothyronine on the synthesis and degradation of the small subunit of renal cortical (sodium/potassium)-adenosine triphosphate. *J Biol Chem.* 1980;255:2131.

42. Graves SW, Brown B, Valdes RJr. An endogenous digoxin-like substance in patients with renal impairment. *Ann Int Med.* 1983;99:604.

43. Bygrave FL. Mitochondra and the control of intracellular calcium. *Biol Rev.* 1978;53:43.

44. Mitchell P. Possible molecular mechanisms of the protonmotive function of cytochrome systems. *J Theor Biol.* 1976;62:327.

45. Crane RK, Malathi KP, Fairclough P. Some characteristics of kidney sodium-dependent glucose carrier reconstituted into sonicated liposomes. *Am J Physiol.* 1979;234:E1.

46. Klip A, Grinstein S, Semenza G. The small intestinal sodium alpha-D-glucose cotransporter is inserted in the brush border asymmetrically. *Ann NY Acad Sci.* 1980;358:374.

47. Nellans HN, Frizell RA, Schultz SG. Coupled sodium-chloride influxes across the brush border of rabbit ileum. *Am J Physiol.* 1973;255:467.

48. Frizzell RA, Dugas MC, Schultz SG. Sodium-chloride transport by rabbit gall bladder: direct evidence for a couple sodium-chloride influx process. *J Gen Physiol.* 1975;65:637.

49. Karlstad MD, Sayeed MM. Alpha-aminoisobutyric acid transport in rat soleus muscle during endotoxic shock. *Am J Physiol.* 1985;248:R142.

50. Curran PF, Schultz SC, Chez RA, Fuisz RE. Kinetic reactions of the sodium–amino acid interaction at the mucosal brush border of intestine. *Am J Physiol.* 1967;219:1027.

51. Samarzija I, Fronter E. Elective studies on amino acid transport across brush border membrane of rat proximal tubule *in vivo. Pfluger's Arch.* 1975;359:R119.

52. Blaustein MP. The interrelationship between sodium and calcium fluxes across cell membranes. *Rev Physiol Biochem Pharmacol.* 1974;70:33.

53. Van Rossum, GDV. Net movements of calcium and magnesium in slices of rat liver. *J Gen Physiol.* 1977;55:18.

54. Orlov SN, Pokudin NI, Kavtsov K. On the mechansim of the regulation of intracellular calcium distribution in adipose tissue [in Russian]. *Biokhimiya.* 1980;45:408.

55. Scheid CR, Honeyman TW, Fay RS. Mechanism of beta-adrenergic relaxation of smooth muscle. *Nature.* 1979;277:32.

56. Aickin CC, Thomas RC. An investigation of the ionic mechanism of intracellular pH regulation in mouse soleus muscle fibers. *J Physiol.* 1977; 273:295.

57. Moore RD. Stimulation of sodium–hydrogen exchange by insulin. *Biophys.* 1981;33:203.

58. Proverbio F, Conresen-Guidi M, Whittembury

G. Ouabain sensitive sodium stimulation of a magnesium-dependent ATPase in kidney tissue. *Biochem Biophys Acta.* 1975;394:281.

59. Whittam R. *Enzymatic Aspects of the Sodium Pump.* New York: Wiley–Interscience; 1970.

60. Soling HD, Kleineke J. Species dependent regulation of hepatic gluconeogenesis in higher animals. In: Hansen RW, Mehlman MA, eds. *Gluconeogenesis: Its Regulation in Mammalian Species.* New York: John Wiley; 1976:369–462.

61. Pestka S. Protein Biosynthesis: mechanisms, requirements and potassium dependency. In: Bittar, EE, ed. *Membranes and Transport,* Vol 3. London: Wiley–Interscience; 1971: chap 9.

62. Gomes-Puyou A, Sandoval F, Pena A, Charez E, Tuena A. Effect of sodium and potassium on mitochondrial respiratory control, oxygen uptake and adenosine triphosphatase activity. *J Biol Chem.* 1969;244:5339.

63. Pallotta BS, Magleby KL, Barret JN. Single channel recordings of calcium-activated potassium currents in rat muscle cell culture. *Nature.* 1981;293:471.

64. Thompson SH. Three pharmacological distinct channels in molluscan neuron. *J Physiol.* 1977; 265:465.

65. Matteson DR, Deutch C. Potassium channel in T lymphocytes: a patch clamp study. *Nature.* 1984; 307:468.

66. Charest R, Blackmore PF, Berthon B, Exton JH. Changes in free cytosolic calcium in hepatocytes following alpha-adrenergic stimulation. *J Biol Chem.* 1983;258:8769.

67. Joseph SK, Thomas AP, Williams RJ, Irvine RF, Williamson JR. Myo-inositol 1,4,5-triphosphate: a second messenger for the hormonal mobilizattion of intracellular calcium in liver. *J Biol Chem.* 1984;259:3077.

68. Hems DA, Whitton PD. Stimulation by vasopressin of glycogen breakdown and gluconeogenesis in the perfused rat liver. *Biochem J.* 1973;136: 705.

69. Nicholls DG. The regulation of extramitochondrial free calcium ion concentration by rat liver mitochondria. *Biochem J.* 1978;176:463.

70. Fiskum G, Lehninger AL. The mechanism of regulation of mitochondrial calcium transport. *Fed Proc.* 1980;39:2432.

71. Chien KR, Farber JL. Microsomal membrane dysfunction in ischemic rat liver cells. *Arch Biochem Biophys.* 1977;180:191.

72. Chien KR, Reeves JP, Buja M, Bonte F, Parkey RW, Willerson JT. Phospholipid alterations in canine ischemic myocardium. *Circ Res.* 1981;48: 711.

73. Hinshaw LB. Overview of endotoxin shock. In: Cowley RA, Trump BF, eds. *Pathophysiology of Shock, Anoxia, and Ischemia.* Baltimore: Williams & Wilkins; 1982: chap 16.

74. Chaudry I, Baue AE. Overview of hemorrhagic shock. In: Cowley RA, Trump BF, eds. *Pathophysiology of Shock, Anoxia, and Ischemia.* Baltimore: Williams & Wilkins; 1982: chap 15.

75. Opie LH. Calcium antagonists. In: *The Heart.* London: Grune & Stratton; 1984: chap 18.

76. Fuhrman FA, Crimson JM. Early changes in the distribution of sodium, potassium and water in rabbit muscles following release of tourniquest. *Am J Physiol.* 1951;166:424.

77. Fuhrman FA. Electrolytes and glycogen in injured tissues. In: Stoner HB, Threfal CJ, eds. *The Biochemical Response Injury.* Springfield, Il: Charles C Thomas; 1960:5–21.

78. Calkins D, Taylor IM, Hastings AB. Potassium exchange in the isolated rat diaphragm: effect of anoxia and cold. *Am J Physiol.* 1954;177:211.

79. Paul DH. The effect of anoxia on the isolated rat phrenic-nerve–diaphragm preparation. *J Physiol.* 1961;155:358.

80. Flink EB, Hastings AB, Lowry JK. Changes in potassium and sodium concentrations in liver slices accompanying incubations *in vitro.* Am J Physiol. 1950;163:598.

81. Petterson S, Gelin LE, Jonsson O, Schersten T. Effect of warm ischemia on the potassium and sodium contents and on the sodium–potassium–ATPase activity in the dog kidney slices. *Eur Surg Res.* 1974;6:330.

82. Croker BP, Saladino AJ, Trump BF. Ion movements in cell injury: relationship between energy metabolism and pathogenesis of lethal injury in the toad bladder. *Am J Pathol.* 1970;59:247.

83. Sahaphong S, Trump BF. Studies of cellular injury in isolated kidney tubule of the flounder, V: effect of inhibiting sulfhydryl groups of plasma membrane with the organic mercurials, pCMB (parachloromercuro benzoate) and pCMBS (parachloromercurobenzene sulfonate). *Am J Pathol.* 1971;63:277.

84. Saladino AJ, Hawkins HK, Trump BF. Ion transport and cell injury: alterations in ultrastructure and ion flux in toad bladder epithelium after treatment with surfactants. *Lab Invest.* 1968; 18:833.

85. Haljamae H. Effects of hemorrhagic shock and treatment with hyhpothermia on the potassium content and transport of single mammalian skeletal muscle cells. *Acta Physiol Scand.* 1970; 78:189.

86. Haljamae H. Shock reactions in skeletal muscle: a method for determining the potassium content of single skeletal muscle cells. *Acta Chir Scand.* 1967;133:259.

87. Haljamae H. Sampling of nanoliter volumes of mammalian subcutaneous tissue fluid and ultramicro flame photometric analyses of the potassium and sodium concentrations. *Acta Physiol Scand.* 1970;78:1.

88. Campion DS, Lynch LJ, Rector FC Jr, Carter N, Shires GT. Effect of hemorrhagic shock on transmembrane potential. *Surgery.* 1969;66:1051.

89. Trunkey DD, Illner H, Wagner IY, Shires GT. The effect of hemorrhagic shock on intracellular muscle action potentials in the primate. *Surgery.* 1973;74:241.

90. Trump BF, Laiho KU, Berezesky IK. The role of ion movements in cell injury and shock. *Circ Shock.* 1979;6:182.

91. Sayeed MM, Adler RJ, Chaudry IH, Baue AE. Effect of hemorrhagic shock on hemorrhagic

transmembrane potentials and intracellular electrolytes, *in vivo. Am J Physiol.* 1981;240:R211.

92. Sayeed MM, Baue AE. Sodium–potassium transport in rat liver slices in hemorrhagic shock. *Am J Physiol.* 1973;224:1265.

93. Sayeed MM, Connors JP, Henton BJ. Sodium fluxes in rat liver slices in hemorrhagic shock. *Physiologist.* 1976;19:354.

94. Sayeed MM, Wurth M, Chaudry IH, Baue AE. Cation transport in the liver in hemorrhagic shock. *Circ Shock.* 1974;1:195.

95. Russo MA, Van Rossom GDV, Galeotti T. Observations on the regulation of cell volume and metabolic control *in vitro*: changes in the composition and ultrastructure of liver slices under conditions of varying metabolic and transporting activity. *J Membr Biol.* 1977;31:267.

96. Mullins LJ, Noda K. the influence of sodium-free solutions on membrane potential of frog muscle fibers. *J Gen Physiol.* 1963;47:117.

97. Flear CTG. Electrolyte and body water changes after trauma. *J Clin Pathol.* 1970;23(suppl. 4):16.

98. Flear CTG, Bhattacharya SS, Singh CM. Solute and water exchanges between cells and extracellular fluids in health and disturbances after trauma. *J Parenter Enter Nutr.* 1980;4:98.

99. Silver IA. Some effects of *E. Coli* endotoxin on cells in culture. In: Maida J, Person RJ, eds. *Pathophysiological Effects of Endotoxins at the Cellular Level.* New York: Alan R Liss; 1981:81–95.

100. Gibson WH, Cook JJ, Gatipon G, Moses ME. Effect of endotoxin shock on skeletal muscle cell membrane potential. *Surgery.* 1977;81:571.

101. Trunkey DD, Illner H, Wagner IY, Shires GT. The effect of septic shock on skeletal muscle action potentials in the primate. *Surgery* 1979;85:638.

102. Illner HP, Shires GT. Membrane defect and energy status of rabbit skeletal muscle cells in sepsis and septic shock. *Arch Surg.* 1981,116:1302.

103. Sayeed MM. Alterations in hepatic sodium–potassium transport during endotoxemia in rats. *Am J Physiol.* 1984;247:R465.

104. Claret M, Mazet JL. Ionic fluxes and permeabilities of cell membrane in rat liver. *J Physiol.* 1972;223.279.

105. Katz B. *Nerve, Muscle, and Synapse.* New York: McGraw-Hill; 1966.

106. Williams JA, Withrow CD, Woodbury DM. Effect of ouabain and diphenylhydantoin on transmembrane potentials, intracellular electrolytes and cell pH of rat muscle and liver. *J Physiol* 1971;212:101.

107. Lambotte L. Effect of anoxia and ATP depletion on the membrane potential and permeability of dog liver. *J Physiol.* 1977;269:53.

108. Freedman N, Dambach G. Antagonistic effect of insulin on glucagon-evoked hyperpolarization: a correlation between changes in membrane potential and gluconeogenesis. *Biochim Biophys Acta.* 1980;596:180.

109. Haylett DG. Effect of sypathomimetic amines on ^{45}Ca efflux from liver slices. *Br J Pharmacol.* 1976;57:158.

110. Jenkinson DH, Koller K. Interaction between the effects of alpha- and beta-adrenoreceptor agonists and adenine nucleotides on membrane potential of cells in guinea pig liver slices. *Br J Pharmacol.* 1977;59:163.

111. Peterson OH. The effect of glucagon on liver cell membrane potential. *J Physiol.* 1974;239:647.

112. Lambotte L. Effect of activation of alpha- and beta-adrenergic receptor on hepatic cell membrane potential in perfused dog liver. *J Physiol.* 1973;323:181.

113. Tolbert MEM, Fain JN. Studies in the regulation of gluconeogenesis in isolated rat liver. *J Biol Chem.* 1974;249:1162.

114. Sayeed MM, Baue AE. Mitochondrial metabolism of succinate, beta-hydroxybutyrate, and alpha-ketoglutarate in hemorrhagic shock. *Am J Physiol.* 1971;220:1275.

115. VanRossum GDV. The relationship of sodium and potassium ion transport to respiration and adenine nucleotide content of liver slices treated with inhibitors of respiration. *Biochem J.* 1972;129:427.

116. Goldstein I. Lysosomes and their relation to the cell in shock. In: *The Cell in Shock.* Kalamazoo, Mich.: A Scope publication of the UpJohn Co.; 1974:30.

117. Morrison DC. Endotoxin-cell membrane interactions leading to transmembrane signaling. In: Inman D, Mandy JM, eds. *Current Topics in Molecular Immunology.* New York: Plenum; 1981:187.

118. Majde JA, Person RJ. *Pathophysiological Effects of Endotoxin at the Cellular Level.* New York: Alan R Liss; 1981.

119. Morrison DC, Ulevitch RJ. The effects of bacterial endotoxins on host mediation systems. *Am J Pathol.* 1978;93:527.

120. Conner J, Fine J, Kusano K, Panas MJT, Prosser CL. Potentiation by endotoxin of responses associated with increased in calcium conductance. *Proc Natl Acad Sci.* 1973;70:3301.

121. Hess ML, Briggs FN. The effect of Gram-negative endotoxin in the calcium uptake activity of sarcoplasmic reticulum. *Biochem Biophys Res Comm.* 1971;45:917.

122. Hess ML, Soulsby ME, Davis JA, Briggs FN. The influence of venous return on cardiac mechanical and sarcoplasmic reticulum function during endotoxemia. *Circ Shock.* 1977;4:143.

123. Nelson KM, Spitzer JA. Alteration of adipocyte calcium homeostasis by *Escherichia Coli* endotoxin. *Am J Physiol.* 1985;248:R331.

124. Sayeed, MM. Alterations in cellular calcium regulation in the liver in endotoxic shock. *Am J Physiol.* 1986;250:R884.

125. Claret-Berthon B, Claret M, Mazet JL. Fluxes and distribution of calcium in rat liver cells: kinetic analysis and identification of pools. *J Physiol.* 1977;272:259.

126. Giese A. *Cell Physiology.* Philadelphia: WB Saunder; 1968:247.

127. Lotersztajn S, Hanoune J, Pecker E. A high affinity calcium-stimulated magnesium-dependent ATPase in rat liver plasma membrane. *J Biol Chem.* 1981;256:11209.

128. Kraus-Friedmann N, Biber H, Murer H, Carafoli E. Calcium uptake in isolated hepatic plasma-membrane vesicles. *Eur J Biochem.* 1982;129:7.

129. Burgess GM, McKinney JS, Fabiato A, Leslie BA, Putney JW Jr. Calcium pools in saponin-permeabilized guinea pig hepatocytes. *J Biol Chem.* 1983;258:15336.

130. Murphy E, Coll K, Rich TL, Williamson JR. Hormonal effects on calcium homeostasis in isolated hepatocytes. *J Biol Chem.* 1980;255:6600.

131. Chen JL, Babcock DF, Lardy HA. Norepinephrine, vasopressin, glucagon, and A23187 induce efflux of calcium from an exchangeable pool in rat hepatocytes. *Proc Natl Acad Sci USA.* 1978;75:2234.

132. Kimura S, Kugai N, Tado R, Kojma I, Abe K, Ogata E. Sources of calcium mobilized by alpha-adrenergic stimulation in perfused rat liver. *Horm Metabol Res.* 1982;14:133.

133. Exton JH. Mechanisms involved in alpha-adrenergic phenomenon: role of calcium ions in actions of catecholamines in liver and other tissue. *Am J Physiol.* 1980;238:E5.

134. Jones SB, Romano J. Plasma catecholamines in the conscious rat during endotoxicosis. *Circ Shock.* 1984;14:189.

135. Nayler WA, Poole-Wilson DA, Williams A. Hypoxia and calcium. *J Mol Cell Cardiol.* 1979;11:683.

136. Hulsmann WC, Lamers JMJ, Stam H, Breeman WAP. Calcium overload in endotoxemia. *Life Sci.* 1981;29:1009.

137. Schane FAX, Kane AB, Young EE, Farber JL. Calcium dependence of toxic cell death: a final common pathway. *Science.* 1979;206:700.

3

Mechanisms of Complement-Mediated Membrane Damage

John C. Papadimitriou
Moon L. Shin
Getrud M. Hänsch

INTRODUCTION

The discovery of the bactericidal potency of cell-free serum by Buchner in 1889 initiated the studies of the complement system.[1,2] The attempts to identify the factors responsible for the killing of bacteria have contributed to the development of *in vitro* model systems utilizing targets such as erythrocytes, tumor cells, or artificial membranes. With the use of cytolytic assays that are endpoint measurements of the complement action, cytolytic assays of the complement proteins, their interactions with target membranes, and their regulation during activation have been uncovered.

The complement system consists of 14 proteins, excluding inhibitors and regulatory enzymes. These proteins are found in blood or other body fluids as precursor forms. When the complement system is activated, it serves as a nonspecific effector system with various biological functions. In the inflammatory process, peptides that have been cleaved from complement proteins during the activation participate as chemical mediators by increasing vascular permeability, by attracting leucocytes, and by promoting phagocytosis of invading particles such as bacteria.[3] The production of these biologically important mediators and

the generation of the cytolytic activities require activation of the complement system. The activation of late complement proteins (C5 through C9) results in the formation of a heteropolymeric complex inserted into the target cell membrane, causing lysis. This chapter is concerned with this membrane attack activity of the complement system.

ACTIVATION OF THE COMPLEMENT SYSTEM

The complement proteins can be activated by two routes: the classical and the alternative pathways (see Figure 3-1). The classical pathway is comprised of the proteins Ca (made up of three subunits, Cl1, Clr, and Cls), C4, C2, and C3 (this listing corresponds to the reaction sequence); the alternative pathway proteins are C3, B, D, and P.

The classical pathway is activated when globular regions of Clq combine with Fc_2 domains of the antibody in an immune complex.[4] Clq can also be activated by choline-bound C-reactive protein, certain enveloped viruses, cardiolipin-a mitochondrial lipid, and central nerve myelin.[5-10] The mechanism of Clq activation is not well understood, but it results in the cleavage of Cls,

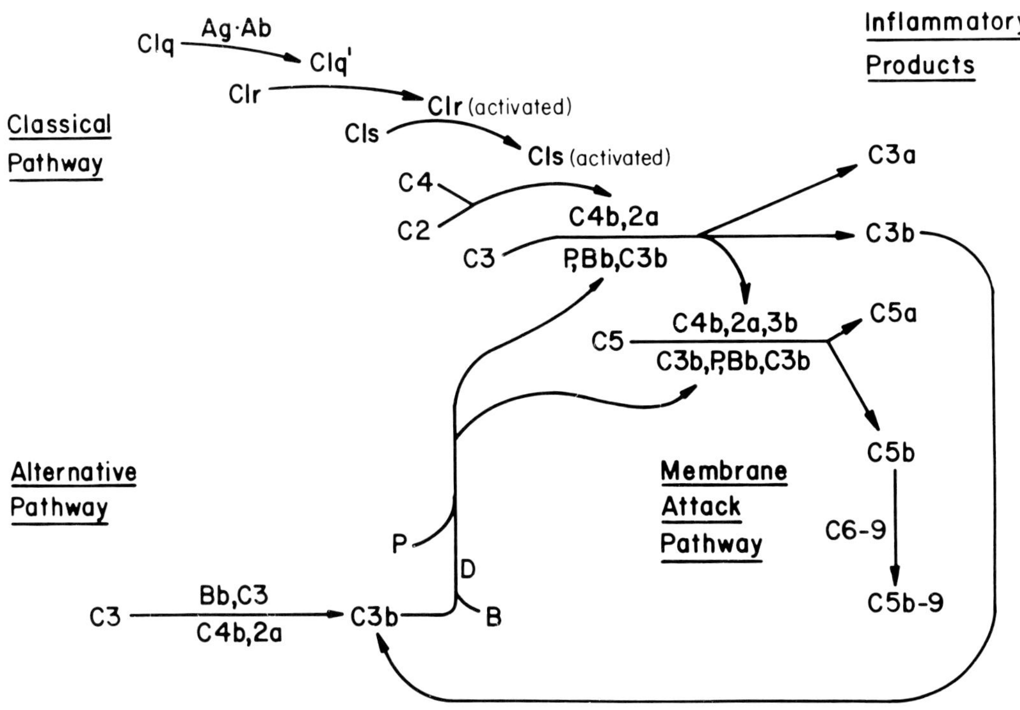

Figure 3-1. Schematic presentation of the classical and alternative pathways of the complement activation.

generating the Cls enzyme. The Cls enzyme then activates C4 and C2 by cleavage, forming the C4b,2a enzyme complex, the so-called C3 convertase. This C3 convertase, bound to the cell membrane or an immune complex, in turn cleaves C3 into C3a and C3b. The catalytic site is on C2a, while C4b supplies the binding site for C3. When C3 is activated, a C5 cleaving enzyme, C4b2a3b, is generated.[11] Both C4b and C3b bind convalently either to hydroxyl or amino groups on the surface of membrane or immune complexes.[12-15] C3b provides a binding site for C5, while C2a cleaves C5.[11] Thus, the activation reactions of Clo through C5 involve the cleavage of peptide bonds in precursor molecules. By contrast, the reactions of C6 through C9 do not appear to require peptide cleavage.

Only certain classes and subclasses of antibody can activate the classical pathway. This restriction may depend upon the ability of their Fc segments to interact with Clq. The juxtaposition of two Fc segments is nec-

essary for activation to ensue. Therefore, only one IgM molecule bound to a cell surface is necessary to activate complement, whereas two IgG molecules critically close to one another are necessary to cause the same reaction.[12] Thus, formation of IgG doublets is influenced by the antigen density on the cell surface and also by factors that affect the mobility of antigens within the plane of the cell membrane.[16]

The alternative pathway is activated when factor D, present in the blood as an active enzyme, cleaves factor B in a C3,B or a C3b,B complex, thus generating a C3 cleaving enzyme, C3Bb or C3bBb, respectively. These enzyme complexes are subject to spontaneous dissociation and can be inactivated by control proteins H and I. However, preperdin (P) can add stability by binding to C3bBb to form the trimolecular enzyme complex C3bBbP. The catalytic site is on B, while C3b binds covalently to the microbial surface or immune complex.[17]

The alternative pathway is activated efficiently by a variety of substances such as microbes or microbial products, certain mammalian cells, aggregated IgA, and peripheral nerve myelin, to list a few.[18-22] Sheep erythrocytes activate the alternative pathway after removal of sialic acid from their cell surface.[23] The presence of sialic acid appears to be important for the proper function of factor H, by permitting C3 convertase inactivation by H and I. The C3 convertase, generated by either the classical or alternative pathway activation, can also cleave C5 when in the presence of an additional C3b serving as a binding site for C5.[24]

Once C5 is cleaved into C5a and C5b, the subsequent activation of C6 to C9 runs its course. The five proteins—C5, C6, C7, C8, and C9—belong to the final common pathway that mediates the cytolytic function of complement by attacking the membrane lipid bilayers of targets such as bacteria, protozoa, fungi, and cells of higher organisms. In addition to the membrane attack action, activation of the complement system, via either the classical or the alternative pathway, generates fragments of C2, C3, and C5, which mediate a variety of biological activities, notably opsonization (C3b, C3bi, C4b), chemotaxis (C5a), anaphylatoxin activities (C3a and C5a), kininlike activities (C4a, C2b), activation of both B-lymphocytes and macrophages (C3b), release of arachidonic acid–derived inflammatory mediators (C3a, C3b, C5b), and regulation of T- and B-cell activation (C3a, C5a).[25]

CYTOTOXIC ACTIVITIES OF COMPLEMENT

Activation of the Late Components, C5 through C9

The activation of late components C5 through C9 is initiated by the cleavage of C5 to generate an "active form" of C5, designated C5b. C5b can bind to C6 to form a stable C5b6 complex that can react reversibly with the cell membrane.[26-28] Since C5b6 complexes can be isolated and stored, they are frequently used to initiate interaction with biological membranes, thus bypassing the earlier activation phase.[29,30] Functionally active C56 complexes can also be generated by pH-shift or by freezing and thawing a mixture of C5 and C6.[31-34]

Subsequent binding of C5b6 with C7 results in the formation of a metastable C5b-7 complex that inserts into the lipid bilayer of a target when the reaction occurs close to the surface of the membrane. In the fluid phase, the membrane reactive properties are readily lost due to self-aggregation or by reacting with serum inhibitors, in the form of S-protein or lipoproteins.[29,35] The binding of C8 to a membrane-associated C5b67 occurs via its beta-subunit of C8 (MW 64,000 daltons), which is noncovalently linked to the alpha–gamma portion of the C8.[36,37] The alpha-chain of C8 in a C5b-8 complex inserts into the lipid bilayer and provides the C9 binding site.[38]

Interaction of Components C5 through C9 with Target Membrane: The Doughnut Hypothesis

Following the generation of C5b6, the sequential binding of late components C7, C8, and C9 on the surface of the target membrane can cause the cytolysis of the target cell. The demonstration that the lipid bilayer of the cell membrane is the target of C5b-9 and the visualization of ultrastructural craterlike complement lesions led Mayer to introduce a theoretical model for the mechanism of C5b-9–mediated membrane damage.[39-43] This model was also based on the fluid mosaic principle of the cell membrane and the one-hit theory of complement lysis. The latter postulates that a single lesion at the site of fixation of the complement protein complex C5b-9 is responsible for the lysis of an erythrocyte. According to the proposed model, the complement lesion is a rigid, hollow structure, shaped like a doughnut. It transverses the lipid bilayer and has a hydrophobic exterior that faces the phospholipid bilayer of membrane, and a hollow hydrophilic core channel through which salt and water can exchange freely between the interior of the cell and the extracellular environment. It has been pointed out that the late-acting complement protein complex C5b-9 is the most probable source of the structural elements of such a transmembrane channel.

Exposure of Hydrophobic Domains in Complement Proteins

The transmembrane channel hypothesis involves the assumption that interactions among the late-acting complement proteins lead to the exposure of hydrophobic peptides that can be inserted into the lipid bilayer. Through this reasoning, renewed attention was drawn to several prior publications that demonstrated phospholipid release from bacteria, liposomes, erythrocytes, and guinea pig hepatoma cells following complement attack.[41,44-48] Quantitative studies by Shin and co-workers showed that approximately one-third of the phosphatidyl-choline release from liposomes occurred after C5b67 formation, and the remainder on reaction of C8 and C9 with C5b67; no lipid release was observed at C5b6 stage (Figure 3-2).[41] The mechanism by which phospholipids are released is not yet clear. A displacement of the membrane lipids occurring as a consequence of C5b-9 insertion into the membrane is improbable since only selected phospholipids, but not cholesterol,

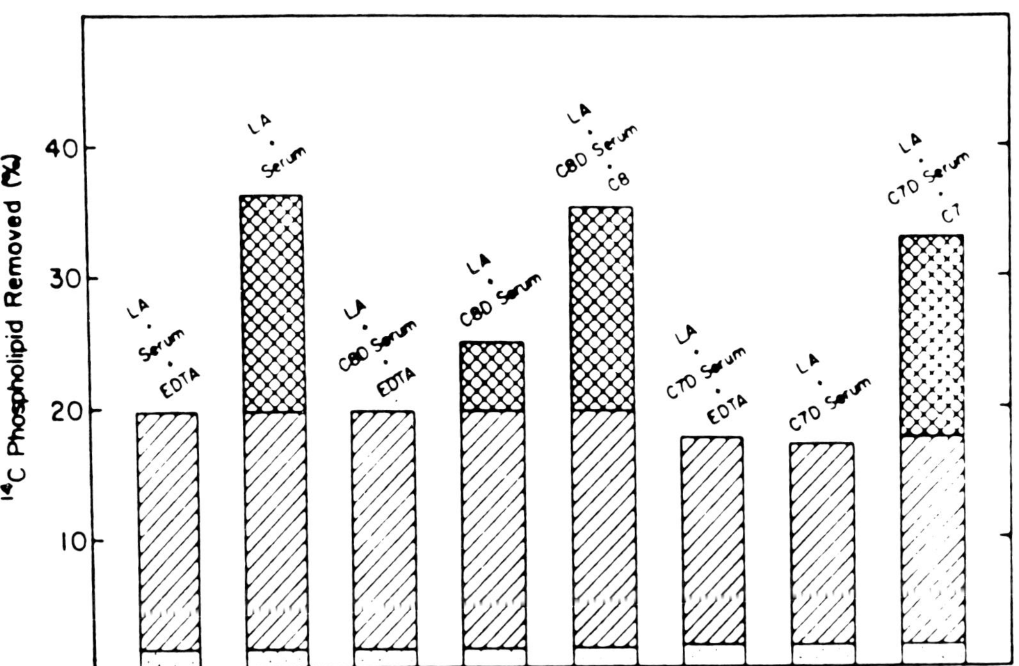

Figure 3-2. Studies of ^{14}C-phospholipid removal from antibody-coated liposomes (LA) by C7-deficient (C7D) and C8-deficient (C8D) human sera. The dotted areas represent release of ^{14}C-PC from LA incubated with buffer alone. The striped areas represent controls in whch LA were treated with normal human serum + EDTA, C8D serum + EDTA or C7D serum + EDTA (EDTA prevents complement activation). The degree of marker release by LA + C7D serum, also shown as a striped bar, is equal to those of the controls, indicating no ^{14}C-PC release attributable to activation of the complement sequence through C6. The cross-hatched areas represent the release of marker attributable to C activation by LA + normal human serum, by LA + purified C8 or by LA + C7D human serum + purified C7. (From Shin et al., 1977.)[41]

were released.[41,49] The binding of C5b-9 to phospholipids leading to the release of C5b-9 phospholipid complex as comicelle appears to be the most probable cause of the release. Such structures have been considered for the interaction of the hydrophobic segment of the membrane protein, cytochrome-b5, with phosphatidylcholine liposomes.[50] The exposure of hydrophobic domains on activated C5b-9 was also studied by quantitative immunoelectrophoresis, demonstrating the binding of radiolabeled Triton X-100 to C5b-9, and by direct measurement of ^{14}C-phospholipid bound to the terminal complement intermediates.[29,51,52] This lipid-binding property of the membrane attack complexes acquired during sequential activating stages was also associated with a transformation of the complement proteins from hydrophilic to an amphiphilic state, which can be demonstrated by methods such as charge-shift electrophoresis.[51]

The most direct evidence for the insertion of complement peptides C5 through C9 comes from experiments by Hu and co-workers.[28] By utilizing a photoreactive glycolipid compound, it was demonstrated that all the complement proteins of the C5b-9 complex are sequentially inserted into the bilayer of liposomes during the formation of functional channel (Figure 3-3).

The Structure of the Complement Lesion

It is important to notice that the typical ultrastructure of C5b-9 complexes failed to appear on erythrocyte membranes when the C9:C5b-8 ratio was less than G (Figure 3-4a).[53] These data, together with the observation that C9 has the capacity to polymerize spontaneously and to form structures morphologically identical to the ultrastructure of C5b-9, which is formed with excess C9 (Figure 3-4b), led to the hypothesis that tubular poly-C9 within and a C5b-9 complex constitute the structural bases of C5b-9

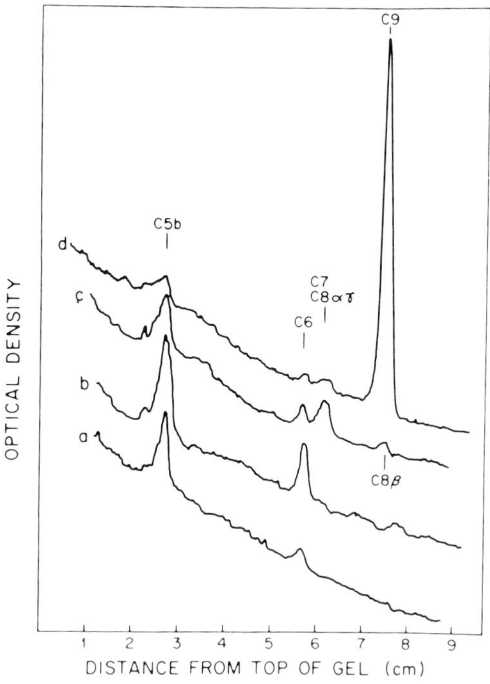

Figure 3-3. Photolabeling of C components as a function of complex assembly. Lipid vesicles containing ^{14}C-labeled, photoreactive glycolipid probes (12 APS-GlcN) were treated with (a) C5b,6, (b) C5b-7, (c) C5b-8, or (d) C5b-9. The samples were irradiated (360 nmol, 2 minutes), the proteins were separated on SDS–PAGE. The densitometric scans of fluorogranes of the gel shown here indicate the interaction of C proteins with the probe in hydrophobic milieu. (From Hu et al., 1981.)[28]

channels.[54,55] However, direct participation of C8, and possibly of C5b-C7, is also likely, because C8 alpha-chain is found in the lipid domain of membrane-bound C5b-9 and functional channels can be formed with as little as one molecule of C9 in a C5b-9 complex, and apparently without C9.[56,57]

In pioneering electron microscopic studies, Humphrey and Dourmashkin have shown that complement lesions formed on target erythrocytes as a characteristic crater-like lesion of an outer diameter of 20 nm, with an inner crater, filled with staining material, of approximately 10 nm in diameter.[42] A numerical correlation was found between the number of "rings" and the predictable number of functional lesions.

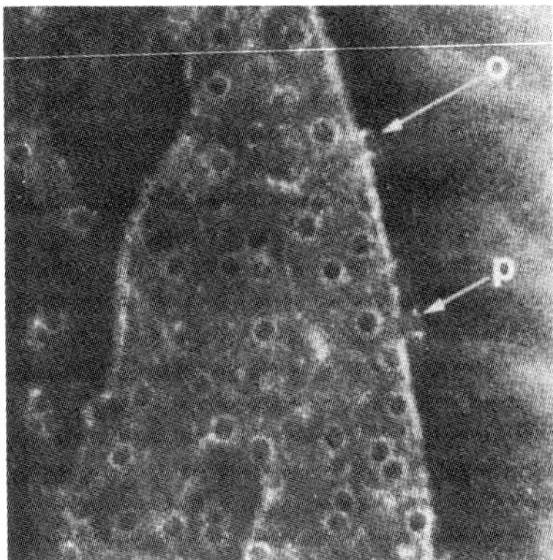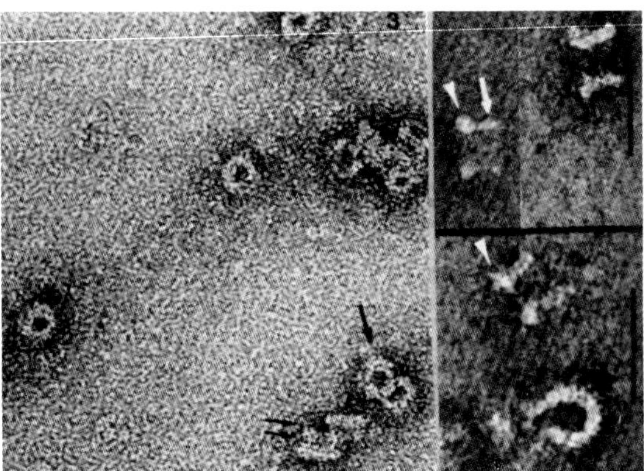

Figure 3-4. Ultrastructure of complement lesions. (*a*) Sheep erythrocytes were lysed with antibody and complement. Complement lesions in membrane-ghost stained with phosphotungistic acid are seen in side projections (arrow) as well as in top view. Scale bars indicate 100 μm. (From Bhakdi S, Tranum-Jensen J. *J Immunol.* 1984;133:1453.)[53] (*b*) Ultrastructure of poly-C9. (From Podack ER, Tschopp J. *Proc Natl Acad Sci USA.* 1982;79:574–578.)

When assembled by adding C5b6, C7, C8, and C9 sequentially, the typical lesion was seen only after the addition of C9.[42] In profile view, the ringlike structures looked like funnels with a hollow core.[58,59]

Further studies of these complement lesions, subunit analyses, and functional and morphological studies following reincorporation into the liposomes, showed that they are made of one molecule of C5b, C6, C7, and C8, and six or more of C9. The lesion appeared as a hollow, thin-walled cylinder of 10-nm internal diameter and 15-nm height, and with an anulus of from 3 to 5 nm. Subsequent studies on the three-dimensional structure of C5b-9 in lipid vesicles gave essentially similar results, conclusively showing centrally located hydrophilic passage.[55]

Function of C5b-9 Channels

Besides the morphological evaluation of the C5b-9 complexes in the membrane, extensive functional studies have been performed to characterize the complement-mediated membrane damage. In earlier work, planar lipid bilayers (BLM) were used since they possess many characteristics that are similar to those of the biological membranes and because kinetics and quantitative changes in membrane permeability to ions can be measured electrically.

Increased ion flow across BLM was observed initially by Barfort and co-workers after treatment of BLM with antigens, antibodies, and fresh serum as a source of complement.[60] Detailed experiments concerning the assembly and properties of the complement channels were studied by Michaels and co-workers with isolated complement proteins.[61,62] Sequential treatment of 60-Å thick BLM with C5b6,C7 produced no conductance changes, whereas further addition of C8 produced a modest increase, suggesting that the transmembrane channels are already formed on C5b6-8 stage. Subsequent addition of C9 greatly amplified the ion flux (see Figure 3-5).

Additional observations were that C5b,6 complexes have the capacity of producing conductance increments in membranes up to 40-Å thick and conversion to the C5b6,7 complex can also produce increased conductance in membranes up to 45-Å thick.[62] These observations suggest progressively deeper insertion of complement components during sequential assembly.

The earliest information on measurement of complement channels in mammalian cells emerged from Donnan effect blocking

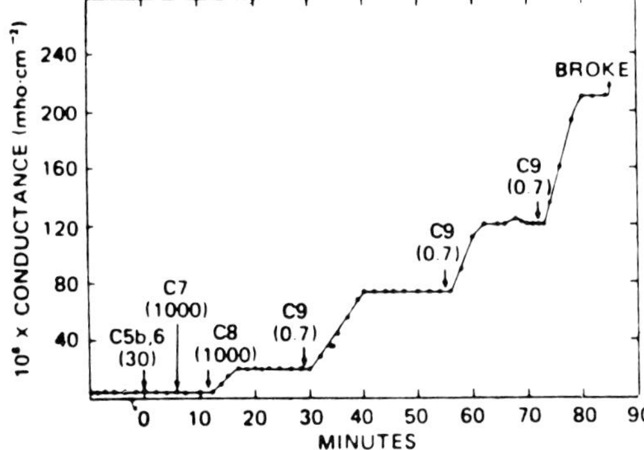

Figure 3-5. Sequential action of complement proteins on lecithin BLM. The aqueous phase contained 150 mmol KCl and 5 mmol histidine buffered at pH 7.0. The BLM was formed from egg lecithin in n-decane (membrane thickness = 6.0 nm). The vertical arrows mark the time of each protein added. The total units of activity added in each instance are given in parentheses. The abscissa is elapsed time in minutes; the ordinate is the specific membrane conductance. (From Michaels et al., 1976).[61]

experiments by Green and co-workers in 1959 and by Sears and co-workers in 1964. These suggested that the effective diameter of the complement channel is not less than 40 Å.[63,64] Molecular sieving experiments using resealed erythrocyte ghosts carrying C5b-9 with maximum number of C9 have further refined this estimate between the range of 55 Å and 100 Å.[48,65]

Soon, it became evident that the 55-Å channel might not be the only kind of channel formed by C5b-9. In measuring electrical conductance of a single complement channel in BLM, coexistence of channels varying in size was observed, indicating a functional heterogeneity of C5b-9, a fact also confirmed in erythrocytes.[57,66-68] The increase in channel size in relation to the C multiplicity has been elegantly shown by methodical kinetic and endpoint analysis of double-marker sieving experiments from resealed erythrocyte ghosts with C9 titration.[57] A size heterogeneity of C5b-9 complexes is further supported by data on C5b-9 extracted from complement-lysed membranes in which C5b-9 complexes of various molecular sizes, ranging from 23S to 34S, were

identified.[69] Thus, it appears that C9-multiplicity is the cause of channel-size heterogeneity.

With regard to the role of C9 in lysing the target cell, an observation of special interest was the formation of poly-C9 channels that were morphologically indistinguishable from C5b-9 complexes, as discussed earlier. This tubular poly-C9 is SDS (sodium dodecyl sulfate-detergent)-resistant, whereas multiple C9 of less than six monomers existing in C5b-9 is SDS-susceptible.[53-55]

The C9 polymerization does not appear to represent an experimental artifact. The role of C9 polymerization by C5b-6 is about 10,000-fold greater than spontaneous poly-C9 formation.[70] These findings provided a new hypothesis regarding channel structure: That a polymerized form of C9 forms the transmembrane channel, with C5b-8 functioning as the polymerizing agent.[54] However, the formation of lytic channels in the absence of SDS-resistant poly-C9 and the evidence of small channels formed with one or few C9 indicate that tubular poly-C9 is not the only functional unit of C5b-9. The fact that C9 cleaved by alpha-thrombin can lyse cells while failing to polymerize also supports the hypothesis that formation of poly-C9 is not necessary to provide a functional channel (Figure 3-6).[71]

Recently, the presence of C5, C6, C7, and C8 in a complex containing SDS-resistant poly-C9 in a 2.5% agarose gel further strengthened the concept that the functional and chemical unit of membrane attack complex could be a heteropolymer consisting of one molecule of C5b, C6, C7, C8 and one or more (up to 18) molecules of C9.[72] Other reports have suggested C9 numbers up to 21 molecules.[54,55]

Factors that Modulate the Efficiency of Complement-mediated Membrane Damage

Complement-mediated lysis by C5b-9 is a multiple-step event that primarily occurs on and in the membrane. Thus, the binding of

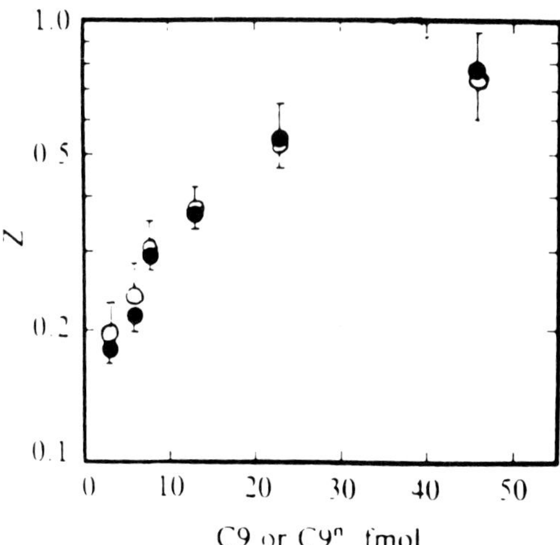

Figure 3-6. Hemolytic activity of C9 and α-thrombin treated C9. EACl-8 cells with varying concentrations of human C9 or alpha-thrombin treated C9 (C9ⁿ). Z plot of the hemolytic activities C9 of C9ʰ as measured under single-hit conditions is given (from Dankert et al., 1985).[71]

activated components to the membrane and the subsequent insertion and channel formation by C proteins could be greatly influenced by properties of the membrane surface as well as of the lipid bilayer *per se.*

A variety of substances could affect the binding of C5b67 to the membrane. Besides the specific interaction of S-protein and LDL with C5b-7, other substances such as polyanions, polycations, myelin basic protein, or lysin-rich histones can interfere with the binding of C5b67 to the target membranes.[29,35,73–75]

Treatment with proteases or neuraminidase also enhances the susceptibility of erythrocytes or nucleated cells to C5b-9 attack.[76–78] It is not clear whether this lytic enhancement is due to removal of the surface charges. However, such an effect would be in line with the fact that binding of C5b6 to the target cell is greatly dependent on the ionic strength of the medium.

The efficiency of insertion and/or channel formation could also be affected by the lipid properties of the membrane. As shown by Shin and co-workers, the marker release by C5b-9 or C5b-8 from liposomes decreased progressively with increasing acyl-chain length as well as with the degree of saturation of the fatty acids.[79] Furthermore, the membrane cholesterol concentration affects marker release considerably. With increasing cholesterol concentration, a decrease in marker release was observed.[80–82] By increasing the membrane fluidity as it is achieved by altering the ratio of lecithin to sphingomyelin of the erythrocyte membrane, the lysis by C5b-9 is increased in tandem with membrane fluidity.[82] However, membrane fluidity is not the only factor that influences the lytic efficiency of complement attack. Studies with *Acholeplasma laidlawii* demonstrated that organisms incorporated with unsaturated phospholipids had a higher membrane fluidity but were less susceptible to complement attack than control organisms.[83] Thus, besides the overall fluidity of the membrane, focal changes in membrane lipid disorder might also alter the efficiency of complement attack. Incorporation of A2C, a fluidizing and fusogenic agent, into membranes increased their susceptibility to complement attack while affecting the overall fluidity of the membrane only slightly.[84] A similar effect was seen by incorporation of myristoleyl alcohol, the cis-isomer of a C14:1 aliphatic alcohol. Both A2C and myristoleyl alcohol have a pronounced bend in the middle of the acyl chain that could affect the acyl-chain packing, which in turn could promote the lysis.[84–86] In contrast, agents that tighten the acyl-chain packing should decrease the susceptibility toward the complement attack. The inhibitory effect of cholesterol probably can be attributed to that effect.

Effect of Complement Species on Lytic Efficiency

The efficiency of erythrocyte lysis varies greatly with the species of complement.[87–89] Thus, lysis by C5b678 formed on sheep erythrocytes is higher with guinea pig C9 than with human C9.[90] On the other hand, guinea pig C8/C9 is inefficient in lysing human erythrocytes.[91,92] Comparing the lysis of erythrocytes of five different species by C5b6,7 and C8/C9 from various species, it was found that lysis was least efficient when C8/C9 were homologous with regard to the target cell.[93] The different efficiency was due neither to an incompatible interaction between the late components nor to an inefficient binding on C5b67, but rather to a reduced insertion of homologous C9.[32,88,90,94,95]

Factors regulating the lysis in a homologous system have recently been identified. An integral membrane protein isolated from human erythrocytes and called "Decay Accelerating Factor" (DAF) by Nicholson-Weller and co-workers appears to be responsible for the increased sensitivity of erythrocytes to lysis by homologous C.[96,97] This effect is demonstrated in the case of patients with paroxysmal nocturnal hemoglobinuria (PNH), whose erythrocytes are devoid of DAF.[98,99] Decay Accelerating Factor primarily modulates the C3/C5 convertase formation on homologous erythrocytes, thus reducing the C3 and C5 cleavage; it has no effect on lysis of E (erythrocytes) by C5b-9 or on C5b67 reaction with C8/C9.[100,101] Thus, an additional defect existing in these PNH–E responsible for the observed enhanced lysis by C5b-9 has to exist. Recently, another membrane protein in human erythrocytes, distinct from DAF, was found to have the capacity to bind human C8, thus inhibiting the lysis by human C5b-9 and preventing C9 polymerization.[101,102] This so-called C8-binding protein is also absent in erythrocytes of patients suffering from PNH.[103] The PNH-erythrocytes are highly susceptible to lysis by human C5b-9, which can be inhibited by C8-binding protein.[103]

The biological importance of such species restriction is evident. It minimizes self-inflicted cell damage by the membrane attack component during a specific or nonspecific inflammatory process.

THE COMPLEMENT-MEDIATED KILLING OF NUCLEATED CELLS (NC)

Most of the data on the interaction of C5b-9 with membranes are derived from studies employing erythrocytes or artificial bilayers as targets. However, complement-mediated killing of bacteria, protozoa, and nucleated cells (NC) such as tumor cells is of special interest. Although the complement-activation mechanism is similar, the killing processes could differ significantly since these targets are metabolically active and possibly able to defend themselves by complex cellular mechanisms, including active membrane mobilization.

The observation that the extent of lysis of NC by C did not necessarily correlate with the extent of C activation or fixation prompted the investigation of various parameters influencing the ability of C5b-9 to lyse NC.[104–107] One extensive series of studies examined the effect of several metabolic modulators on the lytic efficient of C.[49,108–120] Metabolic inhibitors such as puromycin and actinomycin D, or these agents plus mitomycin, adriamycin, 5-fluorouracil, and others, increased the sensitivity of certain tumor cells to lysis by C.[109,111–113] In contrast, an increase in resistance was observed after guinea pig hepatoma cells were treated with various anabolic hormones such as hydrocortisone, insulin, and epinephrine.[115,116] Although the data derived from studies with these metabolic modulating agents suggest that the resistance of NC to cytolysis by C can be associated with metabolic activities, it is difficult to ascertain specific site(s) of action since most of the agents used could affect many metabolic pathways and endproducts. However, in spite of this nonspecificity, a correlation has been made between the effect of these drugs on the resistance of cells to C-mediated cytolysis and lipid metabolism.[117,119,121,122] An important corollary to this association between lipid synthesis and resistance to lysis by C can be derived from the fact that modulation of the lipid composition of the membrane lipid bilayer will alter the susceptibility of liposomes and erythrocytes, microorganisms, and NC to C-mediated lysis, most likely by enhancing the efficiency of C attack due to modulation of C channel formation.[79,83,123,124]

It is important to note that lysis of cells by C can be separated into two distinct phases: The first phase pertains to the requirement for activation of C with subsequent formation of functional C channels in the plasma membrane (PM). The second phase involves the ability of C channels in the PM of NC to mediate cytolysis. There is reason to believe that C channels in the PM of NC, unlike erythrocytes, need not necessarily lead to cytolysis, and that NC are capable of surviving a limited C attack due to a metabolic defense response mechanism(s) that interferes with the ability of C channels in the PM to mediate cytolysis. The first direct evidence demonstrating that lysis of NC by C exhibits multiple-hit characteristics in the sense that NC remain viable in the presence of few channels was shown by Koski and co-workers.[125] In her experiments (Figure 3-7), erythrocytes and NC coated with excess antibody were exposed to excess C6-deficient rabbit serum reconstituted with varying dilutions of C6; cell lysis was measured at the kinetic endpoint. The results revealed that the NC remained viable in the presence of a limited number of $^{86}Rb^+$-releasing C channels and required multiple C channels for cytolysis, in contrast to the single channel requirements for erythrocytes.

For NC to survive in the presence of a limited number of C channels in the PM, C channels in the PM would have to be functionally destroyed or removed from the cell surface at a rate sufficient to overcome the lethal ion shifts mediated by C channels. The first direct evidence showing that C5b-9 channels were functionally unstable on NC was derived from intracellular marker release experiments in which the half-life span of the channels was found to be 1 minute at 37°C in contrast to the stable nature of C

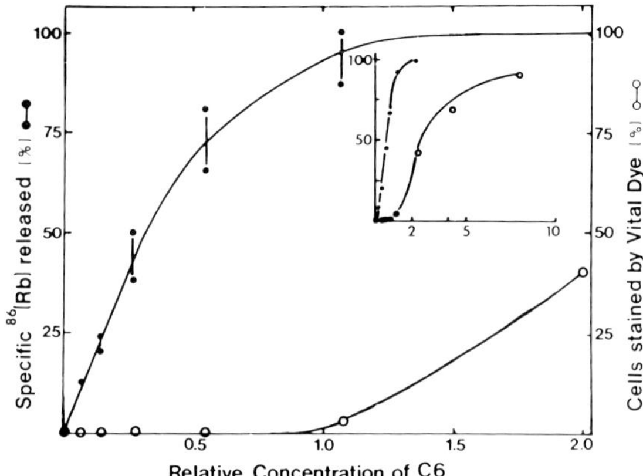

Figure 3-7. Multiple-channel requirement of NC death by C5b-9. Antibody-sensitized, ^{86}Rb-loaded Molt 4 cells at 37°C were treated with excess C6-deficient rabbit serum and limiting human C6. Cell death, measured by vital dye uptake, was measured at 90 minutes, and the marker release at 20 minutes. The cell death curve was sigmoidal, while the marker release was monotonic with respect to C6. (From Koski et al., 1983).[125]

channels on erythrocytes.[126] The half-life for C5b-9 on polymorphonuclear leukocytes, as determined by measuring radiolabeled anti-C9 binding to the remaining C5b-9 on the cell surface, was 3-1/2 minutes.[127]

In experiments by Carney and co-workers, the stability of various terminal C complexes (TCC) in the PM of Ehrlich cells was measured.[128] The disappearance was observed to be a temperature-dependent process, and the initial half-lives for the various terminal C complexes were 30 minutes for C5b-7, 20 minutes for C5b-8, and 10 minutes for C5b-8 containing limited sublytic doses of C9 (C5b-8,9^{lim}) (Figure 3-8). Thus, it is evident that the functional properties of the TCC, especially channel size, might be responsible for differences in the rate of disappearance since TCC known to have larger pore sizes disappeared more rapidly than those with a smaller pore size. The mechanisms responsible for the different rates of disappearance for the various TCC are unknown. To study the effect on disappearance of ion flux, such as Ca^{2+}, through these complexes, the disappearance rate was

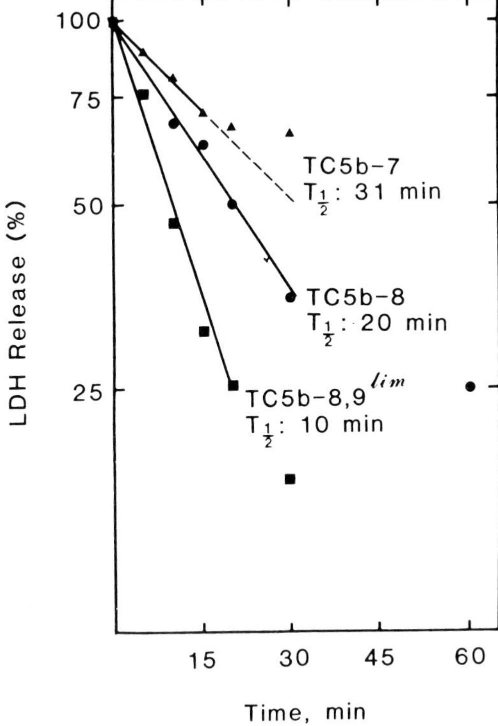

Figure 3-8. Disappearance rate of terminal C complexes in the plasma membrane of EATC. The results in Figure 3-4 were expressed on a semilog arithmic graph. The initial T 1/2 for C5b-7 complexes is 31 minutes, 20 minutes for C5b-8, and 10 minutes for C5b-8 in the presence of a sublytic number of C5b-9. (From Carney et al., 1985).[128]

measured in the presence of various concentrations of extracellular Ca^{2+}, $[Ca^{2+}]_o$.[129] It was found that the $[Ca^{2+}]_o$ influenced the rate of disappearance and the T 1/2 of C5b-8,9^{lim} was 5-1/2 minutes with 1.5 mmol Ca^{2+}, as compared with a T 1/2 of 10 minutes in the presence of 0.15 mmol Ca^{2+} (Table 3-1).

Immunoelectron microscopic tracing of C5b-8 in NC showed that they are rapidly endocytosed by "coated vesicles" and eventually found in multivesicular bodies of the cells (see Figure 3-9).[129] Other authors have reported TCC elimination from the PM by membrane shedding or by a combination of endocytosis and exocytosis.[130,131]

The rapid removal of a limited number of C5b-9 from the surface of NC, such as Ehrlich cells and neutrophils, is stimulated by the Ca^{2+} signal generated by these complexes.[129,132] A weaker Ca^{2+} signal is generated by C5b-8 complexes,[129] through the smaller pore size and less stable channels created by these latter complexes.[60,133] This signal is believed to be responsible, at least in part, for the C5b-8 elimination.[129] The mechanism by which C5b-7 are eliminated is not clear because these complexes do not exist in the form of a channel nor do they generate a Ca^{2+} signal.[134] Recently, Carney and co-workers investigated signal-messenger generation by TCC in addition to Ca^{2+}, as well as the differential effect of Ca^{2+} on signal messengers involved in regulating the elimination process in Ehrlich cells.[135] Both C5b-8 and Cb5-9 were shown to stimulate protein kinase C (PKC) activity that was

TABLE 3-1

Effect of $[Ca^{2+}]_o$ on the Initial T 1/2 (minutes) for TCC on Line-2 Ehrlich Cells^a

TCC	Assay	$[Ca^{2+}]_o$ 1.5 mmol	0.15 mmol	0.015 mmol	Relative Increase
C5b-8	Functional	—	6.0	14.1	2.35
	FACS	—	12.5	26.2	2.10
C5b-8-9^{lim}	Functional	0.97	1.0	—	1.96
	FACS	1.3	2.7	—	2.07

^{a}Results represent the average values derived by analysis of at least three time-points run in duplicate or triplicate from at least two experiments.

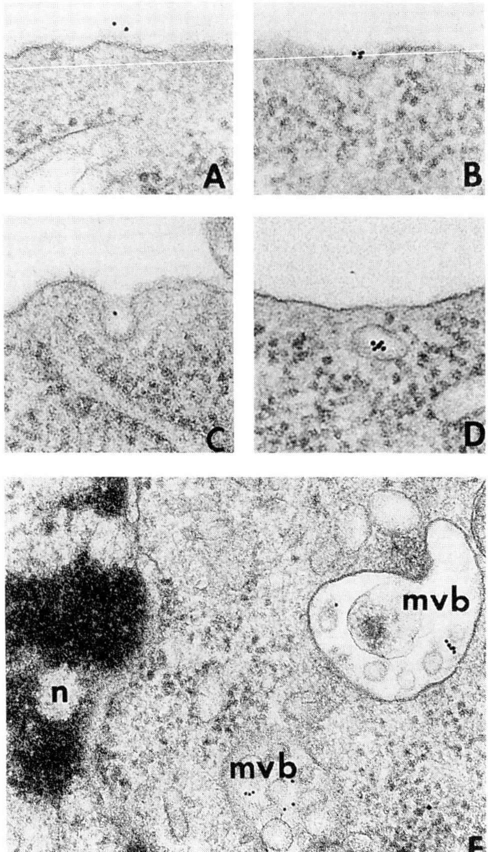

Figure 3-9. The fate of C5b-8 complexes on the cell surface of EATC studied by immunoelectron microscopy. TAC8 treated with goat anti-C5 IgG at 0°C were incubated with colloidal gold particles conjugated to rabbit antigoat IgG at 0°C. Cells were fixed immediately or after incubation at 37°C for 7 minutes or 30 minutes. Colloidal gold particles were exclusively associated with the membrane surface when the cells were maintained at (a) 0°C, and (b and c) were frequently seen in areas of membrane invagination, or (d) occasionally in intracellular vesicles near the cell surface after 7 minutes at 37°C. After incubation of 37°C for 30 minutes, colloidal gold particles were (e) most frequently seen in intracellular multivesicular bodies (n, nucleus; mvb, multivesicular body). (From Carney et al., 1985.)[128]

only partially mediated by the Ca^{2+} signal generated by these complexes. The PKC activity appears to be mediating TCC elimination since H-7—a PKC inhibitor—partially inhibited this process. C5b-7 and Cb5-8 were shown to cause generation of

cAMP. Unlike PKC activation, however, this messenger (cAMP) did not stimulate the elimination process.

OTHER BIOLOGICAL ACTIVITIES OF THE LATE COMPLEMENT COMPONENTS

In numerous studies, the effect of complement on membrane lipid hydrolysis has been described.[136-138] While some of the catabolic events might be related to repair mechanisms of NC, other events appear to be related to destructive processes. The phospholipid degradation accompanied by the generation of lysolecithin has been suggested as a destructive process in the cell.[139]

In addition, Sandberg and associates described a complement-dependent stimulation of prostaglandin synthesis with resultant bone resorption that required activation of late C components.[140] Subsequent studies by others have shown conclusively that mobilization of arachidonic acid (AA) and conversion into various metabolic intermediates occur following C5b-8 and/or C5b-9 assembly on a target cell surface.[141-144]

The mechanism of AA release through C5b-9 has been investigated in oligodendrocytes. It appears that a Ca^{2+}-dependent phospholipase A_2 activation is involved, which is further regulated by protein kinase mechanisms.[145] Other authors have attributed AA release to inhibition of its reacylation to lysolecithin by C5b-9.[146]

Protein kinase C, an enzyme that is activated by the receptor-mediated hydrolysis of inositol phospholipids, relays information from a variety of extracellular signals across the membrane.[147] It has been shown to be activated by sublytic TCC in the PM of Ehrlich cells and platelets.[135,148] Production of cyclic AMP (cAMP), another important intracellular messenger, was also found to be stimulated by TCC.[148]

Other PM-associated enzyme systems are also affected by TCC. The reacylation of lysolecithin is inhibited, and the membrane lipid transmethylation is enhanced.[146,149] The former effect could contribute to AA release and, secondarily, to the inflammatory response, whereas the latter could be involved in the cellular defense against C. Additional manipulation of these enzyme systems by inhibitors increases the cell susceptibility to C by different mechanisms: Inhibition of the lysolecithin reacylation causes an impaired elimination of TCC from the cell surface, whereas inhibition of membrane lipid transmethylation causes a more efficient deposition of TCC on the cell surface.[150]

Other effects of TCC interaction with cell membranes include stimulation of production of oxygen radicals by neutrophils, activation of platelet prothrombinase activity, release of IL-1 from glomerular mesangial cells, and hydrolysis of myelin basic protein.[151-154]

Thus, a complex and dynamic interaction of TCC with the PM emerges from these studies. Even in erythrocytes, which were considered up to now to be passive C-targets, it was recently shown that exposure to low concentration of C induces a large but reversible increase in membrane permeability that does not lead to lysis.[155] Thus, erythrocytes could theoretically possess defense mechanisms against C, including elimination processes as has been shown in platelets, another nonnucleated cell.[130]

To add into the complexity of the PM–TCC interaction, it has been recently shown that TCC interacts with cytoskeletal components: The lateral diffusion coefficient of the bulk membrane proteins decreases with the assembly of TCC on the PM at low C doses, whereas at high C doses the mobility of membrane proteins increase.[156]

From this, it becomes evident that TCC can interact with cells in multiple ways, leading to multiple outcomes. The most crucial question is, of course, How does C kill the target cell? It is supposed that erythrocytes are killed through colloid osmotic deregulation.[63] On the other hand, NC appear to be killed by a mechanism(s) independent of cell swelling.[157] Different mechanisms could be operative, including unregulated Ca^{2+} influx and/or mitochondrial injury.

Some circumstances of cell death have been attributed to permanent impairment of calcium regulation.[158] Formation of C5b-8 or C5b-9 on target cells induces a rapid increase in cellular Ca^{2+}.[129,159] In evaluating the effect of extracellular calcium ($[Ca^{2+}]_o$) on cell death when the number of C5b-9 channels is not limiting, it was found that increased $[Ca^{2+}]_o$ is associated with increased rate of cell death.[160]

There are many theoretical reasons why calcium overload may be detrimental to cells. An increase in free cytosolic calcium could activate calcium-dependent phospholipases, resulting in hydrolysis of membrane lipid with generation of free fatty acids and lysophospholipids.[161] Elevations in cytosolic Ca^{2+} will also increase calcium ATPase activity, resulting in enhanced depletion of ATP. (For review see Cheung et al.[162]) Elevated $[Ca^{2+}]_i$ also causes uncoupling of oxidative phosphorylation in mitochondria, thus further compromising the energy reservoir of the cell.[163]

Complement attack induces lipid turnover in the target NC with stimulation of synthesis and release of specific lipids. (For review see Ohanian and Slager.[119]) The synthesis and release of cardiolipin (CL), a mitochondrial lipid,[164] appeared to be the major component of the released lipid.[49] Thus, enhanced intracellular lipid metabolism in addition to the hydrolysis of cytoplasmic membrane lipid is thought to be involved in humoral immune killing.[119] Also, ultrastructural changes of mitochondria, including swelling, have been observed in lymphocytes following C-attack.[165]

In preliminary studies, we have examined kinetically ATP levels in Ehrlich ascites tumor cells bearing lytic C5b-9 lesions in proportion to LDH release (the latter serves as a marker of cell death). A much faster increase in ATP depletion than LDH release was observed. This indicates a primary effect of the C-attack on the energy metabolism of the cell.

RECAPITULATION

The reaction of activated C5 (C5b), together with C6, C7, C8, and C9, results in exposure of hydrophobic domains on the complement proteins. If the reaction takes place in the immediate vicinity of an appropriate target membrane, insertion of the complement peptides into the membrane lipid bilayer may ensue.

Insertion of C peptides is followed by assembly of transmembrane channels made of C5b-8,9n complexes. The lipid-binding capacity of the amphiphilic complement protein also allows reorientation of the lipid bilayer.

Recently, great progress has been made toward classification of the precise molecular structures of C5b-9 channels with central pores made of C proteins. These channels are effective in lysing both erythrocytes and NC, but with marked differences among these cell types. The C5b-9 attack could also modulate the membrane-associated enzyme activities, possibly resulting in increased cell defense, cytotoxicity, or promotion of the inflammatory reaction. The biological significance of C5b-9–mediated cell stimulation is beginning to be appreciated.

REFERENCES

1. Buchner H. Ueber die bakterientoetende Wirkung des zellfreien Blutserums. *Zentralbl. Bakteriol.* 1889;6:1.
2. Buchner H.Ueber die naehere Natur der bakterientoetenden Sunstanz im Blutserum. *Zentralbl. Bakteriol.* 1889;6:561.
3. Vogt W. Activation, activities and pharmacologically active products of complement. *Pharmacol Rev.* 1974;26:125.
4. Shelton E, Yonemasu K, Stroud RM. Ultrastructure of the human complement component, Clq. *Proc Natl Acad Sci.* 1972;69:65.
5. Volanakis JE, Kaplan MH. Specificity of C-reactive protein for choline phosphate residues of pneumococcal C-polysaccharide. *Proc Soc Exp Biol Med.* 1971;136:612.
6. Volanakis JE, Kaplan MH. Interaction of C-reactive protein complexes with the complement system, II: Consumption of guinea pig complement by CRP complexes; requirement of human Clq. *J Immunol.* 1974;113:9.
7. Cooper NR, Jensen FC, Welsh RM, Oldstone MBA. Lysis of RNA tumor viruses by human serum: direct antibody-independent triggering of the classical complement pathway. *J Exp Med.* 1976;144:970.
8. Bartholomew RM, Esser AF, Muller-Eberhard HJ. Lysis of oncornaviruses by human serum: isolation of the viral complement (Cl) receptor and identification as PL5E. *J Exp Med.* 1978;147:844.
9. Kovacsovics T, Tschopp J, Kress A, Isliker H. Antibody-independent activation of Cl, the first component of complement, by cardiolipin. *J Immunol.* 1985;135:2695.
10. Vanguri P, Silverman B, Koski L, Shin ML. Complement activation by isolated myelin: activation of the classical pathway in the absence of myelin specific antibodies. *Proc Natl Acad Sci USA.* 1982;79:3290.
11. Shin HS, Pickering RJ, Mayer MM. The fifth component of the guinea pig complement system, II: mechanism of SAC 1, 4, 2, 3, 5b formation and C5 consumption by SAC 1, 4, 2, 3. *J Immunol.* 1971;106:473.
12. Ishizaka T, Ishizaka K, Borsos T, Rapp HJ. Cl fixation by human isoagglutinins: fixation of Cl by G and M but not by A antibody. *J Immunol.* 1966;97:716.
13. Law SK, Lichtenberg NA, Levine RP. Evidence for an ester linkage between the labile binding site of C3b and receptive surfaces. *J Immunol.* 1979;123:1288.
14. Law SK, Lichtenberg NA, Holcombe FH, Levine RP. Interaction between the labile binding sites of the fourth (4) and fifth (5) human complement proteins and erythrocyte cell membrane. *J Immunol.* 1980;125:634.
15. Sim RB, Twose TM, Patterson DS, Sim E. The covalent binding reaction of complement C3. *J Biochem.* 1981;193:115.
16. Edidin M, Henney C. The effect of capping H-2 antigens on the susceptibility of target cells to humoral and T-cell–mediated lysis. *Nature New Biol.* 1973;246:47.
17. Fearon DT. Activation of the alternative complement pathway. *CRC Crit Rev Immunol.* 1979;1:1.
18. Gewurz H, Shin HS, Mergenhagne SE. Interactions of the complement system with endotoxin lipopolysaccharides: consumption of each of the six terminal complement components. *J Exp Med.* 1968;128:1049.
19. Platts-Mills TAE, Ishizaka K. The activation of the alternative pathway of human complement by rabbit cells. *J Immunol.* 1974;113:348.
20. Fearon DT, Austen KF. Activation of the alternative complement pathway with rabbit erythrocytes by circumvention of the regulatory action of endogenous control proteins. *J Exp Med.* 1977;146:22.
21. Sandberg AL, Osler AG, Shin HS, Oliveira B.

Biologic activities of guinea pig antibodies, II: mode of complement interaction with gamma 1 and gamma 2 immunoglobulin. *J Immunol.* 1970;104:329.

22. Koski CL, Vanguri P, Shin ML. Activation of the alternative pathway of complement by human peripheral nerve myelin. *J Immunol.* 1985;134:1810.

23. Fearon DT. Regulation by membrane sialic acid of BLH-dependent decay dissociation of the amplification C3 convertase of the alternative complement pathway. *Proc Natl Acad Sci USA.* 1978;75:1971.

24. Shin HS, Pickering RJ, Mayer MM. The fifth component of the guinea pig complement system, III: dissociation and transfer of C5b and the probable site of C5b fixation. *J Immunol.* 1971; 106:480.

25. Rother K, Till GO, eds. *The Complement System.* Berlin: Springer-Verlag; 1988.

26. Thompson RA, Rowe DS. Reactive hemolysis: a distinct form of red cell lysis. *Immunology.* 1968;14:745.

27. Goldlust MB, Shin HS, Hammer CH, Mayer MM. Studies of complement complex C5b,6 eluted from EAC-6: reaction of C5b,6 with EAC4b,3b and evidence on the role of C2a and C3b in the activation of C5. *J Immunol.* 1974;113:998.

28. Hu V, Esser A, Podack E, Wisniesky B. The membrane attack mechanism of complement: photolabeling reveals insertion of terminal proteins into target membrane. *J Immunol.* 1981; 127:380.

29. Podack ER, Muller-Eberhard HJ. Binding of deoxycholate, phosphatidyl choline vesicles, lipoprotein or S-protein to complexes of terminal complement components. *J Immunol.* 1978; 121:1025.

30. Yamamoto K, Gewurz H. The complex of C5b and C6: isolation, characterization and identification of a modified form of C5b consisting of three polypeptide chains. *J Immunol.* 1978;120:2008.

31. Rother U, Hansch GM, Rauterberg EW, Jungfer H, Rother K. Deviated lysis; lysis of unsensitized cells by complement, V: generation of the activity by low pH or low ionic strength. *Z Immunforsch.* 1978;155:118.

32. Hansch GM, Hammer C, Mayer MM, Shin ML. Activation of the fifth and sixth component of the complement system: similarities between C5b6 and C (56)[a] with respect to lytic enhancement by cell bound C3b or A2c, and species preferences of target cell. *J Immunol.* 1981;127:999.

33. Hammer CH, Hansch GM, Gresham H, Shin ML. Activation of the fifth and sixth component of the human complement system, C6 dependent cleavage of C5 in acid and the formation of a bimolar lytic complex C56[a]. *J Immunol.* 1983;131:892.

34. Dessauer A, Rother U, Rother K. Freeze thaw activation of the complement attack phase, I: separation of two steps in the formation of the active C56 complex; II: comparison of the convertase generated C56 with C56 generated by freezing

and thawing. *Acta Pathol Microbiol Immunol Scand.* 1984 (C; Suppl 284)92:75–88.

35. Podack ER, Kolb WP, Muller-Eberhard HJ. The SC5b-7 complex: formation, isolation, properties, and subunit composition. *J Immunol.* 1977; 119:2024.

36. Monahan JB, Sodetz JM. Binding of the eight component of human complement to the soluble cytolytic complex is mediated by its subunit. *J Biol Chem.* 1980;255:10579.

37. Monahan JB, Sodetz JM. Role of the alpha-subunit in interaction of the eighth component of human complement with the membrane-bound cytolutic complex. *J Biol Chem.* 1981;256:3258.

38. Steckel EW, York RG, Monahan JB, Sodetz JM. The eighth component of human complement. *J Biol Chem.* 1980;255:11997.

39. Haxby JA, Kinsky CB, Kinsky SC. Immune response of a liposomal model membrane. *Proc Natl Acad Sci USA.* 1968;61:300.

40. Deleted.

41. Shin ML, Paznekas WA, Abramovitz AS, Mayer MM. On the mechanism of membrane damage by C: exposure of hydrophobic sites on activated C proteins. *J Immunol.* 1977;119:1358.

42. Humphrey HJ, Dourmashkin RR. The lesions in cell membranes caused by complement. *Adv Immunol.* 1969;75:115.

43. Mayer MM. Mechanism of cytolisis by complement. *Proc Natl Acad Sci USA.* 1972;69:2954.

44. Wilson L, Spitznagel J. Characteristics of complement-dependent release of phospholipid from *Escherichia coli. Infect Immunol.* 1971;4:23.

45. Inoue K, Kinoshita T, Okada M, Akiyama Y. Release of phospholipids from complement-mediated lesions on the surface structure of *Escherichia coli. J Immunol.* 1977;119:65.

46. Kinoshita T, Inoue K, Okada M, Akiyama Y. Release of phospholipids from liposomal model membrane damaged by antibody and complement. *J Immunol.* 1977;119:73.

47. Ohanian SH, Schlager SI, Borsos T. Molecular interactions of cells with antibody and complement. In: Reisfeld R, Inman FP, eds. *Contemporary Topics in Molecular Immunology.* New York: Plenum; 1978:7.

48. Giavedoni EB, Chow YM, Dalmasso AP. The functional size of the primary complement lesion in resealed erythrocyte membrane ghosts. *J Immunol.* 1979;122:240.

49. Schlager SI, Ohanian SH, Borsos T. Identification of lipid synthesized and released by tumor cells under complement attack by antibody and complement. *J Immunol.* 1978;120:1644.

50. Robinson NC, Tanford C. The binding of deoxycholate, Triton X-100, sodium dodecyl sulfate and phosphatidyl choline vesicles to cytochrome b5. *Biochemistry.* 1975;14:369.

51. Bhakdi S, Bjerrum OJ, Bhakdi-Lehnen B, Tranum-Jensen J. Complement lysis: evidence for an amphiphilic nature of the terminal membrane C5b-9 complex of human complement. *J Immunol.* 1978;121:2526.

52. Podack ER, Biesecker G, Muller-Eberhard HJ. Membrane attack complex of complement: gen-

eration of high-affinity phospholipid binding sites by fusion of five hydrophilic plasma proteins. *Proc Natl Acad Sci USA.* 1979;76:897.

53. Bhakdi S, Tranum-Jensen J. On the cause and nature of C9-related heterogeneity of terminal complement complexes generated on target erythrocytes through the action of whole serum. *J Immunol.* 1984;133:1453.

54. Podack ER, Tschopp J, Muller-Eberhard HJ. Molecular organization of C9 within the membrane attack complex of complement: induction of circular C9 polymerization by the C5b-8 assembly. *J Exp Med.* 1982;156:268.

55. Podack ER, Muller-Eberhard HJ, Horst H, Hoppe W. Membrane attack complex of complement (MAC): three-dimensional analysis of MAC-phospholipid vesicles recombinants. *J Immunol.* 1982;128:2353.

56. Sodetz JM, Monahan JB. The eighth component of human complement: function of its subunit in interactions with the cytolytic complex of complement. *Fed Proc.* 1981;40:1020.

57. Ramm LE, Whitlow MB, Mayer MM. Size distribution and stability of the transmembrane channels formed by complement complex C5b-9. *Mol Immunol.* 1983;20:155.

58. Bhakdi S, Ey P, Bhakdi-Lehnen B. Isolation of the terminal complement complex from target sheep erythrocyte membranes. *Biochim Biophys Acta.* 1976;419:445.

59. Tranum-Jensen J, Bhakdi S, Bhakdi-Lehnen B, Bjerrum OJ, Speth V. Complement lysis: the ultrastructure and orientation of the C5b-9 complex on target sheep erythrocyte membranes. *Scand J Immunol.* 1978;7:45.

60. Barfort P, Arquilla ER, Vogelhut PO. Resistance changes in lipid bilayers: immunological applications. *Science.* 1968;160:1119.

61. Michaels DW, Abramovitz AS, Hammer CH, Mayer MM. Increased ion permeability of planar lipid bilayer membranes after treatment with the C5b-9 cytolytic attack mechanism of complement. *Proc Natl Acad Sci USA.* 1976;73:2852.

62. Michaels DW, Abramovitz AS, Hammer CH, Mayer MM. Characterization of the complement lesion: the formation of transmembrane channels and their mechanism of assembly. *J Immunol.* 1978;120:1785.

63. Green H, Barrow CP, Goldberg B. Effect of antibody and complement on permeability control in ascites tumor cells and erythrocytes. *J Exp Med.* 1959;110:699.

64. Sears DA, Weed RI, Swisher SN. Differences in the mechanism of *in vitro* immune hemolysis related to antibody specificity. *J Clin Invest.* 1964;43:975.

65. Ramm LE, Mayer MM. Life-span and size of the transmembrane channels formed by large doses of complement. *J Immunol.* 1980;124:2281.

66. Sims PJ, Lauf PK. Steady-state analysis of tracer exchange across the C5b-9 complement lesion in a biological membrane. *Proc Natl Acad Sci USA.* 1978;75:5669.

67. Boyle MDP, Borsos T. Studies on the terminal stages of immune hemolysis, V: evidence that not all complement-produced transmembrane channels are equal. *J Immunol.* 1979;123:71.

68. Boyle MDP, Gee AP, Borsos T. Studies on the terminal stages of immune hemolysis, VI: osmotic blockers of differing stokes radii detect complement induced transmembrane channels of differing size. *J Immunol.* 1979;123:77.

69. Ware CF, Wetsel RA, Kolb WP. Physicochemical characterization of fluid phase (S-C5b-9) and membrane derived (MC5b-9) attack complexes of human complement purified by immunoadsorbent affinity chromatography of selective detergent extraction. *Mol Immunol.* 1981;18:521.

70. Tschopp J, Podack ER, Muller-Eberhard HJ. The membrane attack complex of complement: C5b-8 complex as accelerator of C9 polymerization. *J Immunol.* 1985;134:495.

71. Dankert JR, Esser AF. Proteolytic modification of human complement protein C9, loss of poly (C9) and circular lesion formation without impairment of function. *Proc Natl Acad Sci USA.* 1985;82:2128.

72. Tschopp J, Podack ER. Membranolysis by the ninth component of human complement. *Biochem Biophys Res Comm.* 1981;100:1409.

73. Lint TF, Behrends CL, Gewurz H. Serum lipoproteins and C5,6,7-INH activity. *J Immunol.* 1977;119:883.

74. Baker PJ, Lint TF, McLeod BC, Behrends CL, Gewurz H. Studies on the inhibition of C56-induced lysis (reactive lysis), VI: modulation of C56-induced lysis by polyanions and polycations. *J Immunol.* 1975;114:554.

75. Baker PJ, Lint TF, Siegel J, Kies MW, Gewurz J. Potentiation of C5,6-initiated lysis by leukocyte cationic proteins, myelin basic proteins and lysine-rich histones. *Immunology.* 1976;30:467.

76. Boyle MDP, Ohanian S, Borsos T. Lysis of tumor cells by antibody and complement, VI: enhanced killing of enzyme pretreated tumor cells. *J Immunol* 1976;116:661.

77. Boyle MDP, Ohanian S, Borsos T. Effect of protease treatment on the sensitivity of tumor cells to antibody-GPC killing. *Clin Immunol Immunopath.* 1978;10:84.

78. Lauf PK. Immunological and physiological characteristics of the rapid immune hemolysis of neuraminidase-treated sheep red cells produced by fresh guinea pig serum. *J Exp Med.* 1975;142:974.

79. Shin ML, Paznekas WA, Mayer MM. On the mechanism of membrane damage by complement: the effect of length and saturation of the acyl chains in liposomal bilayers and the effect of cholesterol concentration in sheep erythrocyte and liposomal membranes. *J Immunol.* 1978;120:1996.

80. Hesketh TR, Dourmashkin RR, Payne SN, Humphrey JH, Lachman PJ. Lesions due to complement in lipid membranes. *Nature* 1971;233:620–633.

81. Richards RL, Gewurz H, Siegel J, Alving CR. Interaction of C-reactive protein and complement with liposomes, 11: influences of membrane composition. *J Immunol.* 1979;122:1185.

82. Shin ML, Paznekas WA, Mayer MM. Effect of membrane fluidity on efficiency of sheep erythrocyte lysis by terminal complement proteins. *Fed Proc.* 1979;38:1468.

83. Dahl JS, Dahl CE, Levine RP. Role of lipid fatty acyl composition and membrane fluidity in the resistance of *Acholeplasma laidlawii* to complement-mediated killing. *J Immunol.* 1979;123:104.

84. Shin ML, Hansch GM, Mayer MM. Effect of agents that produce membane disorder on lysis of erythrocytes by complement. *Proc Natl Acad Sci USA.* 1981;78:2522.

85. Kosower EM, Kosower NS, Faltin Z, Diver A, Saltoun G, Frensdorff A. Membrane mobility agents: a new class of biologically active molecules. *Biochim Biophys Acta.* 1974;363:261.

86. Upreti GC, Shirley R, Jain MK. Intrinsic differences in the perturbing ability of alkanols in bilayer: action of phospholipase A$_2$ on the alkanol-modified phospholipid bilayer. *J Membrane Biol.* 1980;55:97.

87. Bordet J. Les serums hemolytiques, les antitoxines et les théories des serums cytolytiques. *Annales de L'Institut Pasteur.* 1900;14:257.

88. Hansch GM, Hammer CH, Vanguri P, Shin ML. Homologous species restriction in lysis of erythrocytes by terminal complement proteins. *Proc Natl Acad Sci USA.* 1981;78:5118.

89. Houle JJ, Hoffman EM. Evidence for restriction of the ability of complement to lyse homologous erythrocytes. *J Immunol.* 1984;133:1444.

90. Kitamura H, Inai S. Molecular analysis of the reaction of C9 with EAC1-8. *J Immunol.* 1974;113:1992.

91. Rosenfeld SI, Packman CH, Jenkins DE, Countryman JK, Leddy JP. Complement lysis of human erythrocytes, III: differing effectiveness of human and guinea pig C9 on normal and paroxysmal nocturnal hemoglobinuria cells. *J Immunol.* 1980;125:2063.

92. Hansch GM, Hammer CH, Jiji R, Rother U, Shin M. lysis of paroxysmal nocturnal hemoglobinuria erythrocytes by acid-activated serum. *Immunobiology.* 1983;164:118.

93. Hänsch GM, Hammer CH, Vanguri P, Shin ML. Homologous species restriction in lysis of erythrocytes by terminal complement proteins. *Proc Natl Acad Sci USA* 1981;78:5118–5121.

94. Yamamoto K. Lytic activity of C5b-9 complexes for erythrocytes from the species other than sheep: C9 rather than the C8-dependent lytic activity. *J Immunol.* 1977;119:1482.

95. Hu VU, Shin ML. Species-related target cell lysis by human complement, complement-lysed erythrocytes from different species differ in their ratio of bound to inserted C9. *J Immunol.* 1984;133:2133.

96. Hoffman EM. Inhibition of complement by a substance isolated from human erythrocytes, I: extraction from human erythrocyte stromata. *Immunochemistry.* 1969;6:405.

97. Nicholson-Weller A, Burge J, Fearon DT, Weller PF, Austen KF. Isolation of a human erythrocyte membrane glycoprotein with decay accelerating activity for C3 convertases of the complement system. *J Immunol.* 1982;129:1984.

98. Nicholson-Weller A, March JP, Rosenfeld SI, Austen KF. Affected erythrocytes of patients with paroxysmal nocturnal hemoglobinuria are deficient in the complement regulatory protein, decay accelerating factor. *Proc Natl Acad Sci USA.* 1983;80:5066.

99. Pangburn MK, Schreiber RD, Trombold JS, Muller-Eberhard HJ. Paroxysmal nocturnal hemoglobinuria: deficiency in factor H-like functions of the abnormal erythrocytes. *J Exp Med.* 1983;157:1971.

100. Medof MF, Kinoshita T, Nussenzwerf V. Inhibition of complement activation on the surface of cells after incorporation of decay-accelerating factors (DAF) into their membranes. *J Exp Med.* 1984;160:1558.

101. Shin ML, Hansch G, Hu V, Nicholson-Weller A. Membrane factors responsible for homologous species restriction of complement-mediated lysis: evidence for a factor other than DAF operating at the stage of C8 and C9. *J Immunol.* 1986;136:1777.

102. Schonermark S, Rauterberg EW, Shin ML, Loeke S, Roelcke D, Hansch GM. Homologous species restriction in lysis of human erythrocytes: a membrane-derived protein with C8-binding capacity function as an inhibitor. *J Immunol.* 1986;136:1772.

103. Schonermark S, Roelcke D, Hansch GM. Homologous species restriction in complement mediated lysis: Identification of a regulatory membrane protein. *6th Intl Congress of Immunology.* 1986. Abstract.

104. Pellegrino MA, Ferrone S, Cooper NR, Dierich MP, Reisfeld RA. Variation in susceptibility of a human lymphoid cell line to immune lysis during the cell cycle. *J Exp Med.* 1974;140:578.

105. Ohanian SH, Borsos T. Lysis of tumor cells by antibody and complement, II: lack of correlation between amount of C4 and C3 fixed and cell lysis. *J Immunol.* 1975;114:1292.

106. Boyle MDP, Ohanian SH, Borsos T. Studies on the terminal stages of antibody-complement-mediated killing of a tumor cell, I: evidence for the existence of an intermediate, T*. *J Immunol.* 1976;116:1272.

107. Burakoff SJ, Martz E, Benacerraf B. Is the primary complement lesion insufficient for lysis? Failure of cells damaged under osmotic protection to lyse in EDTA or at low temperature after removal of osmotic protection. *Clin Immunol Immunopath.* 1978;4:108.

108. Ebbesen P, Arnus KM. Enhancement of the dye-exclusion cytotoxic test by insulin and inhibition by cyclic adenosine 3:5-monophosphate and theophylline. *Transplantation.* 1973;16:476.

109. Ferrone S, Pellegrino MA, Dierich MP, Reisfeld RA. Effect of inhibitors of macromolecular synthesis on HLA antibody mediated lysis of cultured lymphoblasts. *Tissue Antigens.* 1974;4:275.

110. Kaliner M, Austen KF. Adenosine 3'5'-Monophosphate: inhibition of complement-mediated cell lysis. *Science* 1974;183:659.

111. Segerling MS, Ohanian SH, Borsos T. Effect of metabolic inhibitors on killing of tumor cells by antibody and complement. *J Natl Cancer Inst.* 1974;53:1411.

112. Segerling MS, Ohanian SH, Borsos T. Chemotherapeutic drugs increase killing of tumor cells by antibody and complement. *Science.* 1975; 188:55.

113. Segerling MS, Ohanian SH, Borsos T. Enhancing effect by metabolic inhibitors on the killing of tumor cells by antibody and complement. *Cancer Res.* 1975;35:3195.

114. Boyle MDP, Ohanian SH, Borsos T. Studies on the terminal stages of antibody-complement–mediated killing of a tumor cell, II: inhibition of transformation of T* to dead cells by 3'5'cAMP. *J Immunol.* 1976;116:1276.

115. Boyle MDP, Ohanian SH, Borsos T. Studies on the terminal stages of antibody-complement–mediated killing of a tumor cell, III: effect of membrane active agents. *J Immunol.* 1976; 117:106.

116. Schlager SI, Ohanian SH, Borsos T. Inhibition of antibody-complement–mediated killing of tumor cells by hormones. *Cancer Res.* 1976;36:3672.

117. Schlager SI, Ohanian SH, Borsos T. Correlation between the ability of tumor cells to resist humoral immune attack and their ability to synthesize lipid. *J Immunol.* 1978;120:463.

118. Lo TN, Boyle MDP. Relationship between the intracellular cyclic adenosine 3':5'-monophosphage level of tumor cells and their sensitivity to killing by antibody and complement. *Cancer Res.* 1979;39:3156.

119. Ohanian SH, Schlager SI. Humoral immune killing of nucleated cells: mechanisms of complement-mediated attack and target cell defense. *CRC Crit Rev Immunol.* 1981;1:165.

120. Ramm LE, Whitlow WB, Mayer MM. Complement lysis of nucleated cells: effect of temperature and puromycin on the number of channels required for cytolysis. *Molec Immunol.* 1984; 21:1015.

121. Schlager SI, Ohanian SH. A role for fatty acid composition of complex cellular lipids in the susceptibility of tumor cells to humoral immune killing. *J Immunol.* 1979;123:146.

122. Schlager SI, Ohanian SH. Tumor cell lipid composition and sensitivity to humoral immune killing, II: influence of plasma membrane and intracellular lipid and fatty acid content. *J Immunol.* 1980;125:508.

123. Yoo TJ, Chin HC, Spector AA, Whitaker RJ, Denning CM, Lee NF. Effect of fatty acid modification of cultured hepatoma cells on susceptibility to complement-mediated cytolysis. *Cancer Res.* 1980;40:1084.

124. Ohanian SH, Schlager SI, Saha S. Effect of lipids, structural precursors of lipids and fatty acids on complement-mediated killing of antibody sensitized nucleated cells. *Mol Immunol.* 1982;19:535.

125. Koski CL, Ramm LE, Hammer CH, Mayer MM, Shin ML. Cytolysis of nucleated cells by complement: cell death displays multi-hit characteristics. *Proc Natl Acad Sci USA.* 1983;80:3816.

126. Ramm LE, Whitlow MB, Koski CL, Shin ML, Mayer MM. Elimination of complement channels from the plasma membranes of U937, a nucleated mammalian cell line: temperature dependence of the elimination rate. *J Immunol.* 1983;131:1411.

127. Morgan BP, Campbell AK, Luzio JP, Hallett MB. Recovery of polymorphonuclear leucocytes from complement attack. *Biochem Soc Trans.* 1984; 12:779.

128. Carney DF, Koski CL, Shin ML. Elimination of terminal complement intermediates from the plasma membrane of nucleated cells: the rate of disappearance differs for cells carrying C5b-7 or C5b-8 or a mixture of C5b-8 with a limited number of C5b-9. *J Immunol.* 1985;134:1804.

129. Carney DF, Hammer CH, Shin ML. Elimination of terminal complement complexes in the plasma membrane of nucleated cells: influence of extracellular Ca^{2+} and association with cellular Ca^{2+}. *J Immunol.* 1986;137:263.

130. Sims PJ, Wiedmer T. Repolarization of the membrane potential of blood platelets after complement damage: evidence for a Ca^{2+}-dependent exocytotic elimination of C5b-9 pores. *Blood* 1986;68:556–661.

131. Morgan BP, Dankert JR, Esser AF. Recovery of human neutrophils from complement attack: removal of the membrane attack complex by endocytosis and exocytosis. *J Immunol.* 1987;138:246.

132. Morgan BP, Campbell AK. A recovery of human polymorphonuclear leucocytes from sublytic complement attack is mediated by changes in intracellular free calcium. *Biochem J.* 1985;231:205.

133. Ramm LE, Whitlow MB, Mayer MM. Size of the transmembrane channels produced by complement proteins C5b-8. *J Immunol.* 1982;129:1143.

134. Ramm LE, Michaels DW, Whitlow MB, Mayer MM. In: August JT, ed. *Biological Response Mediators and Modulators.* New York: Academic Press; 1983:117–139.

135. Carney DF, Shin ML. *Terminal complement complexes (TCC) generate multiple signal messengers: protein kinase C but not cAMP is involved in rapid elimination of TCC from the surface of Ehrlich cells.* Unpublished manuscript, 1989.

136. Smith JK, Becker EL. Serum complement and the enzymatic degradation of erythrocyte phospholipids. *J Immunol.* 1968;100:459.

137. Guttler F. Phospholipid synthesis in HeLa cells exposed to immunoglobulin G and complement. *J Biochem.* 1972;128:953.

138. Giavedoni EB, Dalmasso AP. The induction by complement of a change in KSCN-dissociable red cell membrane lipids. *J Immunol.* 1976;116:1163.

139. Fischer H. Lysolecithin and the action of complement. *Ann NY Acad Sci.* 1964;116:1063.

140. Sandberg AL, Raisz LG, Goodson MJ, Simmons HA, Mergenhagen SE. Initiation of bone resorption by the classical and alternative pathways and its mediation by prostaglandins. *J Immunol.* 1977;119:1378.

141. Imagawa DK, Osifchin NE, Paznekas WA, Shin

ML, Mayer MM. Consequences of cell membrane attack by complement: release of arachidonate and formation of inflammatory mediators. *Proc Natl Acad Sci USA.* 1983;80:6647.

142. Betz M, Hansch GM. Release of arachidonic acid: a new function of the late complement components. *Immunobiology.* 1984;166:473.

143. Hansch GM, Seitz M, Martinotti G, Betz M, Rauterberg EW, Gemsa D. Macrophages release arachidonic acid, prostaglandin E$_2$, and thromboxane in response to late complement components. *J Immunol.* 1984;133:2145.

144. Shirazi Y, Imagawa DK, Shin ML. Release of leukotriene B$_4$ from sublethally-injured oligodendrocytes by terminal complement proteins. *J Neurochem.* 1987;48:271.

145. Shirazi Y, McMorris FA, Shin ML. Arachidonic acid mobilization and phosphoinositide turnover by the terminal complement complex, C5b-9, in rat oligodendrocytes C$_6$ glioma cell hybrids. *J Immunol.* In press, 1989;142:4385–4391.

146. Hansch GM, Gemsa D, Resch K. Induction of prostanoid synthesis in human platelets by the late complement components C5b-9 and channel forming antibiotic nystatin: inhibition of the reacylation of liberated arachidonic acid. *J Immunol.* 1985;135:1320.

147. Nishizuka Y. Studies and perspectives of protein kinase C. *Science.* 1986;233:305.

148. Wiedmer T, Ando B, Sims PJ. Complement C5b-9-stimulated platelet secretion is associated with a Ca^{2+}-induced activation of cellular protein kinases. *J Biol Chem.* 1987;262:13674.

149. Hansch GM, Betz M, Shin ML. Cytolysis of nucleated cells by complement: inhibition of membrane-transmethylation enhances cell death by C5b-9. *J Immunol.* 1984;132:1440.

150. Papadimitriou JC, Carney DF, Shin ML. Effects of metabolic inhibitors on the nucleated cell resistance to complement (C) attack. In preparation, 1989.

151. Roberts PA, Morgan BP, Campbell AK. 2-chloroadenosine inhibits complement-induced reactive oxygen metabolite production and recovery of human polymorphonuclear leucocytes attacked by complement. *Biochem Biophys Res Commun.* 1985;126:692.

152. Wiedmer T, Esmon CT, Sims PJ. On the mechanism by which complement proteins C5b-9 increase platelet prothrombinase activity. *J Biol Chem.* 1986;261:14587.

153. Lovett DH, Hansch GM, Goppelt M, Resch K, Gemsa D. Activation of glomerular mesangial cells by the terminal membrane attack complex of complement. *J Immunol.* 1987;138:2473.

154. Vanguri P, Shin ML. Hydrolysis of myelin basic protein induced in human myelin by terminal complement complexes. *J Biol Chem.* 1988;263:7228.

155. Halperin JA, Nicholson-Weller A, Brugnara C, Tosteson DC. Complement induces a transient increase in membrane permeability in unlysed erythrocytes. *J Clin Invest.* 1988;82:594.

156. Liu Z-Y, Sanders ME, Hu VW. Effect of complement on the lateral mobility of erythrocyte membrane proteins: evidence for terminal complex interaction with cytoskeletal components. *J Immunol.* 1989;142:2370.

157. Kim S-H, Carney DF, Papadimitriou JC, Shin ML. Effect of osmotic protection on nucleated cell killing by C5b-9: cell death is not affected by the prevention of cell swelling. *Mol Immunol.* 1989;26:323.

158. Duvall E, Wyllie AH. Death and the cell. *Immunol Today.* 1986;7:115.

159. Campbell AK, Daw RA, Luzio JP. Rapid increase in intracellular free Ca^{2+} induced by antibody plus complement. *FEBS Lett.* 1979;107:55.

160. Kim S-H, Carney DF, Hammer CH, Shin ML. Nucleated cell killing by complement: effects of C5b-9 channel size and extracellular Ca^{2+} on the lytic process. *J Immunol.* 1987;138:1530.

161. Chien KR, Abrams J, Serroni A, Martin JT, Farber JL. Accelerated phospholipid degradation and associated membrane dysfunction in irreversible ischemic liver cell injury. *J Biol Chem.* 1978;253:4809.

162. Cheung JM, Bouventre JV, Malis CD, Leaf A. Calcium and ischemic injury. *N Engl J Med.* 1986;26:1670.

163. Lehninger AL. Mitochondria and calcium ion transport. *Biochem J.* 1970;119:129.

164. Marsh P. Spectroscopic studies of membrane structure. In: Campbell PN, Aldridge WN, eds. *Essays in Biochemistry,* Vol. XI. New York: Academic Press; 1975:139.

165. Lambertenghi-Deliliers G, Ferrone S, Ranzi T, Sirichia G. Immune lysis of normal and AET-treated lymphocytes: a study by transmission and scanning electron microscopy. *Blood.* 1971; 38:759.

4

Plasma Membrane and Mitochondrial Changes in Cell Injury

Wolfgang J. Mergner
Michael Costa
John B. Classen
Howard J. Leventhal

IN VITRO CELL INJURY STUDIES: SCOPE AND PURPOSE

The Organelles in Cell Injury

In cell injury, all organelles of the cell are affected. Some organelles cease to provide essential functions for the cell and thus jeopardize its survival. Organelles display variable sensitivity to ischemic injury. A greater sensitivity means, in this context, that morphological changes can be seen within a short(er) interval after onset of injury. Such an organelle is the nucleus, which ultrastructurally shows early clumping of nuclear chromatin. Detailed studies of the isolated nucleus as a function of ischemic time are not available, although it is assumed that metabolic stress in cell injury inhibits synthetic functions of the nucleus. Little is known about nuclear recovery from injury. Indirect evidence from inhibitor studies indicates that even in the absence of nuclear function cell equilibrium can be maintained for a short period of time. Thus, sensitive organelles may not be the primary contributors to cell death. In contrast, if the plasma membrane is severely altered, such as in complement injury, the regular cell functions deteriorate rapidly (see Chapter 3).

Scanning electron micrographs of such cells reveal a jumble of filaments and ruptured organelles. Products released from such injured cells and detected in the lymph, particularly those of mitochondrial and sarcoplasmic reticulum origin, may complex with humoral factors such as human C1q *in vivo.*[1]

The question may be raised, does the plasma membrane or any other intracellular membrane develop sufficient autolysis during early ischemia to explain the advent of irreversible cell injury and cell necrosis? In this chapter, we describe studies on isolated organelles obtained from parenchymal cells or studies on organelles injured *in vitro* after isolation in order to elucidate the structural and functional changes that mark irreversible injury and early necrosis. It is important to understand that these studies search for function of damaged cell particles in a reconstituted system. In this normalized system, the studies aim to describe the functional deficit that has occurred by the autocatalytic processes during ischemia and/or other forms of cell injury.

Following ischemia, the ATP levels drop dramatically; but most cells can recover, in spite of this early drop in ATP, if the depression does not last beyond certain time limits. Since a very poor correlation exists in most parenchymal cells between the fast initial drop of ATP and later cell death, several hypotheses have been advanced to explain the mechanism of cell death under those circumstances. One such is the Energy Hypothesis.[2] The energy hypothesis has been proposed under several different guises. Jennings et al. have postulated that a direct relationship exists between low and decreasing levels of high energy phosphate and lethal injury in dog myocardium.[3] Reimer et al. suggested that the loss of adenine nucleotides and deficient resynthesis may cause cell death.[4] Pine et al. proposed that the cellular osmotic load, due to paralysis of volume regulation in the absence of ATP, is one key element in cell killing.[5] Others believe that acute damage of mitochondrial membranes or of plasmalemma is caused by activated phospholipases, reorganization of phospholipids by lateral phase separation, activated proteases, accumulated lysophosphoglycerides, leukotriens, acyl esters, protons, oxygen radicals, calcium, or destruction of the ATP synthetase machinery—in particular, the proton gradient and uncoupling of mitochondria.[6-18]

Normal Organelles Exposed to Damaging Agents

External Phospholipases

Incubation of isolated whole mitochondria with externally added phospholipases C produced damage to ion transport and oxidative phosphorylation, while respiration remained unimpaired.[19] In contrast, phospholipase C added to submitochondrial particles alters their respiration.[20] Submitochondrial particles change the orientation of their membranes in the process of their preparation; that is, they turn inside-out. This would mean that the location of the phospholipid damage by phospholipase C on the inside or the outside of the mitochondrial membrane determines the functional damage. Early studies already suggested that NADH-dependent pathways of electron transport were more sensitive to the action of phospholipases than succinate oxidation. In subsequent studies it became apparent that, in particular, the NADH-Q_{10} segment is vulnerable to lipolysis and phospholipolysis by either phospholipase A or C. Albumin in the suspension medium, whose action may be the capture of lysis products such as fatty acids, was not effective in protecting mitochondria.[19,20]

Addition of Fatty Acids and Arachidonic Acid to Isolated Mitochondria

Addition of free arachidonic acid to brain mitochondria by Hillered and Chan caused a dose-dependent increase in substrate-supported respiration (state 4 respiration).[21] Uncoupling occurred, and a concomitant decrease of substrate, phosphate-, and adenosine-diphosphate–stimulated respiration (state 3 respiration). Albumin in the suspension medium reduced the uncoupling but did not affect the depression of state 3 respiration. Based on the experimental data, it is suggested that arachidonic acid inhibits mitochondrial ATP synthesis in cerebral ischemia and may also interfere with postischemic recovery.

Aging of Organelles as Model of *in Vivo* Degradation in Injury

Many organelles, such as mitochondria and endoplasmic reticula, prove to be labile and are found to degrade quickly if removed from cells. This process of degradation of isolated organelles is called *aging*. The phenomenon of aging has been studied as a model for autolysis. It is assumed that the same forces of degradation that are influencing isolated extracellular organelles are also destroying organelles *in situ* during autolysis. The best experiments of organelle aging have used mitochondria, but aging of other organelles has been described. The aging-induced degradation process in mitochondria can be separated into two phases:

A time period when respiratory control is lost and a time period when ATPase activity increases.[22]

Underlying these functional alterations are morphological changes such as swelling of mitochondria, flocculent densities in the matrix space, and disruption of mitochondria. Luzikov proposed a sequence of events to explain the aging process.[22] The aging-related autolysis is initiated by alteration of oxidative phosphorylation. Altered oxidative phophorylation leads to increased permeability toward low molecular substances. If this change progresses, increased permeability toward proteins and macromolecules occurs, and high amplitude swelling ensues.

Alteration of Oxidative Phosphorylation

Oxidative phosphorylation coupled with NAD oxidation is a complex process requiring multiple enzyme steps. Alteration of this synthetic function of mitochondria occurs during the early postischemic degradation and in aging.[23] Other functions that decline early include the stimulation of respiration upon addition of ADP, uncoupler-stimulated respiration using dinitrophenol (DNP), and ADP-dependent calcium transport.[24] Simultaneously, ATPase activity is increased. Increased ATPase may rapidly exhaust the supply of endogenous ATP. This may not be the case in an acid suspension medium.

Permeability toward Low Molecular Weight Substances

Membrane alterations during aging cause increased permeability to low molecular weight components. For example, Siekevitz and Potter showed that aging liver mitochondria rapidly lose NAD, NADP, ADP, ATP, and ITP.[25] There appears to be a direct relationship between loss of small molecular substances and organelle swelling.[26] Significant mitochondrial swelling starts when the level of endogenous ATP falls below 20%. Swelling has a major role in aging of isolated mitochondria because it

leads to activation of previously latent enzymes. For example, enzymes that depend on pyridine dinucleotide substrate may be activated. Exogenous pyridine nucleotides can be incorporated into aging mitochondria.

Permeability toward Proteins

Membrane alterations cause increased permeability toward soluble proteins. Since matrix proteins are tightly contained within the confines of the inner compartment (or matrix), appearance of proteins outside the matrix indicates severe membrane damage. Malate dehydrogenase, for example, exists in two isoenzymes, one in the cytosol and one in mitochondrial matrix. Mitochondrial damage is indicated by the appearance of malate dehydrogenase isoenzyme in the serum. Estraad-O investigated the exodus of soluble matrix enzymes from aging liver mitochondria incubated at $37°C$.[27] The appearance of glutamate dehydrogenase and aspartate aminotransferase in the suspending medium were indicators of severe mitochondrial damage since the appearance of enzymes in the medium was linked with evidence of phospholipid dissolution of mitochondrial inner membranes.[28]

High Amplitude Swelling

High amplitude swelling is the most severe form of swelling and the last stage of mitochondrial degeneration in aging. Morphological changes of swelling are observed in mitochondria during the normal cycle and are also an early sign of injury. A number of agents will cause mitochondrial swelling. Wojtczak and Lehninger divided the agents that cause mitochondrial swelling into two groups.[29] The first group includes agents whose effects can be inhibited by external albumin, the so-called U-factor. Such agents include calcium and thyroxine. The second group of agents, is like phosphate, insensitive to external albumin. The high amplitude swelling in aging cannot be inhibited by external albumin. Mitochondrial swelling is caused by active or passive trans-location of ions, with water following passively.[30] Aging mitochondria may swell as a result of deficient permeability control caused by membrane alteration and supported by the high protein content of the matrix space. Thus, the corresponding morphological change would be caused by dilution of the matrix space and appears as electron lucency in electron micrographs.

The Mechanisms of Organelle Aging

Autocatalytic processes have been implicated in the degradation of isolated organelles. These processes result from activation of catalytic enzymes in the membranes or within the proteinaceous core of these organelles. Autocatalytic dissolution is aggravated by the absence of maintenance functions and detoxification of noxious stimuli. If functions are preserved, organelles can retard the autocatalytic destruction. Enzymes or substances that are critical for autocatalytic dissolution include phospholipases, free radicals, and proteases.

Phospholipase A_1 and A_2

Nachbaur et al. and Waite and Sisson showed that phospholipase A_1 and A_2 are present in the plasma membrane; phospholipase A_2 in mitochondria.[31,32] These phospholipases are dependent on calcium cations. Phospholipase-related deterioration in aging either reduces membrane phospholipids or produces lysis products that disturb the membranes. Both possibilities seem to have been detected in aging mitochondria. The first possibility was developed by Rossi et al., who showed that mitochondrial aging is accompanied by a decrease in content of mitochondrial phospholipids, such as phosphotidyl-choline, phosphotidyl-ethanolamine, and phosphotidyl-serine.[33] Even mitochondria isolated from ischemic tissue showed a decline in cardiolipin.[34] The decline in phospholipids can be inhibited by local anesthetics such as Nupercaine, an inhibitor of phospholipase. The role of lysis products was shown in experiments using serum albumin, which was partially protective.[35] Serum albumin serves as a carrier protein or scavenger of free fatty acids. The ameliorating effects of both phospholipase inhibitor and albumin may be indicative of some of the molecular events that lead to organelle destruction. Further evidence for the role of fatty acids in aging, with and without ATP, is derived from experiments of Parce et al., who added oleate as a test agent in the incubation medium of freshly isolated aging mitochondria.[36] These mitochondria take up oleate in the presence of ATP. Oleate can be recovered and detected in cardiolipin, a specific phospholipid of mitochondria. Similarly, glycerol-3-[P^{32}]-phosphate is incorporated into phospholipids upon addition of ATP and ATP-induced contraction of pre-swollen aging mitochondria. In the absence of ATP or ADP, mitochondria become altered. It is assumed that fatty acids added to isolated mitochondria disturb the order of the inner membrane and act as an uncoupler. The experiments of Parce et al. have shown that phospholipase-mediated transacylation reactions are highly dependent on ATP levels.[36] A decline in ATP levels may therefore render the organelle incapable of maintaining phospholipid structure and simultaneously expose the organelle to the destructive effects of the degradation products.

Lipid Peroxidation

Mitochondria form free radicals normally and also contain powerful systems that can destroy free radicals. Abnormal mitochondria may form lipoperoxides if free radicals are not disposed of. According to Hanstein and Hatefi, the lipid-sulfur centers, part of the electron transport complexes, conduct the accumulation of labile lipid sulfides.[37] Iron-sulfur centers are in all probability not the only initiators of lipid oxidation in mitochondrial membranes. Other elements of the electron transport chain such as cytochrome aa3 and cuproteins have been detected as a source of free radicals. If, as has been proposed, the electron transport chain itself acts as an initiator of lipid oxidation and superoxide radical formation, powerful systems must exist that normally protect the organelle from the destructive effects. Some elements of this machinery have been identified in mitochondria, for example, superoxide dismutases.[38] A matrix-bound system of superoxide dismutase is characterized by the presence of manganese. A superoxide dismutase located in the intercristal space contains copper and zinc.[39,40] However, the role of free radicals for aging, autolysis, or ischemia is not entirely clear.

Proteases

Specific neutral proteases in mitochondria mediate catalytic reactions under neutral pH conditions. The release of amino acids has been observed in isolated aging mitochondria by Bartley and Birt.[41] Ferdinand et al. showed that isolated mitochondria at 40°C, at a pH of 7.2 and in 0.25 M sucrose, release amino acid at a rate of 0.144 nmol per mg dry weight.[42] That amino acids are released is an indication that autolysis involves protein digestion. Freshly isolated mitochondria do not release amino acid into the supernatant.

Stabilization of Organelles

Luzikov reviewed the idea and presented mechanisms of stabilization of mitochondria *in vitro*.[22] Stabilization occurs and is linked to oxidative phosphorylation. Romani et al. and Goldblatt and Romani showed that animal and plant mitochondria can maintain their respiratory control index for from 3 to 4 days at 25°C if the medium contains substrate (malate, pyruvate, and succinate), NAD, thiamine pyrophosphate, coenzyme A, albumin, and oxygen.[43-46]

Preservation was dependent on the presence of ADP and, probably, on the activity of oxidative phosphorylation. In the external medium, ATP was unable to maintain mitochondria. Parce et al. proposed that the presence of intramitochondrial ATP is critical.[47] These authors have suggested that as levels of ATP decrease inside them, mitochondria lose their ability to reacylate lysophosphatides and to metabolize fatty acids by beta oxidation.[47] The decay is marked by a drastic increase of lysophosphotidylethanolamine and free fatty acids. Luzikov interprets these data as meaning that oxidative phosphorylation is an indispensable prerequisite of preventing mitochondrial disintegration.[22]

STUDIES USING THE *IN VITRO* ISCHEMIA MODEL

The Plasma Membrane

Purification of the Plasma Membrane

The acquisition of pure fragments of plasma membrane from a single cell type has proven to be extremely difficult, particularly after injury to tissues and cells. Investigators such as Franson and Weglicki have resorted to studying isolated cell systems, such as isolated adult heart cells.[48] The overall yield of plasma membrane reported by various investigators is low. Nevertheless, studies of sarcoplasma from whole organs allow some general insight into membrane changes by several approaches. These approaches utilize the preparation of density gradient–derived lipid vesicles. Marker enzymes, such as 5′-nucleotidase or Na^+/K^+–ATPase, adenyl cyclase, and K-stimulated p-nitrophenylphosphatase assist the tracing of the purification steps.[49-52] These approaches have proven that the plasma membrane or sarcolemma is a resilient system that can be well prepared and isolated. But ischemia and reperfusion severely lower the yield of membrane vesicles and also alter the membrane-bound enzyme functions.[52] Some of the sarcolemmal changes occur early and may involve signal transmission, receptor functions, and cell volume regulation. The consequence of severe sarcolemmal injury is irreparable destruction of the barrier function of the membrane. These observations give validity to the dye exclusion tests and enzyme release tests for viability.

Jennings and Hawkins reported that functional and structural derangement occurred in the sarcolemma of ischemic myocytes, with coarse breaks and interruptions seen in irreversible injury. Corr et al. postulated that lysophospholipids causing membrane disorder rather than membrane disruption are responsible for cardiac dysrrhythmias. Nayler et al., in contrast, did not observe holes in the sarcolemma of cardiac myocytes at the end of the ischemic period prior to reperfusion.[17] The large amount of calcium entry at the beginning of reperfusion, according to Nayler et al., does not represent entry through membrane leaks, but is controlled by membrane proteins and is highly pH sensitive. Calcium transfer may be promoted by voltage-activated slow channels. These data argue against the concept that the large influx of calcium upon reflow is a consequence of disruption of the plasmalemma. In fact, the low pH during ischemia may have a protective effect just by blocking the slow channels in myocytes, which recover during reperfusion and then contribute to the calcium overload.[55]

The Enzymes of the Plasma Membrane

Studies on the enzyme system of the plasma membrane have revealed a diversity of responses to ischemia. Some enzymes were inactivated early, even during the reversible phase of injury, while others persisted even after cell disintegration had progressed. Smith et al. investigated the time course of degradation of marker enzymes in liver and heart and found that Na/K–ATPase was even activated in ischemic kidney (2 hours ischemia; Figure 4-1), while

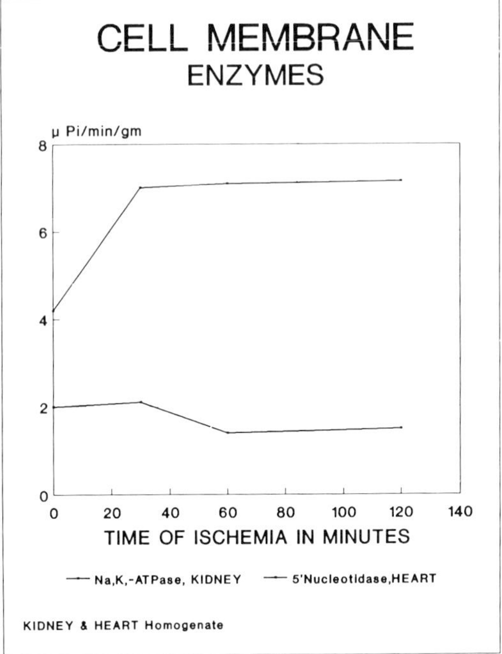

Figure 4-1. Enzyme activity in the first homogenate of rat renal and cardiac tissue. *In vitro* ischemia. (*In vitro* ischemia describes the incubation of organs wrapped and sealed at 37°C. Organelles are then isolated from these organs after homogenization. At no time was the tissue rinsed, washed, or perfused prior to homogenization.)

5′-nucleotidase in ischemic heart was only moderately depressed at 2 hours of ischemia. Significant membrane-bound enzyme changes were observed by Vatner et al., who observed a decline in adenyl cyclase after 1 hour of ischemia.[51] This change could affect the effector system for beta-receptors and could be associated with a relative lack of responsiveness by the heart to adrenergic agents. Ischemia of the guinea pig heart from 60 to 90 minutes caused severe depression of Na^+/K^+-ATPase activities. Reperfusion injury and ischemic injury may follow different but related pathways. Calcium plays a significant role in reperfusion injury, as outlined earlier.[16]

Studies on Phospholipids

Dissolution of the phospholipid layers by phospholipases was one of the early proposals for plasma membrane alteration by Weglicki and by Chien et al.[57,58] These authors could demonstrate phospholipase activity in various pH ranges and the release of fatty acids or arachiodonic acid. A close correlation was demonstrated in liver tissue between accelerated phospholipid degradation and associated membrane dysfunction after irreversible ischemic injury had occurred.[59] A detergent function of amphophilic compounds such as lysophosphatides and long-chain acyl-CoA was postulated by Corr to be associated with plasma membrane dysfunction in the heart and to have initiated the arrhythmias.[60] Despite the early enthusiasm,

more refined studies on the heart failed to find a significant, consistent decrease of sarcolemma phospholipid content; this is in possible contrast to other tissues.[61-63] This was confirmed in a recent study by Suyatna et al., who compared the composition of phospholipids of isolated highly purified sarcolemmal vesicles from normal and autolytic rat myocardium and compared normal and ischemic rat heart of 30 and 60 minutes ischemia.[64] At 60 minutes ischemia, most sarcolemmal membranes have large discontinuities in rat heart (stage 5 morphological changes; Figure 4-2).

Although enrichment of 5'-nucleotidase

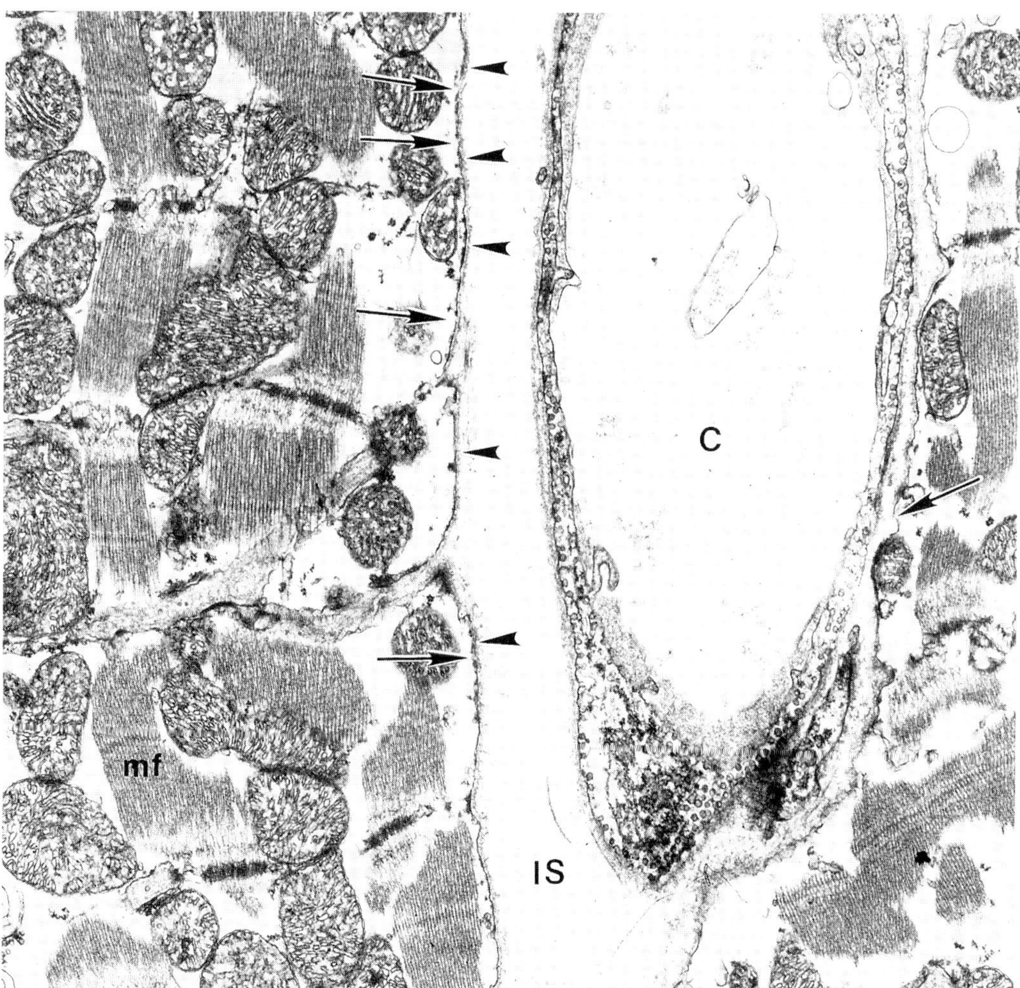

Figure 4-2. Electron micrograph of myocardium with severe ischemic injury. Mitochondria show flocculent densities. There is interstitial edema and extravasation of red blood cells. The basement membrane is intact but the plasma membrane is interrupted in many places. CB = Contraction bands; FD = flocculent densities.

in this study was 50 times, the content of lysophospholipid was not consistently increased in contrast to the reports by other authors such as Sobel et al.[60,61,65] Therefore, even though much is known about phospholipids and phospholipases, the exact mechanism of membrane alteration and the character of the membrane defect have remained a mystery. Recent data support the hypothesis that after ischemia phospholipid reorganization and bilayer destabilization occurred. This destabilization progresses faster after reperfusion.[7] Ultrastructural studies of freeze fractured membranes have demonstrated aggregation of membrane particles and extrusion of multilamellar membrane structures. This apparent dissolution of the plasma membrane has been termed *lateral phase separation* by Post et al.[7]

In Vitro Studies on Mitochondria

Flocculent Densities in Mitochondrial Matrix

Ever since pathologically altered tissues have been examined by electron microscopy, it has been noted that lethal injury is associated with the appearance of flocculent densities in the matrix of mitochondria (Figure 4-3). This association has been so firmly established that this ultrastructural marker of matrix densities has been accepted as the most significant sign of irreversible cell injury. Studies of Itkonen and Collan have convincingly shown that these flocculent densities consisted of proteins that could be digested by proteases.[66] The flocculent matrix densities of mitochondria could be isolated from the ischemic mitochondria if certain conditions were observed. It was proposed that these densities represent products of coagulation of matrix proteins, possibly associated with the depression of the matrix pH in the absence of proton transport. The functional significance of this change is not known and may require detailed future studies. One wishes to know

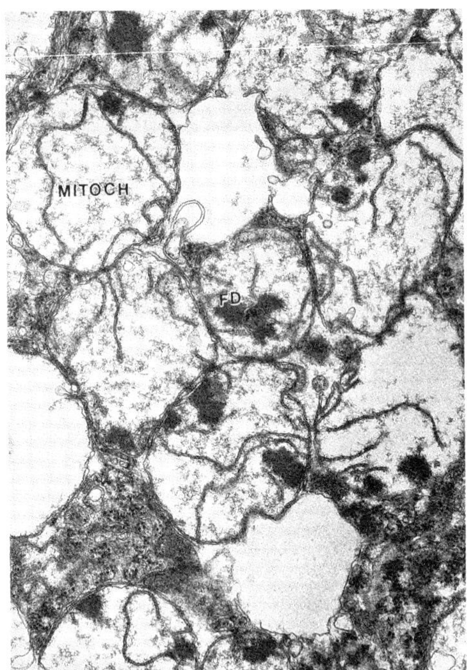

Figure 4-3. Electron micrograph of necrotic rat kidney tubular cell. Mitochondria are swollen (MITOCH). Note the flocculent densities (FD). *In vitro* ischemia.

the association of all matrix-associated functions with the appearance of flocculent densities of the matrix. As shown by Costa and Mergner, flocculent densities occur at a time when the ability to resynthesize ATP ceases and when the proton gradient collapses (see Figure 4-4 and Figure 4-10, p. 99).

Direct Measurement of ATP Synthesis by Isolated Mitochondria Damaged by *in Vitro* Ischemia

The development of conditions sufficiently grave to cause cell death requires time. The duration of viability during ischemia varies with temperature, cell type, cellular activity, and prior oxygen consumption. Since the measurements of viability are relatively coarse, ischemic tolerance is an "unknown" for most cell systems. In an attempt to search for parameters that would correlate with the advent of cell death — such as uncoupling of mitochondria, dye

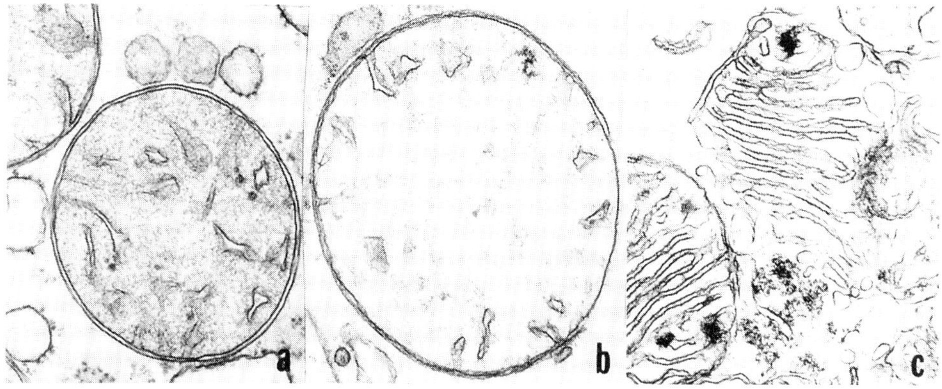

Figure 4-4. Isolated mitochondria from rat liver. (A) Normal mitochondria. (B) Thirty minutes of ischemia. (C) Sixty minutes of ischemia.

uptake and dye transport (in renal cells)— Costa and Mergner studied ATP synthesis in mitochondria.[67] The occurrence of flocculent densities in the matrix of mitochondria was chosen as the indicators of cell death. The ability of mitochondria to synthesize ATP from ADP in the presence of substrate and oxygen, and to deliver ATP externally was tested in isolated mitochondria by the luciferin–luciferase luminescence system (Figure 4-5).[70] Luminescence was observed and measured in the photomultiplier of a spectrophotometer. Mitochondria from ischemic rat heart experienced a sharp decline of ATP synthesis within 30 minutes of ischemia. Rat liver mitochondria isolated under the same conditions and with the same precautions showed decline at 45 minutes of ischemia (Figure 4-6). Morphological studies on the heart and liver tissue and on isolated mitochondria revealed that flocculent densities appeared in both liver and heart a short time later, at 45 and 60 minutes, respectively (Figure 4-4).

One may conclude from these studies that the decline of mitochondrial ATP synthesis and/or ATP export occurred at an early enough interval that it has to be considered in the pathogenesis of cell death. The functional decline was in close enough association to be considered as related, even though the morphological change occurred after the

functional change had been observed. The functional decline of ATP synthesis may be the most sensitive indicator of mitochondrial degeneration and irreversible autolysis.

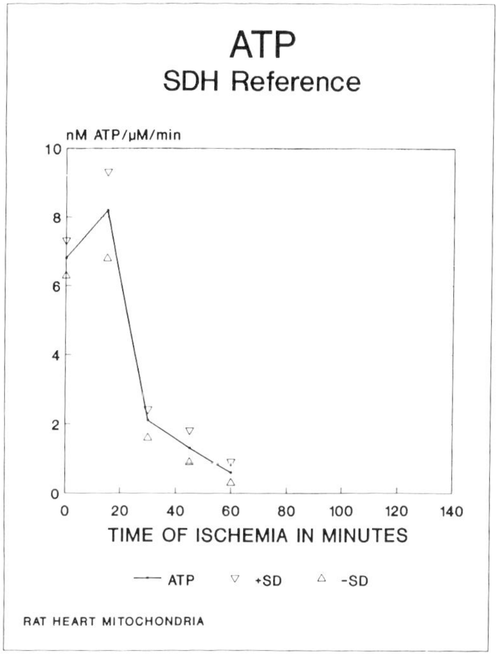

Figure 4-5. Determination ATP synthesis in isolated rat heart mitochondria by the luminescence technique in the luciferin–luciferase system. The technique measures ATP synthesis after injection of ADP/AMP. The enzyme succinate dehydrogenase (SDH) is used as reference and as indicator of the mitochondrial mass. Luminescence is standardized with a known amount of ATP. *In vitro* ischemia. (With permission.)

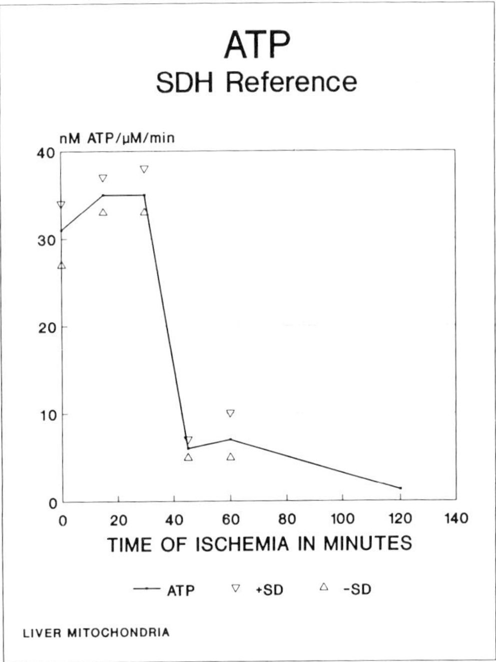

Figure 4-6. Determination ATP synthesis in isolated rat liver mitochondria by the luminescence technique in the luciferin–luciferase system (conditions as in Figure 4-5). *In vitro* ischemia. (With permission.)

Studies on Ischemic Mitochondria in the Presence of ATP

Even short-term ischemia causes ATP levels to remain depressed, even after reperfusion. This may create a threat to cell survival. Mitochondria are injured during short ischemia and can become uncoupled, suggesting that the failure of mitochondria to synthesize sufficient ATP may be connected with the later failure to survive (at least in cells depending on high energy turnover; survival of other cells may be supported by anaerobic glycolysis alone).[68] Classen et al. attempted to determine if this relative lack of energized phosphate delivery is due to accelerated ATP hydrolysis by uncoupled mitochondria or to altered ATP synthesis (Figure 4-7).[68] ATP and/or ADP/AMP were injected into a mitochondrial suspension, and the luminescence of the luciferin–luciferase system of Lemaster and Hackenbrock was observed.[70] This luminescence

system offers researchers the opportunity to observe the kinetics of added ATP or ADP in the presence of normal mitochondria and in mitochondria damaged by ischemia.

For its intended purpose the test system presents advantages over endpoint studies because it is dynamic and allows one to monitor the rate of synthesis and hydrolysis. Injected ATP functions as a probe for mitochondria. In the absence of mitochondria, injected ATP induced a rapidly ascending

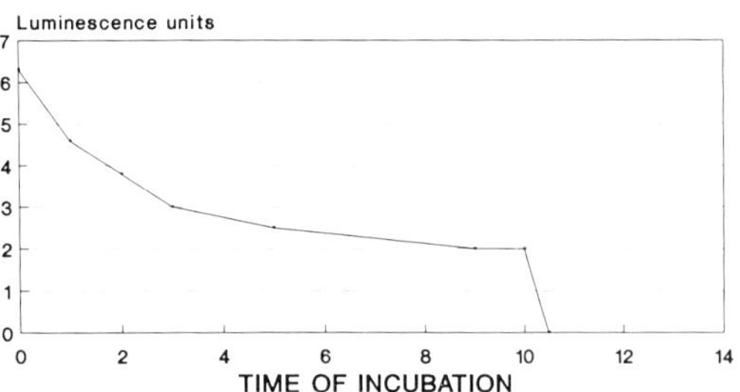

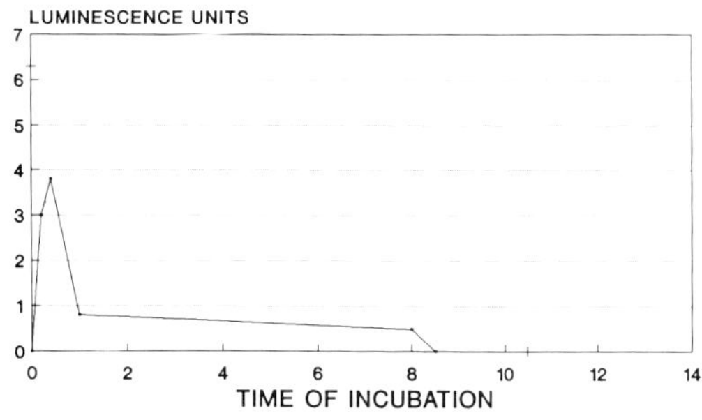

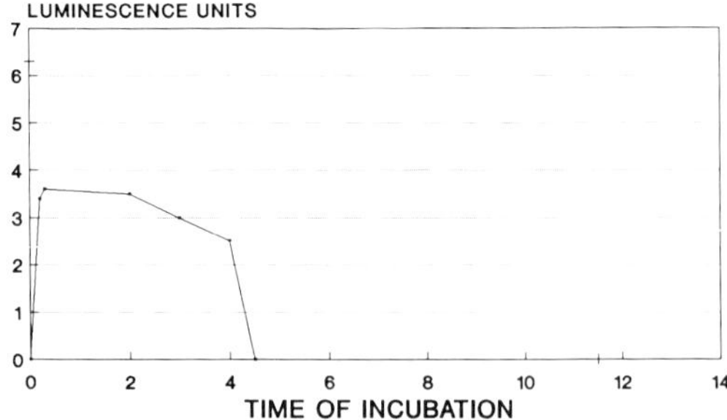

Figure 4-7. Luminescence after addition of ATP or ADP to (A) isolated normal and (B) ischemic mitochondria. (C) Addition of ADP/AMP to normal mitochondria. (With permission.)

peak of luminescence that, after peaking, declined in linear fashion over a period of time. In the presence of normal mitochondria the initial peak was followed by an initial decline, which was captured by a plateau. This plateau lasted for several minutes and subsequently descended abruptly to the baseline. Ischemic rat heart mitochondria (45 minutes) revealed the same initial peak after addition of ATP, but a much more intense early drop developed, followed by a low, slightly above baseline, plateau. Again, a quick drop in luminescence occurred after several minutes. Addition of a second aliquot of fresh ATP to a suspension of ischemic mitochondria obtained additional peaks in response to ATP, but the plateau returned quickly to the previous low level. The results were explained by the assumption that the initial luminescence peak was a direct response to the presence of ATP. The drop in luminescence probably represents ATP hydrolysis. This point was demonstrated by studies using dinitrophenol (DNP), a mitochondrial uncoupler. Uncoupled normal and ischemic mitochondria reveal only the initial peak, without the plateau phase. Work by Rouslin et al. showed that mitochondrial ATPase can, under the right conditions, hydrolyze ATP when heart muscle is made transiently ischemic.[71]

The plateau phase observed in the presence of normal and abnormal mitochondria may represent continued resynthesis of ATP from ADP, while the terminal dive of the luminescence curve may be related to the cessation of ATP resynthesis due to exhaustion of oxygen (or substrate) in the supernatant. This conclusion was supported by experiments with additional substrate and oxygen. Addition of oxygenated supernatant returned the plateau phase. These interpretations are confirmed by luminescence experiments using the addition of ADP/AMP, substrate, phosphate, and oxygen to the suspension of ischemic and normal mitochondria, which revealed a high or low plateau phase, respectively (Figure 4-7C). Based on these studies, it was concluded that under the conditions tested, the synthesis of ATP is greatly reduced and the ATPase catabolic activity remains active.

The Inner Membrane ATPase

The ATPase of the inner membrane of mitochondria is a reverse function of the ATP synthesis. As a catabolic function it can be used to identify the presence of the ATPase enzyme. It cannot be used, however, to test the enzyme for its synthetic function. The synthetic function depends on the coupled interaction between the electron transport system and the ATP synthetase or, more precisely, it depends on the intact inner membrane of mitochondria. The common test system for inner membrane ATPase requires the presence of cofactor, uncoupler, and ATP as substrate. Either ATP lysis or generation of inorganic phosphate serve as the indicator of enzyme activity. One of the important cofactors is magnesium. Mergner et al. tested the ATPase enzyme activity as a function of ischemic time in isolated heart and renal mitochondria of the rat.[72] In both systems an immediate, steady decline of the inner membrane ATPase activity was observed beginning at 5 minutes *in vitro* ischemia (Figure 4-8). Since this inactivation occurred during the early, reversible phase, and since morphological studies by negative staining in these early intervals showed no decline in inner membrane spheres in contrast to later intervals where inner membrane spheres had disappeared, it was assumed that this decline in enzyme activity was due to inhibition rather than destruction of the ATPase enzyme (Figure 4-9). This was recently shown to be a likely mechanism by Rouslin and Pullman, who demonstrated that the ischemic alteration, and particularly the increased acidity, caused reversible binding of the ATPase inhibitor protein described by Pullman and Monroy.[73,74]

The Proton Gradient

The proton gradient of mitochondria is a phenomenon connected to the function of

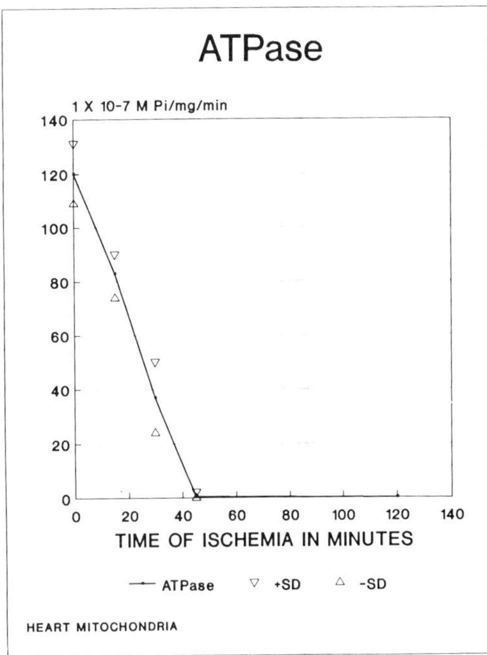

Figure 4-8. ATPase activity in isolated mitochondria from rat heart tissue. *In vitro* ischemia. (With permission.)

charge separation and synthesis of ATP. The proton gradient was originally described by Mitchell, who discovered that metabolizing isolated mitochondria generate a depression of the pH in the suspension medium. The proton gradient is measured in a closed system. Isolated mitochondria are incubated in an anoxic medium. The suspension medium contains electrolytes, substrate, and ADP. The change of pH is measured with a pH electrode. A provision for rapid injection of a measured amount of oxygenated incubation solution is provided. The resulting curve consists of two components: (1) the rapid upstroke, and (2) a shoulder. The shoulder shows the slow dissipation of the pH gradient and is thought to be an expression of the activity of the inner membrane ATPase. This shoulder of the pH curve can be rapidly discharged by the addition of uncoupler such as FCCP. The upstroke is thought to represent the amount of protons released upon addition of oxygen. It gives

the basis for calculation of the H:O ratio. It was determined that in the renal system the proton gradient declined at 60 minutes *in vitro* ischemia, while in the rat heart the proton gradient drops between 15 and 25 minutes of ischemia (Figure 4-10).[76] It has been proposed that this change corresponds to the uncoupling of mitochondria. The time of ischemia related to the collapse of the proton gradient agrees with the just-described reduction of ATP synthesis measured by luminescence studies and with the appearance of flocculent densities in isolated and *in situ* heart mitochondria.

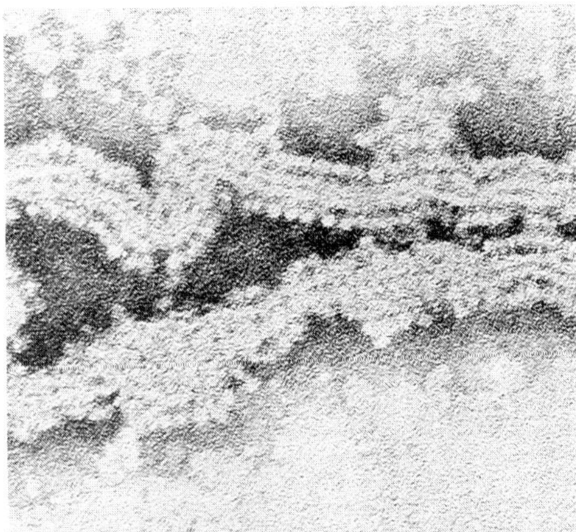

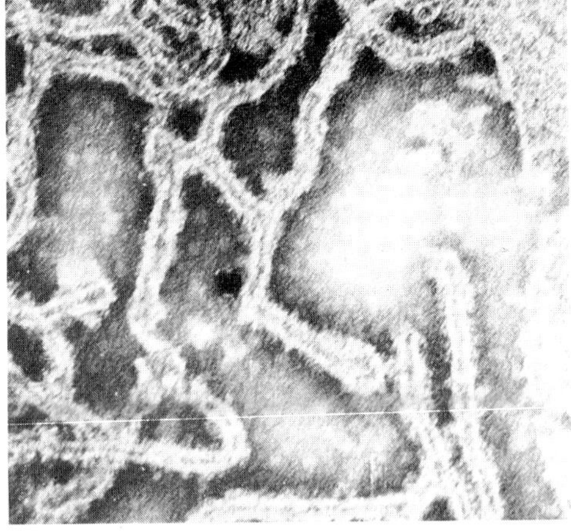

Figure 4-9. Electron micrographs of negatively stained preparations of rat kidney mitochondrial inner membrane by phosphotungstic acid. (A) Two-hour ischemia. (B) Four-hour ischemia. Note the inner membrane spheres visible in *A* and absent in *B*. *In vitro* ischemia. (With permission.)

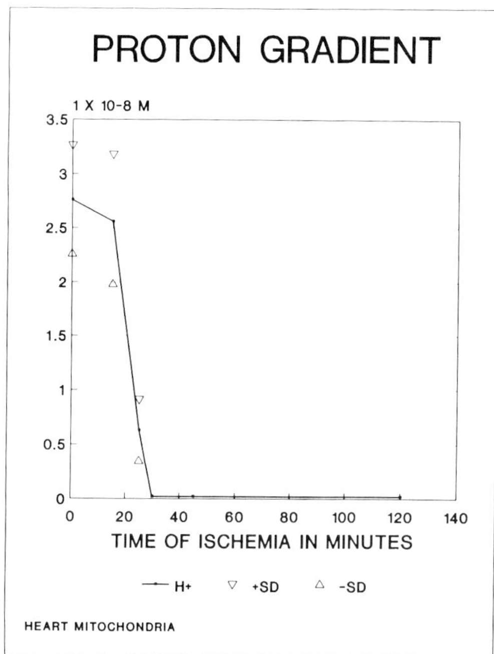

Figure 4-10. Proton gradient observed after injection of a measured amount of oxygenated solution into an anoxic mitochondrial suspension that contained substrate and ADP. *In vitro* ischemia. (With permission.)

Membrane Potential in Cell Injury

The normal membrane potential of mitochondria is 200 mV. It increases in the presence of EGTA and decreases in the presence of ADP to 190 psi. The membrane potential of mitochondria as a function of degeneration due to exposure to penta-chlorodienyl-glutathione, a nephrotoxic conjugate of hexa-chlorobutadiene, was studied by Jones et al.[77] Mitochondrial membrane potential was determined indirectly in terms of TPP activity in the incubation medium by using a TPP selective electrode in combination with a calomel reference electrode. It could be shown that mitochondria were unable to retain calcium at the same time when they were unable to develop and to maintain a membrane potential (Δ psi). No change, however, was noted in anoxic mitochondria. In anoxic mitochondria, changes became apparent when mitochondria were induced to respire. At this point a collapse of the

membrane potential and a release of the calcium occurred. Fuchs et al. appraised mitochondrial membrane potential by surface fluorimetry on DASPMI stained hearts.[78] The fluorescence intensity in living tissue is related to the electrochemical gradient of mitochondria. The author could not find a direct relationship between mitochondrial membrane potential and ATP production under different metabolic conditions. It was assumed that derangement of ATP synthesis occurs first, while the membrane potential is maintained. With the advent of irreversibility the membrane potential will decay.

Studies on Substrate Transport into Mitochondria

Translocation of Substrate in Ischemia. The alteration of substrate translocation in ischemic mitochondria is not well known. Carriers exist for certain substrates. The succinate–malate carrier is highly dependent on pH, which could play a role *in situ* in ischemic tissue with low pH. Glutamate also has a specific carrier. The interactions of a number of carriers create shuttle mechanisms of transport and cotransport that are controlled by the intermediate metabolism in the two compartments, the mitochondrial matrix space and the cytosol.[79]

Carnitine-Acylcarnitine Translocase in Ischemia. Carnitine-acylcarnitine translocase is located in the inner membrane of mitochondria. This enzyme is responsible for the transfer of fatty acids into mitochondria. As such it is an enzyme that is regulated by the metabolic status of the cell. Carbohydrate metabolism is enhanced after coronary ischemia and reperfusion, when fatty acid metabolism is depressed. Pauly et al. observed that the maximal velocity for carnitine exchange of the translocase is reduced 55% after 60 minutes of ischemia in isolated mitochondria from the heart.[80] Based on studies of alteration in matrix glutathione (50% depleted), glutathione disulfide (200% increased) and alteration of the ratio in reduced glutathione to glutathione-disulfate, a modification of protein-sulfhydryl groups was proposed as the mechanism of the inhibition. Even following reflow of 20 minutes the carnitine-acylcarnitine-translocase remained depressed.[80]

Tissue Respiration in the Heart and Kidney

Mergner et al. tested respiration in 1-mm cubes of normal and ischemic rat renal and rat heart tissue, respectively.[66,76,81] The incubation medium contained a buffered iso-osmolar salt solution containing albumin. It was maintained at 37°C and was oxygenated. Glucose was present in one set of experiments but absent in another set of experiments. Respiration was measured either in the Warburg Respirometer, modified by Gilson, or in the Clark oxygen electrode assembly. It was found that respiration proceeded unchanged in resuspended tissue that had previously been exposed to ischemia for 3 hours in the kidney and for 60 minutes in the heart (Figure 4-11). These periods are considered the early phase of ischemia. The respiration declined rather abruptly in renal tissue and more gradually in heart tissue. In the heart this decline continued over the remaining period of the test, that is, 4 hours. Respiration, however, was never zero during the 4 hours of ischemic interval tested. It is important to note that respiration persisted considerably longer after the morphological evidence of necrosis appeared. Necrosis was determined by electron microscopic features of flocculent densities of mitochondrial matrix.

Respiration of Mitochondria Following Cell Injury

In studies by Mergner et al. on respiration of renal mitochondria and on cardiac mitochondria of the rat, changes in state III plus state IV respiration were determined and were influenced by ischemic time. State III respiration is supported by substrate and by ADP; state IV respiration is the respiration that is supported by substrate alone. Two

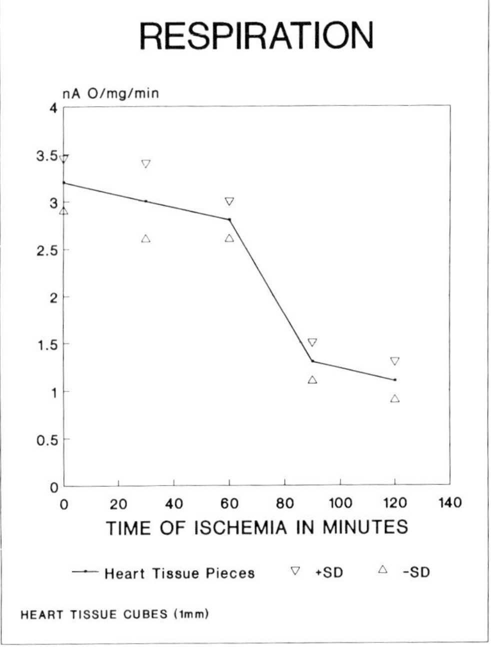

Figure 4-11. Respiration of 1-mm cubes of heart tissue incubated in the Clark oxygen electrode assembly (Yellow Spring). No glucose was present in the suspension medium. *In vitro* ischemia. (With permission.)

categories of substrate were examined: those that require NADH dehydrogenase and those requiring succinate dehydrogenase. The results in rat heart and rat kidney reveal that state III respiration and coupling become depressed during early ischemic intervals, probably a sign of the beginning uncoupling of mitochondria. This relationship was quite clear if glutamate was used as substrate. With glutamate- and succinate-supported respiration, state IV respiration persisted over the entire time interval tested; but some substrates metabolized by complex I had reduced state IV respiration during the early necrotic phase (determined by the appearance of flocculent densities; see also Figure 4-12). The importance of coupling of oxidative phosphorylation and of recovery of state III respiration was recently reported by Weinstin et al. in low flow ischemia and recovery.[83] Recovery of cardiac function determined by *dP/dt* coincided with recovery of state III respiration and coupling of mitochondria.

Respiratory Complexes in Cell Injury

Respiratory complexes are packages of megaenzymes that most likely exist as units within the membrane of mitochondria. They can be isolated as complexes and they can be tested for their ability to either donate or accept electrons. Therefore, either electron donors, hydrogen donors, or electron acceptors are used to test these complexes by spectrophotometry, respiratory activity, or analysis of reduced and oxidized NAD. Oster et al. described the reaction of complex I as a function of ischemic time (Figure 4-13).[84] Complex I is a respiratory complex that by its incorporation of the NAD-dehydrogenase leads the dehydrogenation and electron transport metabolism in mitochondria. Activity of this complex revealed a steady state of slight decline until 45 minutes of ischemia, when a more rapid

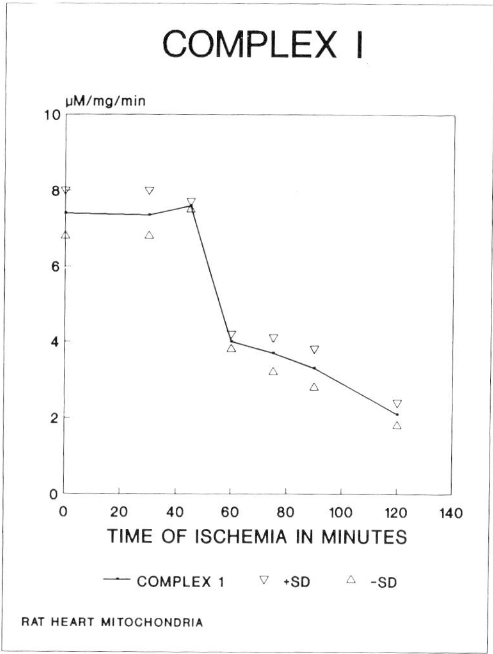

Figure 4-13. Complex I, measured in submitochondrial particles of ischemic rat heart mitochondria by oxidation of an electron donor. *In vitro* ischemia. (With permission.)

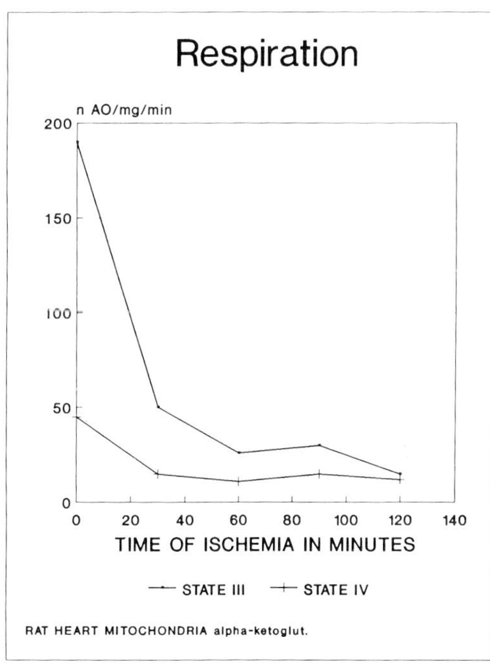

Figure 4-12. Respiration of isolated rat heart mitochondria after the addition of α-ketoglutarate. *In vitro* ischemia. (With permission.)

decline to 50% of the control activity occurred. This overall trend is in agreement with findings by Rouslin et al.[85] The slope was followed by a continuing, but less steep decline over a 4-hour ischemic period (Figure 4-14). Complex I was the only respiratory complex that revealed such a decline; complex II (including the succinate dehydrogenase enzyme), III (including cytochrome b and cytochrome c-1), and IV remained stable during the ischemic interval (4 hours) tested.[71,73,85] Based on these data Smith et al. recommended using complex II, III, and IV related enzyme activity to monitor mitochondrial yield for efficiency of isolation.[56]

Ubiquinone

Ubiquinones are a group of mobile electron and proton carriers. Ubiquinone can be extracted by lipid solvents and quantitated. In preliminary studies, ubiquinone was ex-

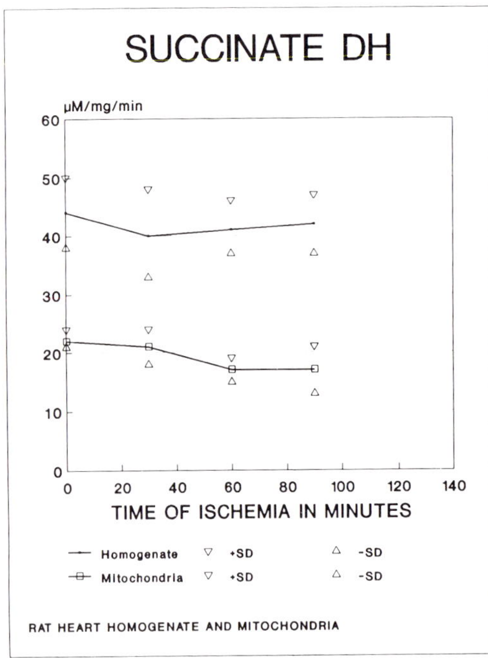

Figure 4-14. Succinate dehydrogenase in the homogenate and in the pellet of isolated rat heart mitochondria. (With permission.)

tracted from ischemic and nonischemic cardiac mitochondria exposed to *in vitro* ischemia. The spectrophotometric analysis revealed a decline in concentration of ubiquinone per milligram protein and per succinate dehydrogenase activity as a function of ischemic time to 50% within a two-hour interval. These data, if confirmed, are potentially interesting because of the association of ubiquinone with complex I. Some supporting evidence has come from studies by Marnbayashi et al., who observed preservation of ischemic rat liver mitochondrial function and liver viability with CoQ_{10}.[86]

Histochemistry and Enzyme Analysis of Ischemic Myocardium

Fredericks et al. studied the relationship of changes in histochemical stains related to mitochondrial enzymes to the occurrence of flocculent densities in the rat liver.[87] It is reported that flocculent densities appeared in the liver after 2 hours of 37°C *in vitro* ischemia. After this time, histochemical staining for glutamate-dehydrogenase, a matrix enzyme, decreased, whereas succinate dehydrogenase histochemical staining—an enzyme of the inner membrane—and monamine oxidase-related histochemical staining—an outer membrane enzyme—persisted.

Nitro-blue-tetrazolium (NBT) converts to formazan by the uptake of two electrons and one proton. The formazan product thus formed is a colored substance. Intracellular redox systems having a lower redox potential than the dye itself can interact with tetrazolium salts. Examples of this system are NAD/NADH, NADP/NADPH, flavoproteins, cytochrome b, cytochrome b5, and cytochrome P_{450}. The system also requires an electron carrier. When free electrons are made available by an enzyme system, the dye can be reduced. Enzyme systems that can perform this task are called *diaphorases*. Nachlas and Schnitka, and Ramkisson have applied the NBT histochemical reaction as a test for tissue dehydrogenases.[88,89] Whether the histochemical test assays the dehydrogenase directly or this reaction is dependent on the presence of certain cofactors, as proposed by Klein and Schaper, has not been settled.[90] If cofactors are important, the following reaction can be visualized:

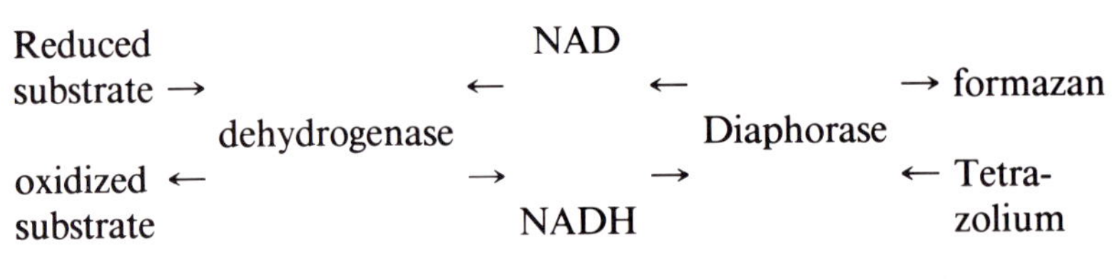

Leventhal examined the relationship between ischemic time and the fading of the NBT reaction in the globally ischemic heart.[91] Three types of staining reactions were found: (1) NBT staining related to pyruvate dehydrogenase declined immediately, (2) NBT histochemical staining supported by malate began to decrease after 60 minutes global ischemia at 37°C (this decline could be retarded, as reported by Klein and Schaper),[90] and (3) NBT histochemical stain supported by succinate as substrate was somewhat reduced but still positive after 24 hours of *in vitro* ischemia. The latter results are in agreement with the data of Klein et al., who argued that substrates that do not require NADH as cofactor will show a staining reaction; that is, declining staining coincides with the loss of NAD in the tissue.[90]

The simultaneous biochemical analysis of the dehydrogenases by Leventhal showed that pyruvate dehydrogenase activity displayed a sharp decline between 45 minutes to 60 minutes of ischemia.[91] This is significantly later than the histochemical reaction (15 minutes), but still the earliest dehydrogenase change of the three dehydrogenases compared. Succinate dehydrogenase was only slightly depressed after 24 hours of ischemia, while malate dehydrogenase activity declined between 3 and 4 hours of *in vitro* ischemia, coinciding with the endpoint of decreasing staining quality. In relation to the morphological hallmark of irreversibility, the decline of pyruvate dehydrogenase activity occurred shortly after flocculent densities had appeared. The histochemical test had no relationship to this event. All changes in histochemical and enzyme tests of malate dehydrogenase and succinate dehydrogenase had no relationship to the morphological point of no return. They also had no relationship to the decline in respiration of tissue cubes.

Summary of Studies on Respiration and Oxidative Phosphorylation

The appearance of flocculent densities as markers of irreversible cell injury indicate that necrosis begins in the globally ischemic rat heart at 37°C at approximately 30 minutes; in the kidney at about 60 minutes; and in the liver at about 45 minutes of ischemia. The time period when flocculent densities appear is identical for ischemic tissue and isolated mitochondria. Simultaneous functional changes of mitochondria are the loss of detectable proton gradient and of detectable active ATP synthesis. Changes that occur prior to the appearance of flocculent densities are the inhibition of the inner membrane ATPase and the fading of pyruvate-related histochemical staining by NBT. Changes in respiration, including complex I respiration, may possibly be early alterations in the phase of necrosis.

Calcium Metabolism in Cell Injury

Mergner et al. examined the ability of renal mitochondria to accumulate calcium after exposure to periods of *in vitro* ischemia (Figure 4-15).[92] The authors found that rela-

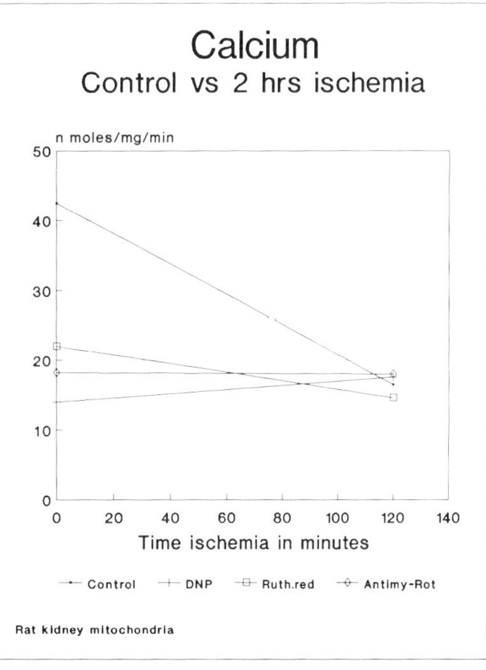

Figure 4-15. Calcium accumulation measured as uptake of [45]Calcium by isolated rat kidney mitochondria. Uptake is supported by substrate and respiration. 0 sub = Absence of substrate; RR = Ruthenium red; DNP = uncoupling of mitochondria by dinitrophenol.

tively short periods of ischemia did not affect the ability of renal mitochondria to accumulate calcium. Mitochondria obtained from ischemic renal tissue of more than 15 minutes showed continued loss of the ability to accumulate calcium either supported by substrate metabolism and respiration, or supported by ATP. Calcium accumulation, however, was not eliminated. This inability of mitochondria to sequester calcium during ischemia and following reflow after prolonged ischemia was confirmed by a number of groups in different organs, particularly in the heart.[2,16,17] It provides one essential point in assigning a leading role to the uncontrolled level of cytosolic calcium in ischemic cell destruction. Levels of cytosolic calcium are directly related to the severity of organelle alteration. Studies that used calcium antagonists such as lidoflazine report partial prevention of loss of mitochondrial oxidative phosphorylation.[93] Arnold et al. determined in the rat kidney that 50 minutes of $37°C$ ischemia caused severe reduction of adenine nucleotides and ATP.[94-96] Mitochondria could not take-up calcium and efflux of mitochondrial calcium occurred. Upon reflow, nucleotides levels recovered, although they remained depressed, and mitochondria regained the ability to accumulate and release calcium within 3 hours of ischemia.

Oxidative Phosphorylation as a Function of pH

In order to simulate the intracellular pH changes in ischemic tissue, Mergner et al. examined mitochondrial function of state 3 and state 4 respiration as a function of the pH of the suspension medium.[97] They found that both extreme alkaline and extreme acid conditions altered state 4 respiration, state 3 respiration, and coupling, while relatively few changes occurred in a pH region ranging from pH 6.5 to 7.8. More recently, Rouslin and Pullman reported that after 20 minutes of myocardial ischemia, mitochondrial ATPase was from 50 to 60%

inhibited.[73] The acid pH favored the linking of an ATPase inhibitor protein reported by Pullman and Monroy.[74] Subsequent studies on submitochondrial particles showed that control ATPase contained 62 percent as much inhibitor as ischemic mitochondria. Submitochondrial particles derived from ischemic myocardium showed an increase of 2.5 percent of inner membrane ATPase. Inhibitor protein could be released from ischemic submitochondrial particles.

Mitochondrial Phospholipids and Phospholipases

Shaikh and Downbar assessed the changes in major phospholipids as a function of ischemic time.[61] The authors found that within an ischemic interval of up to 40 minutes there was no major change in the major phospholipids of the heart. The phospholipids subsequently fell after 8 and 24 hours by 6% and 14%, respectively. Lysophospholipids, however, increased rapidly and reached levels of 60 percent above normal in 8 minutes. The total amount of lysophospholipids, however, remained small (0.6 percent).

The Delivery of Energized Phosphate to Sites of Use in Heart Cells

Modulation of Adenine Nucleotide Translocase Activity

Shug and Subramanian reported decreased adenine nucleotide transferase following ischemia and reperfusion.[98] It was proposed that early mitochondrial swelling occurred in part due to the decreased ADP/ATP transport and depressed oxidative phosphorylation. It is suggested that this process may be linked to free radical formation and lipid peroxidation altering mitochondria and plasma membrane.

The Creatine Kinase–based Energized Phosphate Metabolism

In normoxic hearts the rate of oxidative phosphorylation, the velocity of the creatine kinase reaction, and cardiac performance are closely coupled.[99] Creatine kinase is a phosphotransferase transferring energized phosphate groups from creatine phosphate to ADP:

$$\text{ATP} + \text{creatine} \leftrightarrow \text{ADP} + \text{phosphocreatine}^{100}$$

Phosphocreatine is an example of a high energy phosphate compound, also called *phosphagens*. It is the immediate energy source for contraction in skeletal muscle and nerve tissue, and also plays a crucial role in the myocardium. In the heart the isoenzymes of creatine kinase are found in the mitochondria and the myofilaments. The mitochondrial creatine kinase is of prime importance in the transfer of the newly created high energy phosphate group out of the mitochondria to the site of immediate usage. Phosphocreatine plays an important intermediary role because it is highly diffusible whereas ADP is not.

In ischemic injury, recovery of postischemic function may be limited because of energy production (oxidative phosphorylation), energy utilization (Actomysin ATP), or energy transfer (by creatine kinase).[99] Bittl et al. have shown that both energy production and transfer are deficient in postischemic myocardium.[101,102] The accumulation of inorganic phosphate in ischemic tissue has been implicated for the separation of the creatine kinase from the mitochondrial membrane. Neubauer et al. tested isolated ferret hearts made ischemic for 20, 40, and 60 minutes, then reperfused for 40 minutes. The hearts were compared for myocardial function monitored by the rate pressure product, coronary flow, changes in ATP content, changes in creatine phosphate, inorganic phosphate, cellular pH, and oxygen consumption. It was determined that rate pressure product recovery was depressed in steps of severity beginning at 20 minutes of ischemia and was abolished at 60 minutes of ischemia.

CONCLUSION

We have presented very detailed evidence on ischemia-induced pathological changes of the plasma membrane and mitochondria. The evidence is by no means sufficient to favor one or the other pathogenic mechanisms. In fact, we were unable to present an all-encompassing principle, but were bound to describe a network of many tiles of a mosaic. Studies on aging organelles as a model for autolysis seem to have taught us that persistent function is linked with maintenance of organelle integrity. Mitochondria that are able to synthesize ATP, for example, may stay intact longer.

From the *in vitro* and *in vivo* ischemia studies presented it seems clear that decline of essential membrane-related functions of mitochondria and of the plasma membrane occurs early in ischemia, so early that the decline in these functions may be responsible for the irreversibility of the injury. To our way of thinking, the studies of Nayler et al. on the prolonged time course and the burst of pathological calcium influx after reflow across the plasma membrane are important considerations.[17] From the studies on organelles isolated from ischemic tissue, we have learned that the time course of autolytic alteration may affect some functions earlier than others. Important data demonstrate how small components of the bioenergetic machinery of mitochondria cease to function and may be the indicators of significant alterations in cell metabolism. Data on the decline of the proton gradient deserve, in our opinion, further attention and proof.

ACKNOWLEDGMENT

The outstanding support of Kay Bailey, Librarian at the Armed Forces Institute of Pathology, Washington D.C., is gratefully recognized.

REFERENCES

1. Rossen RD, Michael LH, Kagiyama A, et al. Mechanism of complement activation after coronary artery occlusion: evidence that myocardial ischemia in dogs causes release of constituents of myocardial subcellular origin that complex with human C1q *in vivo. Circ Res.* 1988;62:572.

2. Poole-Wilson PA. What causes cell death? In: Herse DJ, Yellon DM, eds. *Therapeutic Approaches to Myocardial Infarct Size Limitation.* New York: Raven Press; 1984;43–60.

3. Jennings RB, Hawkins HK, Lowe JE, Hill MC, Klottman S, Reimer KAS. Relation between high energy phosphate and lethal injury in myocardial ischemia in the dog. *Am J Pathol.* 1978;92:187–214.

4. Reimer KA, Hill ML, Jennings RB. Prolonged depletion of ATP and of the adenine nucleotide pool due to delayed resynthesis of adenine nucleotides following reversible myocardial ischemic injury in the dog. *J Mol Cell Cardiol.* 1981;13:229–239.

5. Pine MB, Caulefield JB, Bing OHL, Brooks WW, Abelman WH. Resistance of contracting myocardium t swelling with hypoxia and glycolytic blockade. *Cardiovasc Res.* 1979;13:215–224.

6. Weglicki WB, Waite BM, Stam, AC. Association of phospholipase A with a myocardial membrane preparation containing the $(Na^+ + K^+)\text{-}Mg^{2+}\text{-}$ATPase. *J Mol Cell Cardiol.* 1972;4:195–201.

7. Post JA, Ruigrok TJC, Verkleij ASJ. Phospholipid reorganization and bilayer destabilization during myocardial ischemia and reperfusion: A hypothesis. *J Mol Cell Cardiol.* 1988;20:107–112.

8. Ogunro EA, Ferguson AG, Lesch M. A kinetic study of the pH optimum of canine cardiac cathepsin D. *Cardiovasc Res.* 1980;14:254–260.

9. Tones MA, Poole-Wilson PA. Effects of lysophosphatidylcholine on cellular integrity and calcium uptake in rabbit myocardium. *Eur Heart J.* 1983;4(suppl E):73.

10. Tones MA, Letts G, Fleetwood G, Poole-Wilson PA. Failure to demonstrate a pathological role for leukotriene D4 in hypoxic rabbit myocardium. *J Mol Cell Cardiol.* 1983;15(suppl 1):409.

11. Katz AM, Messineo FC. Lipid–membrane interactions in the pathogenesis of ischemic damage in the myocardium. *Circ Res.* 1981;48:1.

12. Altschuld RA, Hostetler JP, Brierly G. Responses of isolated heart cells to hypoxia, reoxygenation and acidosis. *Circ Res.* 1981;49:307.

13. Romson JL, Hook BG, Kunkel SL, Abrams GD, Schork MA, Lucchesi BR. Reduction of the extent of ischemic myocardial injury by neutrophil depletion in the dog. *Circulation.* 1983;67(5):1016–1023.

14. Downey JM, Miura T, Eddy LJ, et al. Xanthine oxidase is not a source of free radicals in the ischemic rabbit heart. *J Mol Cell Cardiol.* 1987;19(11):1053–1060.

15. Papahadjopoulos D. Calcium induced phase changes and fusion in natural and model membranes. In: Porst H, Nicolson GL, eds. *Membrane Fusion: Cell Surface Reviews.* Amsterdam: Elsevier; 1978;765–790.

16. Dhalla NS, Panagia V, Singal PK, et al. Alteration in heart membrane calcium transport during development of ischemia-reperfusion injury. *J Mol Cell Cardiol.* 1988;20(suppl II):3–14.

17. Nayler WG, Panagiotopoulos S, Elz JS, Daly MJ. Calcium-mediated damage during postischemic reperfusion. *J Mol Cell Cardiol.* 1988;20:41–54.

18. Mergner WJ, Chang SH, Trump BF. Studies on the pathogenesis of ischemic cell injury, V: morphologic changes of the pars convoluta (P1 and P2) of the proximal tubule of rat kidney made ischemic *in vitro. Virchows Arch.* 1976;21:211–228.

19. Burstein C, Loyter A, Racker E. Effects of phospholipase on structure and function of mitochondria. *J Biol Chem.* 1971;246:4075–4082.

20. Burstein C, Kandrach A, Racker E. Effect of phospholipase and lipase on submitochondrial particles. *J Biol Chem.* 1971;246:4083–4089.

21. Hillered L, Chan PH. Effects of arachidonic acid on respiratory activities in isolated brain mitochondria. *J Neurosci Res.* 1988;19:94–100.

22. Luzikov VN, Galkin AV, trans., Roodyn DB, ed. *Mitochondrial Biogenesis and Breakdown.* New York: Consultants Bureau; 1985.

23. Mergner WJ, Trump BF. Ultrastructural and functional changes in mitochondria aged *in vitro. Lab Invest.* 1970;22:505.

24. Carafoli E, Gazzotti P. Loss and maintenance of energy linked functions in aged mitochondria. *Biochem Biophys Res Commun.* 1970;39:842–846.

25. Siekovitz P, Potter VR. Adenylate kinase of rat liver mitochondria. *J Biol Chem.* 1953;201:1.

26. Kaufman BT, Kaplan NO. Mechanism of depletion of mitochondrial pyridine nucleotides. *Biochim Biophys Acta.* 1960;39:332–342.

27. Estraad-O S. The release of mitochondrial enzymes by uncoupling agents. *Arch Biochem Biophys.* 1964;106:498.

28. Estraad-O S, Carabez AT, Cabeza AG. Effects of phospholipids on induced enzyme release from mitochondria. Biochemistry. 1966;5:3432–3440.

29. Wojtczak L, Lehninger AL. Formation and disappearance of an endogenous uncoupling factor during swelling and contraction of mitochondria. *Biochim Biophys Acta.* 1961;51:442–456.

30. Brierly GP; Packer L, Gomez-Poyou A, eds. *Monovalent Cation Transport by Mitochondria in Mitochondria Bioengergetics, Biogenesis, and Membrane Structure.* New York: Academic Press; 1976:3.

31. Nachbaur J, Colbeau A, Vignais PM. Distribution of membrane-confined phospholipases A in the rat hepatocyte. *Biochim Biophys Acta.* 1972;274:426–446.

32. Waite M, Sisson P. Utilization of neutral glycerides and phosphoethanolamine by phospholipase A_1 of the plasma membrane of rat liver. *J Biol Chem.* 1973;248:7985.

33. Rossi CS, Lehninger AL. Stoichiometry of respiratory stimulation, accumulation of calcium and

phosphate, oxidative phosphorylation in rat liver mitochondria. *J Biol Chem.* 1964;239:3971.

34. Smith MW, Collan Y, Kahng MW, Trump BF. Changes in mitochondrial lipids of rat kidney during ischemia. *Arch Biochem Biophys.* 1980; 618:192–201.

35. Scarpa A, Lindsey JG. Maintenance of energy-linked functions in rat-liver mitochondria aged in the presence of nupercaine. *Eur J Biochem.* 1972;27:401.

36. Parce JW, Cunningham CC, Waite M. Mitochondrial phospholipase A_2 activity and mitochondrial aging. *Biochemistry.* 1978;17:1634.

37. Hatefi Y, Hanstein WG. On energy conservation and transfer in mitochondria. *Bioenergetics.* 1972;3:129–136.

38. Weisiger RA, Friedovich I. Superoxide dismutase. Organelle specificity. *J Biol Chem.* 1973;248: 3582.

39. Panchenko LF, Brusov O, Gerasimov AM, Lokteva TD. Intra-mitochondrial localization and release of rat liver superoxide dismutase. *FEBS Lett.* 1975;55:84–87.

40. Peeters-Joris C, Vandervoorde AM, Baudhuin P. Subcellular localization of superioxide dismutase in rat liver. *Biochem J.* 1975;134:31.

41. Bartley W, Birt LM. In: Bartley W, Kornberg HL, Quale JR, eds. In: *Essays in Cell Metabolism.* London: Wiley Interscience; 1970:1–44.

42. Alberti KG, Bartley W. The localization of proteolytic activity in rat liver mitochondria and its relation to mitochondrial swelling and aging. *Biochem J.* 1969;111:763–776.

43. Romani RJ, Monadjem A. Mitochondrial longevity *in vitro*: the retention of respiratory control. *Proc Natl Acad Sci USA.* 1970;66:869–873.

44. Ozelkok S, Romani RJ. Restoration of energy linked functions in "aging" rat-liver mitochondria. *Life Sci.* 1974;14:1427–1431.

45. Watanabe F, Kamiike W, Nishimura T, Hashimoto T, Tagawa K. Decrease in mitochondrial levels of adenine nucleotides and concomitant mitochondrial dysfunction in ischemic rat liver. *J Biochem.* 1983;94:493–499.

46. Goldblatt MJ, Romani RJ. Maintenance of respiratory control by beef heart mitochondria incubated at 25 degrees C: response to protective agents and to protective agents and prior stress. *Arch Biochem Biophys.* 1977;183:149–155.

47. Parce JW, Spach PI, Cunningham CC. Deterioration of rat liver mitochondria under conditions of metabolite deprivation. *Biochem J.* 1980;188: 817–822.

48. Franson RC, Pang DC, Towle DW, Weglicki WB. Phospholipase A activity of highly enriched preparations of cardiac sarcolemma from hamster and dog. *J Mol Cell Cardiol.* 1978;10:921.

49. Lloyd D. General methodology for isolation and characterization of mitochondria from microorganisms. *Methods Enzymol* 1979;55:135–44.

50. Mansier P, Charlemagne D, Rossi B, Preteseille M, Swynndgedauw B, Lelievre L. Isolation of impermeable inside-out vesicles from enriched sarcolemma fraction of rat heart. *J Biol Chem.* 1983;258:6628.

51. Vatner DE, Knight DR, Shen Y-T, Thomas JX, Homcy CJ, Vatner SF. One hour of myocardial ischemia in the conscious dog increases beta-adrenergic receptors, but decreases adenyl cylase activity. *J Mol Cell Cardiol.* 1988;20:75–82.

52. Chemnitius JM, Sasaki Y, Burger W, Bing RJ. The effect of ischemia and reperfusion on sarcolemmal function in perfused canine hearts. *J Mol Cell Cardiol.* 1985;17:1139.

53. Jennings RB, Hawkins KH. Ultrastructural changes of acute myocardial ischemia. In: Wildenthal K, ed. *Degradative Processes in Heart and Skeletal Muscle.* Research Monographs in Cell and Tissue Physiology. 1980:295. Monograph 3.

54. Corr PB, Gross RW, Sobel BE. Amphiphatic metabolites and membrane dysfunction. *Circ Res.* 1984;55:135.

55. Sperelakis N. Regulation of calcium slow channels of cardiac muscle by cyclic nucleotides and phosphorylation. *J Mol Cell Cardiol.* 1988; 20:75–106.

56. Smith MW, Wehman N, Mergner WJ. Studies on representative isolation of mitochondria from ischemic myocardium. *Fed Proc.* 1981;40:750.

57. Weglicki WB. Degradation of phospholipids of myocardial membranes. In: Wildenthal K, ed. *Degradative Processes in Heart and Skeletal Muscle, Vol. 3.* Research Monographs in Cell and Tissue Physiology. North Amsterdam: Elsevier; 1980:337.

58. Chien KR, Reeves JP, Buja M, Bonte F, Parkey RW, Willerson JT. Phospholipid alteration in canine ischemic myocardium. *Circ Res.* 1981; 48:711.

59. Chien KR, Abrams J, Serroni A, Martin JT, Farber JL. Accelerated phospholipid degradation and associated membrane dysfunction in irreversible ischemic injury. *J Biol Chem.* 1978;253: 4809.

60. Corr PB, Snyder DW, Lee BI, Gross RW, Keim CR, Sobel BE. Pathophysiological concentrations of lysophosphatides and the slow response. *Am J Physiol.* 1982;234(2):P H187-95.

61. Shaikh NA, Downar E. Time course of changes in porcine myocardial phospholipid levels during ischemia: a reassessment of the lysolipid hypothesis. *Circ Res.* 1981;49:316.

62. Chien KR, Han A, Sen A, Buja LM, Willerson JT. Accumulation of unesterified arachidonic acid in ischemic canine myocardium. *Circ Res.* 1984;54:313.

63. Steenbergen C, Jennings RB. Relationship between lysophospholipid accumulation and plasma membrane injury during total *in vitro* ischemia in dog heart. *J Mol Cell Cardiol.* 1984;16:605.

64. Suyatna FD, van Veldhoven PP, Borgers M, Mannaerts GP. Phospholipid composition and amphiphile content of isolated sarcolemma from normal and autolytic rat myocardium. *J Mol Cell Cardiol.* 1988;20:47.

65. Sobel BE, Corr PB, Robison AK, Goldstein RA, Witkowski FX, Klein MS. Accumulation of lysophosphoglycerides with arrhythmogenic proper-

ties in ischemic myocardium. *J Clin Invest.* 1978;62:546.

66. Itkonen P, Collan Y. Mitochondrial flocculent densities in ischemia: digestion experiments. *Acta Pathol Microbiol Immunol Scand.* 1983; 91:463–468.

67. Costa M, Mergner WJ. ATP synthesis and viability of liver and heart. *Fed Proc.* 1984;43:751.

68. Shimahara Y. Four stages of mitochondrial deterioration in hemorrhagic shock. *Res Exp Med.* 1981;179:23–33.

69. Classen B, Mergner WJ, Costa M. *In vitro* ischemic heart mitochondria using the luciferin–luciferase system as indicator of ATP levels. *J Cell Physiol.* 1989;141:53–59.

70. Lemaster JJ, Hackenbrock CR. Continuous measurements and rapid kinetics of ATP synthesis in rat liver mitochondria, mitoplasts and inner membrane vesicles determined by firefly luminescence. *Eur J Biochem.* 1976;67:1–10.

71. Rouslin W, Erickson JL, Solaro RJ. Effects of oligomycin and acidosis on rates of ATP depletion in ischemic heart muscle. *Am J Physiol.* 1986;250:H503–H508.

72. Mergner WJ, Chang SH, Marzella LL, Kahng MW, Trump BF. Studies on the pathogenesis of ischemic cell injury, VIII: ATPase of rat kidney mitochondria. *Lab Invest.* 1979;40:686–694.

73. Rouslin W, Pullman ME. Protonic inhibition of the mitochondrial adenosine 5′-triphosphatase on ischemic cardiac msucle. Reversible binding of the ATPase inhibitor protein to the mitochondrial ATPase during ischemia. *J Mol Cell Cardiol.* 1987;19(7):661–668.

74. Pullman ME, Monroy GC. A naturally occurring inhibitor of mitochondrial adenosine triphosphatase. *J Biol Chem.* 1963;238:3762.

75. Mitchell P. *Chemiosmotic Coupling and Energy Transduction.* Bodmin, Cornwall: Glynn Research Ltd; 1968.

76. Mergner WJ, Marzella LL, Mergner C, Kahng MW, Smith MW, Trump BF. Studies on the pathogenesis of ischemic cell injury, VII: proton gradient and respiration of renal tissue cubes, renal mitochondria and submitochondrial particles following ischemic cell injury. *Beitr Pathol.* 1977;161:260–271.

77. Jones DP. Renal metabolism during normoxia, hypoxia, and ischemic injury. *Ann Rev Physiol.* 1986;48:33.

78. Fuchs J, Zimmer G, Breiter-Hahn J. A multiparameter analysis of the perfused rat heart: response to ischemia, uncouplers and drugs. *Cell Biochem Funct.* 1987;5:245–253.

79. Chappell JB, Robinson BH. Penetration of mitochondrial membrane by tricarboxylic acid anions. In: Goodwin TW, ed. *The Metabolic Role of Citrate.* New York: Academic Press, 1968;123–133.

80. Pauly DF, Yoon SB, McMillin JB. Carnitine-acylcarnitine translocase in ischemia: evidence for sulfhydryl modification. *Am J Physiol.* 1987; 253:H1557.

81. Mergner WJ, Smith MW, Trump BF. Studies on the pathogenesis of ischemic cell injury, XI: P/O ratio and acceptor control. *Virchows Arch.* 1977;26:17–26.

82. Chance B, Williams GR. The respiratory chain and oxidative phosphorylation. *Adv Ezymol.* 1956;17:65–134.

83. Weinstein ES, Spector ML, Adams EM, Yokum MD, Humphrey RW, Fry DE. Mitochondria and myocardial performance: response of ischemia and reperfusion. *Arch Surg.* 1986;121:324–329.

84. Oster K, Livengood J, Wehman W, Smith MA, Mergner WJ. Studies on respiration of ischemic rat hearts: complex I. *Fed Proc.* 1981;40:751.

85. Rouslin W, Ranganathan S. Impaired function of mitochondrial electron transfer complex I in canine myocardial ischemia: loss of flavin mononucleotide. *J Mol Cell Cardiol.* 1983;15(8):P537–542.

86. Marubayashi S, Dohi K, Ezaki H, Hayashi K, Kawasaki T. Preservation of ischemic rat liver mitochondrial function and liver viability with CoQ$_{10}$. *Surgery.* 1982;91:631–637.

87. Frederiks WM, Marx F, Myagkaya GL. A histochemical study of changes in mitochondrial enzyme activities of rat liver ischemia *in vitro. Virchows Arch [B].* 1986;51:321–329.

88. Schnitka TN, Nachlas MM. Histochemical alterations in ischemic heart muscle and early myocardial infarction. *Am J Pathol.* 1963;42:507–527.

88. Ramkisson RA, Macroscopic identification of early myocardial infarction by dehydrogenase alteration. *J Clin Path.* 1966;19:479.

90. Klein HH, Schaper J, Puschmann S, Nienaber C, Kreuzer H, Schaper W. Loss of canine myocardial nicotinamide adenine dinucleotides determines the transition from reversible to irreversible ischemic damage in myocardial cells. *Bas Res Cardiol.* 1981;76:612–621.

91. Leventhal HJ. *Changes in the Histochemistry and Biochemistry of Dehydrogeneses in Ischemic Myocardium of the Rat: Comparison with Morphology.* Baltimore: University of Maryland; 1987. Thesis.

92. Mergner WJ, Smith MW, Saraphong S, Trump BF. Studies on the pathogenesis of ischemic cell injury, VI: accumulation of calcium by isolated mitochondria of ischemic rat kidney cortex. *Virchow Arch (Cell Pathol).* 1977;26:1–16.

93. Rosenthal RE, Hamud F, Fiskum G, Varghese PJ, Sharpe S. Cerebral ischemia and reperfusion: prevention of brain mitochondria injury by lidoflazine. *J Cereb Flow Metab.* 1987;7:752.

94. Arnold PE, Lumbertgul D, Burke TJ, Schrier RW. Adenine nucleotide metabolism and mitochondrial Ca^{2+} transport following renal ischemia. *Am J Physiol.* 1986;250:F357.

95. Arnold PE, Van-Putten VJ, Lumlertgul D, Burke TJ, Schrier RW. Adenine nucleotide metabolism and mitochondrial Ca^{2+} transport following renal ischemia. *Am J Physiol.* 1986;250:F357–363.

96. Arnold PE, Lumbertgul D, Burke TJ, Schrier RW. *In vitro* versus *in vivo* mitochondrial calcium loading in ischemic acute renal failure. *Am J Physiol.* 1985;248:F845.

97. Mergner WJ, Hickey JM, Chang SH, Beresesky

IK, Weissman HF, Trump BF. Studies on ischemia of the myocardium: stereometry and microprobe analysis. *Proc Int Aca Path XI Int Congress.* 1976.

98. Shug AL, Subramanian R. Modulation of adenine nucleotide translocase activity during myocardial ischemia. *Z Kardiol.* 1987;76:(suppl 5):26.

99. Neubauer S, Hamman BL, Perry SB, Bittl JA, Ingwall JS. Velocity of the creatine kinase reaction decreases in post ischemic myocardium: a 31P-NMR Magnetization Transfer study of the isolated ferret heart. *Circ Res.* 1988;63:1–15.

100. Lehninger A. *Biochemistry.* 2nd ed. New York: Worth Publisher, 1975.

101. Bittl JA, Weisfeld ML, Jacobus WE. Creatine kinase of the heart mitochondria: the progressive loss of enzyme activity during *in vivo* ischemia and its correlation with depressed myocardial function. *J Biol Chem.* 1985;260:208–214.

102. Ambrosio G, Jacobus WE, Bergman CA, Weisman HF, Becker LC. Preserved high energy phosphate metabolic reserve in globally "stunned" hearts, despite reduction of basal ATP content and contractility. *J Mol Cell Cardiol.* 1987; 19:953–964.

5

Cell Metabolism and Microcirculation during Ischemia and Reflow

Wolfgang J. Mergner

INTRODUCTION

The purpose of the circulation is to allow exchange of dissolved gases, substrate, hormones, electrolytes, and fluids between capillaries and the interstitial tissue space (Figure 5-1.) Alteration of circulation and altered perfusion interferes with metabolic exchange. Circulation is not flow alone, but has to provide capillary perfusion. Life in the multicellular organism depends on this commerce of exchange. Interruption of circulation or insufficient return of circulation after an interruption threatens direct survival of cells, tissue, and organs. The exchange across the capillary barrier is considerable, allowing movement of solids, fluids, and gases by diffusion, filtration, and vesicular transport. Gases utilizing the entire endothelial cell surface of $6,000^2$ have a diffusion rate of 55,000 ml/minute in both directions. This applies to oxygen, carbon dioxide, and glucose. Fluids filter passively across the capillary barrier, driven by hydrostatic pressure and by osmotic pressure. If the osmotic pressure is given and steady, the hydrostatic pressure is the most important single determinant of satisfactory capillary perfusion and fluid filtration.

Ischemia is a disorder of circulation and describes any situation in which the critical capillary perfusion of an organ or a tissue is reduced or eliminated. In this chapter, we focus on the vascular bed in ischemia and on the related alteration of cell metabolism. With reduced flow, oxygen becomes deficient, and deficient nutrients are delivered. Accumulation of waste products is also a feature of ischemia.[1-3] All these factors (deficient oxygen and oxidative phosphorylation, deficient substrate and substrate utilization, and waste product accumulation), alone or in combination, induce cell damage. The initial phase of cell damage may be autocatalytic. Cell and interstitial edema, a change in vascular resistance, circulatory regulation and an inflammatory response mediated by adjacent vessels may expand the range of the original tissue injury.

Definitions

Variable scenarios of cell damage result from strokes and from infarcts of kidney, spleen, intestines, heart, and other organs. Frequent causes of ischemia are intrinsic arterial vessel disease, such as atherosclerosis or vasculitis, emboli or thrombi, vascular spasm and compression of vessels by outside forces. Under these conditions, ischemia can be total or partial, temporary or permanent. Ischemia may involve an entire organ, which is called *global ischemia*. It may also only affect a section of an organ, which is called *regional ischemia*.[4] Regional and global ischemia, as well as transient or permanent ischemia, differ in regard to sensi-

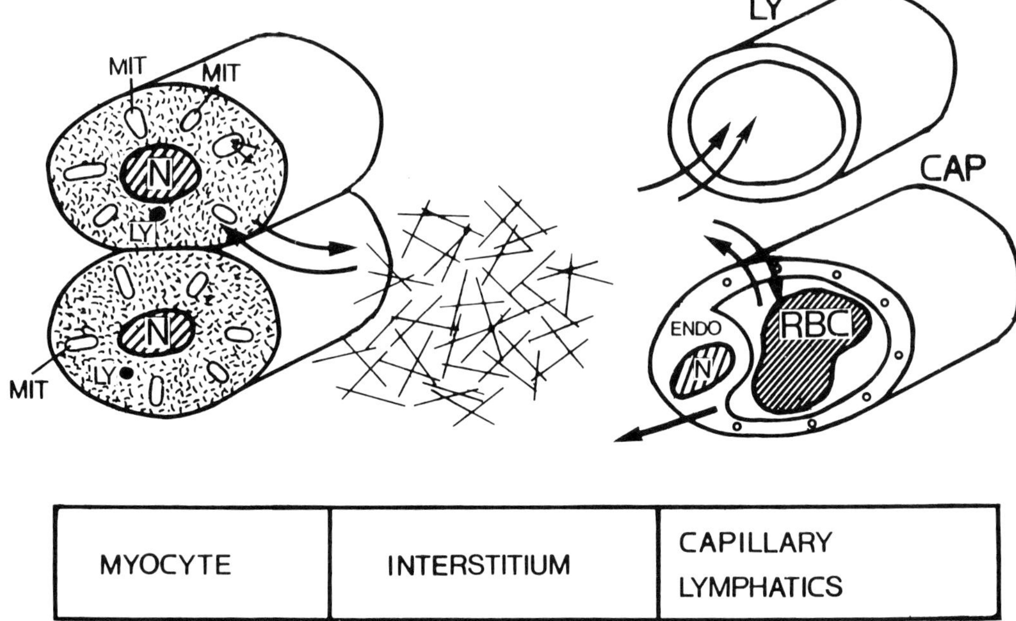

Figure 5-1. Schematic drawing of the tissue spaces in the heart and the exchanges taking place between the tissue spaces. The parenchymal cells are considered suspended in the interstitial space.

tivity of cells and the expected cell damage, the time of cell survival, edema, electrolyte exchange, metabolism, involvement by inflammatory cells, and possibly the predominant mechanism of cell death. Because survival is threatened, ischemia is an adverse condition affecting parenchymal cells, the interstitial space, connective tissue cells, and the local microcirculation. All these histological tissue components play a role in the progression of ischemic injury and have to be considered in an effort to understand the pathophysiological impact of ischemia and to describe or predict outcome.

The topic of this chapter is the role of the vasculature in ischemia during and following regional ischemia. The role of endothelial cells, red blood cells, platelets, and white blood cells in ischemia are discussed and questions are entertained such as: How does the microvasculature contribute to, or ameliorate, the effects of ischemia during the ischemic event? How may damage to the

microvasculature affect or alter reflow and recovery in temporary ischemia?

MICROVASCULATURE: REGULATION AND STRUCTURE

Blood supply is a highly regulated process. Although the regulatory sites, the signals, and the mediators of this regulation have not been completely elucidated, it is apparent that the blood supply is tightly linked to the oxygen requirements and metabolic requirements of the tissues. The level of oxygen consumption at the time of injury—for example, in the heart (MVO_2)—is also a determinant of the speed of necrosis in ischemic tissue.[5] Regional blood supply is related to the perfusion pressure, the volume of blood delivered to the vascular bed, and the resistance of the vascular bed to perfusion. The vascular resistance varies with the local constriction of small arterial vessels and with the tissue pressure. Vascular factors are summarized in the term *vascular tone*.[6] The autonomic nervous system and metabolic products also regulate local resistance and flow.

Regulatory Sites

Resistance is partly an active and partly a passive process, the latter resulting from the vascular structure itself and from compressive forces that act on the tissue. Resistance is also regulated at certain control sites in the regional microvasculature. For example, under physiologic conditions in the normal heart, the resistance component of large epicardial coronary branches is very small in relation to the total coronary resistance and can be neglected.[7] The same is true for other large conducting arteries supplying internal organs and the musculoskeletal system. The conducting arteries may even play a supporting role for regional blood supply. Increased alpha-receptor stimulation increases the tone of extramural coronary criteria and provides sufficient perfusion pressure for subendocardial perfusion.[8] Most of the regulation of flow in the microvasculature depends on resistance in small arteries and arterioles. These vessels account for 93% of the total resistance of the myocardial microvascular bed. They may play a similar role in other tissues.

Arteriovenous bypasses exist in most tissues and have variable importance. An arteriovenous bypass may conduct more perfusion under conditions of high resistance in the microvasculature. Arterioles, the chief site of vascular resistance, have a diameter of less than 100 μm. They branch off larger conduit arteries and approach the subunit of the tissue: the muscle bundle in the heart or skeletal muscle, the glomerulus and tubules in the kidney, the alveolar basket in the lung, the glandular acini in the pancreas. The smaller arterioles (A3 and A4) may reach a diameter of less than 15 μm. The

capillary bed offers a limited amount of resistance under ordinary circumstances. This resistance, however, may increase during injury.

Capillary density and perfusion distance are important determinants for oxygen transport and delivery of nutrients. There are mechanisms in most vascular systems to adjust blood flow to the demand by increasing and decreasing capillary perfusion by increasing the volume of the capillary vascular bed. In many vascular beds, capillaries can be recruited with increased metabolic demands at the regulatory site of precapillary sphincters. Such recruitment increases the capillary density and reduces the distance between cells and the source of oxygen. Such studies have suggested that capillaries can regulate their diameter through contraction of pericytes. Diffusion, distance, amount of flow and/or resistance would be regulated by increased flow through heart capillaries. The meandering shape of capillaries during contraction may also influence intermittent or transient alteration in resistance (Figure 5-2). Such a variable, convoluted course of capillaries is found in the coronary vascular system and in contracting skeletal muscle.

Examples of Structural Arrangements

The microvasculature is specially designed for each organ system. Its three-dimensional structure corresponds exactly to the functional units of the organ.[9] Casts of the capillaries from different organs allow the observer to study capillary and cell interactions in heart, lung, liver, pancreas, adrenals, and kidney.[10-13] Detailed studies also allow morphometric evaluation of the microvasculature and pathophysiological correlation.[13-15] Under some circumstances, cast and perfusion preparations are inadequate for visualization of developing capillaries. Minamikawa et al. reported a new histochemical technique based on vascular-

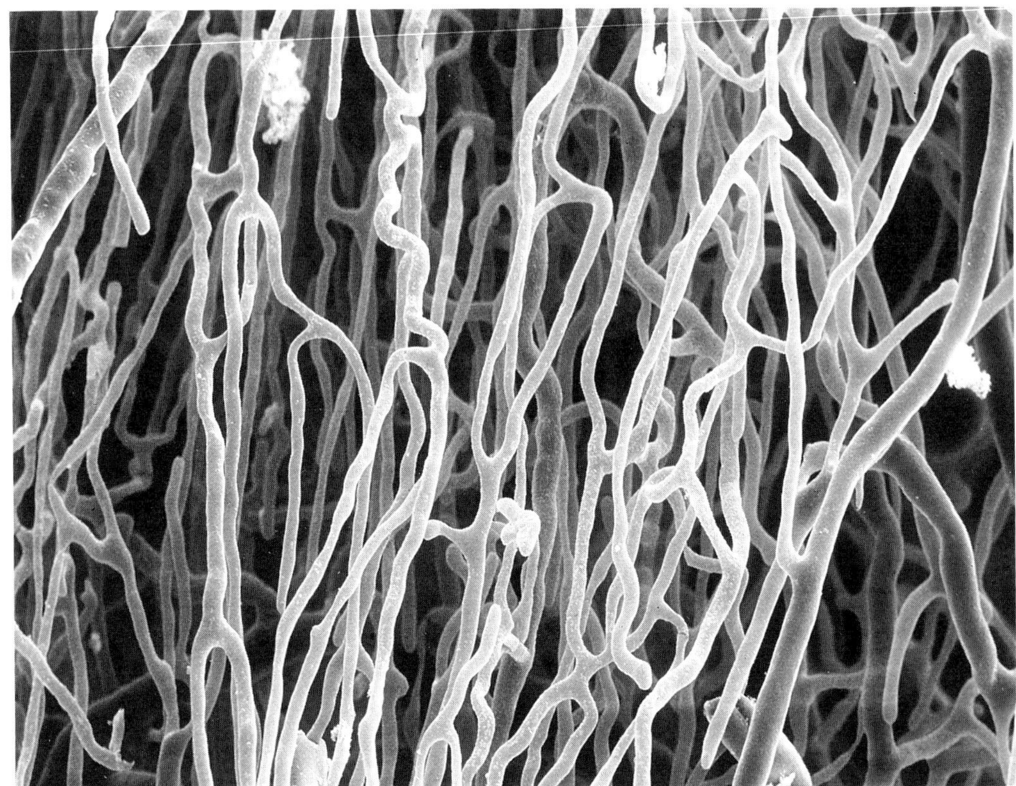

Figure 5-2. Corrosion cast of the capillaries of the heart, seen by scanning electron microscopy. The heart has been relaxed by injecting higher concentration of KCl into the coronary vessels. A low viscosity plastic was subsequently infused, which polymerized within less than 8 minutes. Perfusion continued during the entire time at a mean blood pressure. Corrosion by sodium hydroxide.

specific lectin binding for brain angiogenesis.[16] Of the lectins tested, three (Griffonia simplicifolia agglutinin 1; Ricinus communis agglutinin 1; and Soybean agglutinin) selectively stained wall in brain tissue. With 5-bromo-2-deoxyuridine (BrdU), developing capillaries can be visualized. In some organs, the recognition of such relationships between structure and function have not risen above the level of suggestive evidence, such as the concept of a counter-current system of the capillaries in the myocardium.

Baez reviewed the microvascular terminology.[17] Definitions have been established for the venous segment. According to the proposed definitions, a *venous capillary* is defined as the capillary segment that is formed by the confluence of two or three capillaries. It has a diameter of from 8 to 10 μm. A *postcapillary venule* is a venous vessel of from 8 to 30 μm with an increasing number of pericytes formed by the confluence of several venous capillaries. A *collecting venule* is a microvessel of from 30 to 50 μm with one complete layer of pericytes and a complete layer of veil cells. Smooth muscle cells are present in the wall. A *muscular venule* has a diameter of from 50 to 100 μm and contains a continuous wall of smooth muscle cells. The arterioles of between 50 and 100 μm have more than one smooth muscle layer and are branching into the *terminal arteriole*, which is defined according to Chambers and Zweifach as the final arterial ramification, from 30 to 50 μm in diameter, and having a continuous layer of smooth muscle cells.[18] The *terminal arteriole* may be followed by arteriolar vessel that has a discontinuous smooth muscle layer designated a *metarteriole*. A muscle arrangement around the origin of some capillaries gave rise to the designation *precapillary sphincter*.[19]

Microvasculature of the Heart

A detailed description of the capillaries of the microvasculature of the heart has been presented by Bassingthwaighte et al.[20] Arteries penetrate perpendicularly into the myocardium of the left ventricular wall and give off small arterial branches at right angles. In this system, the endocardium is the last tissue reached by the arteries. The arterioles show a smooth transition to capillary beds. They arise at right angles from the arteries. The capillaries form a dense network around muscle cell strands with multiple branches and anastomoses having a netlike architecture.[21] In corrosion cast preparation, the lumen of capillaries measure between 5 and 7 μm in diameter. There are arteriovenous anastomoses and venous-venous anastomoses. Stoiko and Mergner employed vascular injection by plastic and determined that the arterial and arteriolar branching into capillaries occurred by bifurcation until it reached the capillary level.[22] At the capillary level, H-shaped and Y-shaped capillary anastomoses were seen (Figure 5-2). Many capillaries formed two patterns at the venous confluence. One was a comblike pattern; the other was characterized by a bulblike dilatation at the side of confluence (cauliflower pattern). Stoiko and Mergner showed extensive capillary convolusion during contraction in the myocardial capillaries, similar to the pattern seen in contracted skeletal muscle (Figure 5-3).[22]

Blood flow in the myocardium shows mi-

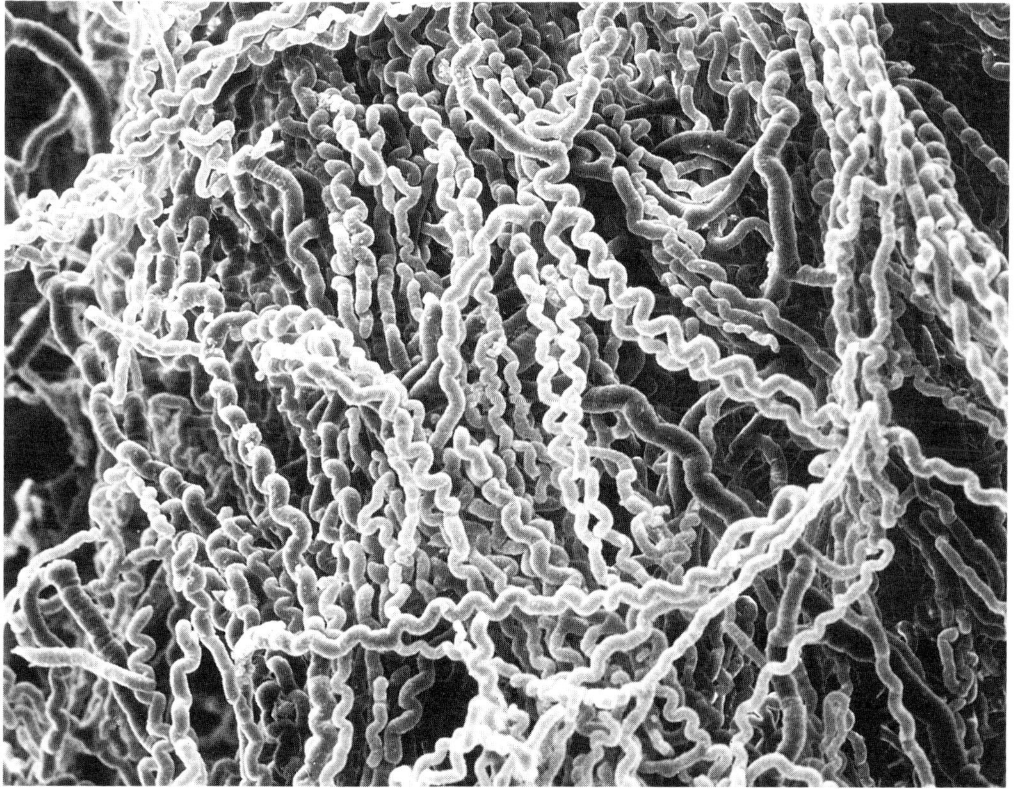

Figure 5-3. Corrosion preparation of the heart, prepared as in Fig. 5-2. The heart was allowed to contract prior to the perfusion and injection of the plastic.

croheterogeneity, with blood flow in the subendocardium 10% higher than that of the outer layers. This higher blood flow results from the higher capillary density of the subendocardial region.[23] The systolic coronary inflow is the result of net forward flow and concealed backflow. The microvasculature of the heart and the endocardium contains dormant capillaries that can be recruited. Porter and Bankstone observed the development of functional components of the microvasculature of the rat by using probes of graded size.[14] Throughout the entire vasculature, there was no leakage of carbon particles. Carbon particles were associated, however, with a few coated vesicles in endothelial cells. Ferritin, in contrast, appeared in luminal plasmalemmal vesicles of endothelial cells and was seen after 60-seconds circulation in the abluminal site of the proximal interstitium. Ferritin was also seen in coated vesicles and multivesicular bodies after 90 seconds of circulation. Reaction product of myoglobin could be detected in all of the cell compartments mentioned and in transendothelial cell channels. It also filled the intercellular space, except the tight junctions. The role of graded small vascular occlusion was tested by gradually increasing the number of microspheres injected into the vasculature. This test increases the reserve of the myocardium. The vascular reserve of the myocardium in sizable (Figure 5-4).

Microvasculature of the Lung

The lung parenchyma is the gas exchange component of the lung and comprises 87% of the lung tissue. The nonparenchymal portion—that is, bronchi, pleura, and subpleural—constitutes the rest. The pulmonary arteries follow the airway system closely. Branches leave the arterial vessels at right angles and approach the prealveolar arterioles. The precapillary vessels have a diameter of from 18 to 20 μm and pass into the intersecting plane of two or three aveolar walls. At this level they commonly divide into the capillary basket of the alveolus. The alveolar capillary bed and alveolar interstitium are organized in such a way as to give the appearance of a continuous sheet of flow around the alveolus, interrupted only by the pillars of connective tissue. These connective tissue pillars are 5 μm in diameter.[9] The venous drainage of the lung arises from the free surface of the alveolus. Venous channels are still located in the alveolar wall and are remote from the airway system.[13,24]

Microvasculature of the Kidney

Horacek et al. described the renal vasculature of Macace monkeys and compared their anatomy to other mammals.[11] The renal microvasculature originates with the afferent arterioles to the renal glomeruli. From the glomeruli, several efferent vascular patterns arise to different regions, such as the subcapsular, midcortical, inner cortical, and medullary regions. The medullary vasculature forms parallel vascular strips toward the medulla, with frequent branching into capillaries forming the plexus of interbundle capillaries. Ascending vessels are seen; they run to the cortico-medullary junction, where these vessels empty into collecting veins. Some veins ascend into the cortex and empty into the proximal one-third of the interlobar veins.

Microvasculature in the Liver

The microvascular unit in the liver consists of a terminal portal venule from which sinusoids arise. Sinusoids form a glomus. The hepatic artery is also part of the hepatic microcirculation. The hepatic artery branches around the terminal bile ducts. The hepatic glomus is drained by two or more hepatic venules.[9] The liver sinusoids are lined by a discontinuous endothelial lining. Such a discontinuous lining may be less than static. Under abnormal conditions it could be shown that the endothelial lining in the liver becomes continuous, a process called *capillarization.*

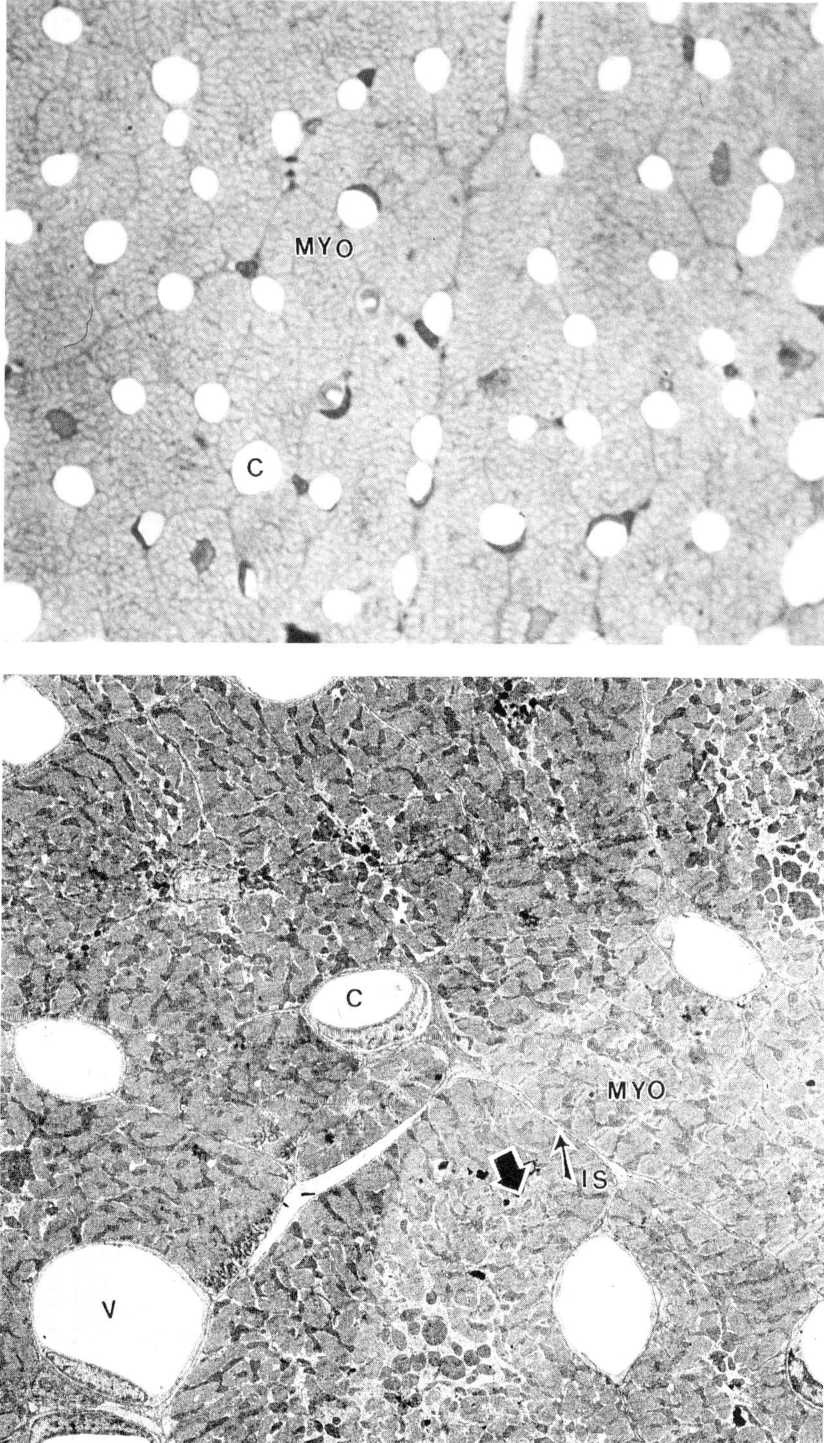

Figure 5-4. Perfusion fixation of myocardium, seen by light and low magnification electron microscopy. Myo: cardiac myocytes; C: capillary lumen; V: venule. Small arrows point to the interstitial space. Large arrow points to the nonedematous myocyte.

Special Functional Arrangements

Countercurrent blood flow has been described for the renal circulation, where it influences the concentrating function of the urine excretion. Since arterial and venous capillaries lie in close apposition, countercurrent diffusional shunting is a possibility.[20] Countercurrent diffusional shunting from arteries to veins has been shown for labeled water and for dissolved hydrogen.[26] Katz and Feigl, however, studied the magnitude of CO_2 countercurrent trapping, describing the net diffusion from intramural coronary venous vessels to coronary arterial vessels, and found only a little carbon dioxide diffusional trapping under normal coronary flow in the coronary circulation.

Tissue Spaces

The capillaries and the parenchymal cells have to be seen as a functional unit of four different spaces: the vascular space, the interstitial space, the lymphatic space, and the cellular space. All four spaces remain in a very close exchange. The capillaries and interstitial space have active exchange through the endothelium, determined by unidirectional transport, diffusion, and reabsorption. The cellular space and the interstitial space exchange is controlled by the plasma membrane of the parenchymal and endothelial cells, respectively. Lymphatic vessels act as an additional regulator of the interstitial space for fluids, larger proteins, and even cells.

The structural stability of the entire unit of vessel and parenchymal cells is guaranteed by basement membranes, basal lamina, and small interlacing bands of collagen fibers binding together parenchymal cells, basement membranes, capillaries, and the interstitial space. These relationships resist expansion of these spaces. If pathological processes such as ischemia, cell injury, and others interfere with the control of these spaces, the volume regulation becomes disturbed. This disturbance leads to accumulation of water in most instances, causing cell edema or edema of the interstitial space. Expansion of the cell space, resulting in cell swelling and edema of the interstitial space, expand the tissue spaces and raise the *tissue pressure*. An increase in tissue pressure may increase the resistance to capillary perfusion.

PHYSIOLOGY OF BLOOD SUPPLY TO TISSUES

Mechanism of Regulation

Marcus has described five basic mechanisms that regulate the blood flow to organs: metabolic regulation, autoregulation, neural modulation, humoral control, and myogenic regulation.[28] The regulated response of the blood flow is altered when the more abnormal conditions, such as *in vivo* regional ischemia, prevail.

Metabolic Regulation

Metabolic regulation may apply to all organ circulations. It has been extensively studied in the myocardium. The dependence of left coronary flow on myocardial oxygen consumption (MVO_2) suggests that tissue such as the heart regulates its own blood supply by causing regional vasodilatation in proportion to the energy expenditure.[6] Under physiologic conditions the tissue metabolism determines tissue blood flow unless superimposed mechanisms stimulate or inhibit flow in response to other regulatory functions, such as perfusion of the skin in the process of temperature regulation of the body. In fact, the linkage of blood flow to metabolic demands is quite tight, and the response in certain vascular beds such as the coronary circulation may occur in seconds. Chemical candidates, such as oxygen, carbon dioxide, pH, potassium, prostaglandins, acetate, and adenosine, have been suggested as the mediators of the vascular response to the metabolic needs.[28-37]

Adenosine is a particularly interesting candidate for the local metabolic control of the microvasculature because it is released as a function of catabolism of adenine nucleotides and it is metabolized by the adenosine deaminase, which converts adenosine to the inactive compound inosine, thus allowing feedback regulation.[38] Although evidence has been presented for the major vasodilatory role of adenosine in the heart, it is not clear whether these factors play a major role or are only modulators to a more general vascular regulatory factor. It was assumed that adenosine is unimportant in controlling coronary blood flow in unstressed dog hearts based on experiments that failed to show reduction of flow following the injection of adenosine deaminase. Therefore, it may be that adenosine is normally below the vasoactive threshold and not important in mediating local metabolic control in the unstressed heart.

Autoregulation and Ischemia

Autoregulation is the capacity of a vascular bed to adjust its resistance so as to maintain a constant blood flow during changes in perfusion pressure. The effect of autoregulation has been recognized since in many organs changes in perfusion pressure are not associated with parallel changes in flow. The autoregulation is effective within a physiologic range that in the heart ranges from 50 to 150 mm Hg according to Boatwright et al. and Guyton et al.[39,40] It is postulated that by its presence or absence autoregulation could contribute to disturbed regional flow downstream from a vascular obstruction (see also the sections entitled "Vascular Waterfall" and "Critical Closing Pressure" that follow).

Neural Control of the Circulation in the Heart

Vascular systems such as the coronary arteries and veins have a rich supply of sympathetic and parasympathetic innervation. There is considerable experimental evidence that the nerve supply influences regional circulation. The coronary arteries and veins contain receptors such as beta-1, beta-2, alpha-1, alpha-2, and muscarinic cholinergic receptors, but the various vascular systems differ in their sensitivity toward activators of these receptors. For example, in the coronary circulation there may be a negligible influence by beta-1 and beta-2 activators, whereas alpha-1-receptors can produce physiologically significant reduction in coronary blood flow.[40]

Beta-stimulation. Under abnormal conditions, a magnification of beta-stimulation has been observed. Such influence has been documented in hypercholesterolemia and following severe coronary obstruction due to atherosclerosis. As discussed later, these altered sensitivities may explain some of the paradoxical effects in cardiovascular shock and in ischemic injury.[41]

Alpha-stimulation. Alpha-stimulation and vasoconstriction is operative during atherosclerotic narrowing of the coronary arteries, during low perfusion (hypoperfusion) of the heart and increased myocardial oxygen extraction. The vasoconstrictor effect following the alpha-stimulation during these conditions does not result in lactate accumulation, indicating satisfactory cellular metabolism.[42] In models using constant pressure perfusion of the heart, a homogenous transmural response of the vasculature was observed during normal perfusion pressure and following low perfusion pressures. During constant-flow hypoperfusion, the steal between endocardium and epicardium was reduced if alpha-stimulation occurred, whereas alpha-blockage increased the steal and caused endocardial injury. Feigl proposed that alpha-receptor–mediated coronary vasoconstriction has a beneficial effect by lessening transmural steal during situations of coronary hypoperfusion, while coronary arteries have a high density of alpha-receptors.[42-48]

Humoral Regulation

Humoral agents released from cells and tissues frequently influence the circulation by stimulating or reducing vascular resistance. Such agents are the catecholamines, aginotensin, vasopressin, several cations, thyroid hormone, adrenal steroids, glucagon, histamine, acetylcholine, serotonin, kinins, substance P, vasoactive intestinal polypeptide (VIP), and prostaglandins. The effect of these agents on different segments of the microvasculature and in different organs is variable. Catecholamines have direct effects on the vascular smooth muscle, where they interact with specific receptors (see earlier). Angiotensin II, the most potent vasopressure known, causes significant vasoconstriction that also involves the coronary vasculature of both large and small coronary vessels. Several local mechanisms also related to angiotensin action blunt the effect of angiotensin on the coronary circulation, however. Vasopressin, a hormone formed in the hypothalamus and stored and released from the posterior pituitary, promotes fluid balance and has a direct vasoconstrictor effect. Circulating cations also have a moderate effect on the vascular resistance. For example, perfusion of coronaries with calcium causes mild coronary constriction, while magnesium opposes the constrictive effect. The effect of potassium is reported to be biphasic.

Of increasing interest in atherosclerotic research has been the altered response of the atherosclerotic vasculature to agents such as acetylcholine, histamine, and serotonin (see following). Some humeral-mediated reactions may be transmitted or modified by the endothelial cells. For example, 5-hydroxytryptamine and noradrenaline response was mediated by endothelial cells.[49-51] Direct endothelial and smooth muscle cell interaction has also been postulated.[52] One major role of response is played by the H2 receptors. Ashton et al. observed that H2 thromboxane, prostacycline antagonists changed the cyclic flow variation in stenosed coronary arteries in dogs.[53]

Myogenic Response in Autoregulation

Autoregulation may also be determined by the myogenic response. Autoregulation introduces several myogenic mechanisms. It has proven difficult to separate the different components that determine autoregulation.[27]

Special Phenomena

Vascular Waterfalls

Vascular waterfall describes a circulation in which tissue pressure rather than venous pressure is the back pressure that opposes arterial pressure.[6] The waterfall exists in vascular systems where a collapsible vessel is surrounded by tissue pressure exceeding venous pressure. In a situation of "waterfall," the flow is determined by the arterial pressure variable and the critical closing pressure, and is not influenced by the venous pressure.

Concealed Systolic Backflow

At systole in the heart and during skeletal muscle contraction, ventricular wall stress and intramuscular pressure rise and the vessels become compressed. As a consequence, arterial flow at peak systole and during maximum skeletal muscle contraction is retrograde, and venous flow rises sharply. The flow reduction in systole may be due to both increased resistance due to increased tissue pressure and to retrograde ejection of blood during contraction. In a case of stenosis, the systolic perfusion pressure may be raised and the diastolic may be lowered. The stenosis, therefore, may resist backflow. If the perfusion pressure is lowered, systolic backflow can be detected.

The Vascular Steal Phenomenon

The steal phenomenon describes a dysfunctioning microcirculation. The transmural steal phenomenon in the heart describes a situation of maldistribution of blood flow across the ventricle. It may occur during episodes of critical flow reduction and represents failure to maintain equal flow distribution. Under those circumstances, the blood may be shifted away from the critical subendocardial region into the epicardium. Nathan and Feigl recently proposed that adrenergic vasoconstriction lessens transmural steal during coronary hypoperfusion in the heart.[54]

Critical Closing Pressure (Stop–Flow)

Critical closing pressure was defined by Burton as the transmural pressure below which active tension caused by smooth muscle causes physical closure of arterioles.[55] This concept may apply to the heart and skeletal muscle. The "opening pressure" consists of at least two components: the arteriolar resistance to filling first described by Burton and the compression of the capillaries described by Guyton.[55,56-58]

Extravascular Compressive Forces

The extravascular compressive forces constitute a small component of total vascular resistance.[58] The extravascular compressive forces can be divided into two components: (1) physical compression and (2) shear caused by twisting of vascular channels as the heart and skeletal muscle contract.[58] If the vascular compressive forces are removed, coronary vascular resistance decreases by between 10 and 25%.[59,60]

Reactive Hyperemia

Hyperemia occurs after short-term occlusion of conducting arteries. During short ischemic intervals, hyperemia is a physiologic reaction. The time of hyperemia is in proportion to the oxygen deficit. Following longer ischemic intervals, hyperemia may become inappropriate, as recently reported by Schaper.[23] Functional hyperemia has been largely viewed as being caused by the interaction between parenchymal cells and the associated microvasculature.[61] Recent evidence demonstrates that functional hyperemia is a complex phenomenon. The hyperemia involves capillaries, arterioles, and small arteries. The interactions between parenchymal cells and vascular segments involves long-range messenger-mediated effects and flow-dependent vasodilatation. The endothelium also plays a role in functional hyperemia.[61]

The Endothelium

The vascular endothelium accounts for a high surface area in most organs. Since the heart is dependent on efficient oxygen delivery, the surface area has been measured as 1,000 cm^2/g of tissue. In the heart, capillary endothelium forms a continuous lining.[61,64] In other and more specialized organs, discontinuous endothelial arrangements have been observed. The luminal plasmalemma of the endothelial cells is supported by a negatively charged glycocalyx that repels neutrophiles and limits adherence. The glycocalyx is also antithrombogenic.[64] The endothelial cells have surface receptors for many vasoactive compounds, such as acetylcholine, histamine, serotonin, catecholamines, bradykinin, and adenine nucleotides. The endothelial cells, particularly the myocardial cells, are metabolically active, secreting plasminogen activator, antithrombin III, prostacyclin (PGI_2) and endothelium derived relaxing factor (EDRF).

A most extensive review of the changes of endothelium in ischemia and reperfusion has been presented recently by Forman et al.[65] Following the pioneering studies of Furchgott et al., the crucial role of the endothelium in controlling the diameter of arteries has been recognized.[66-68] Furchgott tested segments of muscular arteries with and without endothelium in a perfusion chamber, and found that these segments of arteries relax in response to the release of an EDRF. It was found that the EDRF was released in response to a number of nonspecific stimuli such as acetylcholine, thromboxans, histamine, serotonin, and others. The role of endothelial cells in relaxing vascular tone was studied by Vanhoute and co-workers (see following). Pohl and Busse reported that the EDRF inhibits the effects of nitro compounds in isolated arteries.[69] The release of the EDRF depended on the presence of intact endothelium. Regenerated or damaged endothelium was not able to produce relaxing factor. It might even release a constricting factor in response to the same stimuli. Free hemoglobin may inactivate the EDRF. The effect of atherosclerosis on the endothelium has not been clearly established, but a number of studies by Freiman et al. have suggested that hypercholesterolemia itself and the postulated degeneration and regeneration of endothelium may alter the vascular response of muscular arteries.[70]

The Physiology of Collateral Flow

The various vascular systems have different degrees of collateral circulation. In the intestine, large arcuate arteries connect the perfusion field of conduit arteries. In other tissues such as the heart, the collaterals are less prominent. Yet collaterals are found in the heart, although significant species differences exist. Such collaterals are defined as (1) native collaterals or (2) mature collaterals.[28] Native collaterals are tiny vessels, less than 100 mm in diameter, that link the large conduit arteries together. These collaterals exist in the dog and in humans mainly in the subepicardial regions. Their resistance under normal circumstances is 30 to 40 times greater than the minimal resistance of the native coronary vasculature.[71-74] Under abnormal conditions, these collaterals represent pathways into the obstructed vascular bed that offer significant potential for growth while forming mature or developed collaterals; that is, vessels from 1 to 3 mm in diameter.

Recognition of this growth potential and the description of the time phases involved may have been the most significant recent developments in studies on the microcirculation. Resistance in collaterals is determined by extravascular compressive forces and by the perfusion pressure. Maximal flow through mature coronary collaterals may reach between 60 and 75% of normal coronary vasculature. Oxygen supply through collaterals is obviously different for different vascular systems since the resulting blood flow carrying oxygen should represent a balance between pressure and vascular resistance. In the heart a reduced blood flow of 25 ml per minute per 100 g delivers 5 ml of oxygen per minute per 100 g of tissue. It is said that 4 ml of oxygen can be extracted. The nonbeating heart muscle requires only 1.5 ml oxygen per minute per 100 g tissue. Therefore it can be calculated that under certain conditions such as reduced energy demand of the nonbeating heart, there may be enough oxygen delivered by the collateral vascular system to guarantee survival of tissue within the area at risk; that is, within the occluded vascular bed.[71,72,73] These effects may be augmented if the remaining residual perfusion removes metabolic wastes such as protons that inhibit anaerobic glycolysis in ischemic tissue, thus depriving cells of alternate pathways of energy generation.[67]

THE MICROVASCULATURE IN EARLY RESPONSE TO PERMANENT ISCHEMIA

The pressure loss across minor vascular stenoses can be compensated for by a decrease in resistance of the microvascular bed by autoregulation. This compensation is limited to luminal obstructions of 85% or less. Above 85% (single) stenosis, perfusion of the periphery becomes a function of the flow across this stenosis.[7]

Permanent ischemia may be subdivided into several phases. During each phase different structures of the tissue play the predominant role. The initial phase may be considered the hemodynamic phase because tissue reaction and the possibility of mediation depend to a large extent on hemodynamic factors: perfusion and tissue pressure, endothelial leakage and thrombi. At a later phase, activation of inflammatory cells determine the picture and create phenomena that are superimposed on the hemodynamic events. At a still later phase, cell proliferation, phagocytosis and repair processes dominate the histological picture.[75]

Collateral Flow in Early Permanent Ischemia

Permanent ischemia following occlusion of one major artery will not allow flow into the vascular bed supplied unless there are vascular connections from adjacent conduit arteries that can deliver blood into the area at risk. The connecting collateral arteries commonly transmit greatly reduced pressure into the occluded vascular bed due to their high degree of resistance. Following the onset of ischemia, the resistance falls. There are good reasons to believe that ischemic myocardium provides stimulus for near maximal vasodilation of collateral blood vessels.[71] Under these conditions the determinants of collateral blood flow are: (1) the anatomically fixed hydraulic resistance of the collaterals proper; (2) the arterial driving pressure; (3) extravascular resistance, such as radial stress across the ventricular wall and tissue pressure; and (4) the size of the ischemic vascular bed. Schaper et al. calculated that under maximal dilatation of collaterals the resistance of this bed is 3.5 units, which means that a pressure of 350 mm Hg is needed to drive 100 ml of blood through 100 g of tissue per minute.[71] The tissue pressure in the subendocardium creates an unfavorable flow distribution between epicardium and endocardium. This leads to earlier subendocardial damage and to earlier and more severe subendocardial vulnerability and no-reflow phenomenon.

When coronary arteries are occluded for a period of time, then followed by reflow, there is a species-dependent spectrum in the speed and extent of myocardial infarction. Rapid and complete infarction occurred in the rabbit, pig, and rat; however, infarctions proceed at a slower rate in dog and cat. The guinea pig does not exhibit any infarction at all. The final infarct size after reperfusion correlates well with the amount of collateral flow. Collateral flow is low in the rabbit, pig, rat, sheep, pony, baboon, and wild boar. Collateral flow is of medium level in dogs and cats, and is high in guinea pigs. Humans with chronic and slowly progressing coronary artery disease may have considerable collateral flow, similar to dogs and cats (Figure 5-5).[76]

The Border Zone in Regional Ischemia

Sage and Gavin described the microvascular function at the margins of early regional infarctions in the isolated rabbit

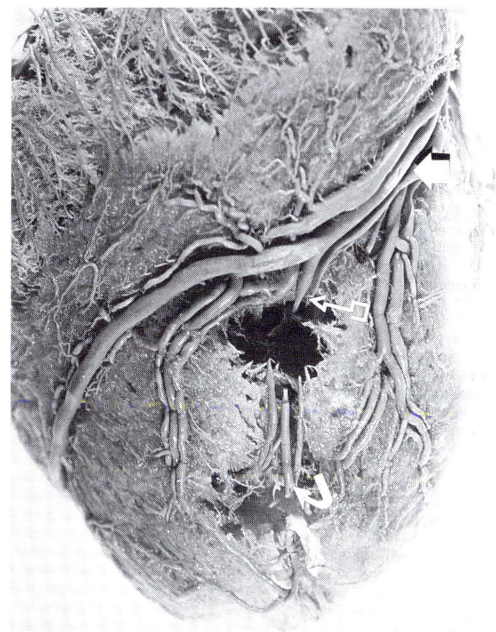

Figure 5-5. Corrosion preparation of a heart, demonstrating the large perfusion field of the left anterior descending coronary artery (big white arrow) and the capillary network. In the center, the second diagonal artery has been ligated (boxed arrow). The remaining diagonal artery is filled, as are the accompanying veins. The pressure in the occluded artery was 45 mm Hg. At the curved arrow a second ligature was applied to isolate the occluded vascular bed, which remained filled and showed blood flow through collaterals.

heart.[77] The microvascular bed adjacent to the regional infarction showed an increase of filled capillaries from 60 to 80 or 90% at 30 minutes occlusion. After 60 minutes of occlusion perfusion defects appeared and only 10% of the vessels were functional. Tissue necrosis produced dilated venous plexuses adjacent to regions of necrosis. These tissue are fed by arterioles entering at right angles. This observation was interpreted as a result of obstruction of the venous effluent by compression through cell and interstitial edema (Figure 5-6).[77]

The Interstitial Space in Ischemia

Following obstruction of a major artery, the interstitial space is affected in several ways, all of which result in increased interstitial pressure and local edema. The interstitial space is the space surrounding parenchymal cells and capillaries. It is a rather fixed space, with significant restriction to fluid movement; it is held together by connective tissue bands. The advent of vascular obstruction will slow down capillary flow and reduce hydrostatic pressure in the vascular system. At the same time, cellular volume increases in response to electrolyte shifts, with cells loosing potassium and gaining sodium and calcium from the interstitial space. Lymphatic drainage and removal of fluids by the venous end of capillaries is decreased. All these events make the tissue pressure rise, causing compression to the microvascular bed and an increased opening pressure. This is particularly prominent in organs (kidney and heart) that allow only limited motion and shearing of parenchymal cells due to a tight capsule and connective tissue connections that restrict motion.

The elevation of the interstitial pressure to from 2 to 3 times normal levels was observed by Hooper and Mergner in the epi-

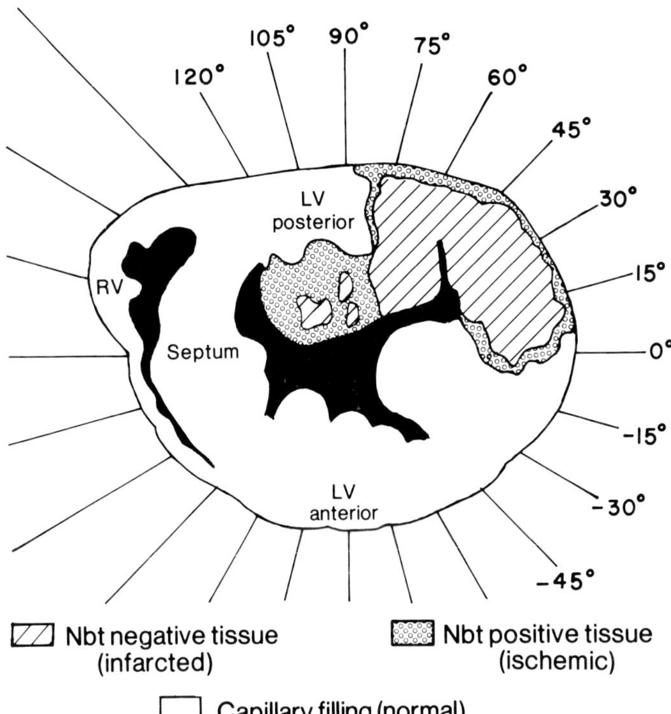

Figure 5-6. Schematic drawing of an ischemic experiment, with the mapping of the NBT positive and NBT negative zones. Capillary filling with white Microfil (R). Note the sectorial orientation of the heart slice. (Courtesy of Dr. G. Mergner, with permission).

cardium of the heart.[78] The interstitial pressure was higher in the center of the infarct. The electrolyte content of the interstitial space also changes, showing a steep rise in potassium, an intracellular electrolyte. The interstitium, therefore, may be a factor that restricts capillary perfusion in the center of the risk area by balancing the reduced perfusion pressure of the microvascular bed with the increased tissue pressure. The latter may increase the opening pressure of the capillary bed. Such a mechanism could explain the nitroblue tetrazodium (NBT) positive zone that surrounds the myocardial infarction in the regional ischemia of the dog myocardium.

Mergner et al. have shown that the area at risk contained two different perfusion levels by microspheres during the period of occlusion.[79] The zone with approximately 45%

reduction of control flow corresponds to the NBT positive zone, while the infarcted myocardium, not staining with NBT, had a flow of from 1 to 10%, (at the lower limit of reliability of the method). Reflow in the NBT positive zone increased to 91% of control level following 90 minutes of occlusion. Reflow in the ischemic tissue reached approximately 87% of control (Figure 5-7). Morphological examination of the reflow in the infarcted tissue showed patches of deficient capillary filling.

Capillary Flow during Total and Partial Ischemia

Braunwald initiated an epoch of research into the role of the border zone and the possibilities of salvaging myocardium.[80] He proposed that reduction in infarct size

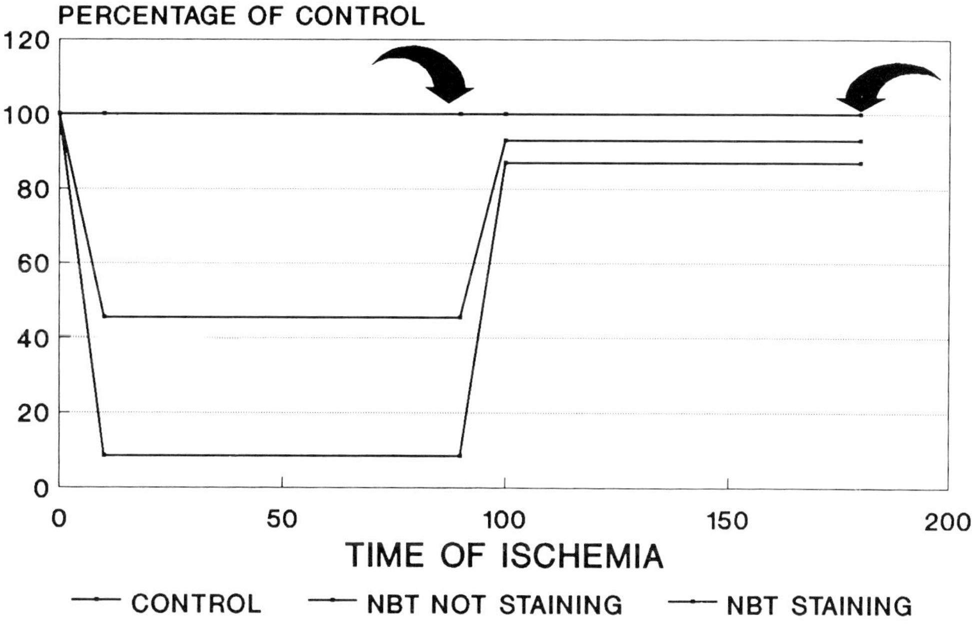

Figure 5-7. Schematic presentation of a reflow experiment. Occlusion of the LAD occurred for 90 minutes, and reflow was allowed for 90 minutes. Flow was determined with radioactive microspheres. Tissue was separately collected from the Microfil staining, the NBT staining, and the NBT nonstaining areas. (Data from Mergner et al. 1988,[79] with permission.)

would improve survival and reduce the morbid consequences of myocardial infarction.

Of the new facts that have found general agreement, one is that only capillary blood flow is nutrient blood flow. Blood flow that bypasses the capillary system or does not perfuse sufficient capillaries causes ischemia. Thus, the final outcome is determined by the adequacy of capillary blood flow. Using the same argument, one may say that it is essential to prove whether capillary perfusion takes place or not in studies of ischemia. The filling of small arteries with contrast material is insufficient evidence of capillary perfusion. Therefore, x-ray examination of contrast-filled vessels is insufficient because of the lack of resolution of vascular structures by x-ray films. Data supported by this method may be unreliable indeed. Tillmans and Kuebler, who studied the capillary bed in ischemia of the heart, observed that within seconds the poststenotic arterioles became dilated.[81] Blood flow velocity in capillaries and venules diminished within the ischemic zone. If the artery was only partially occluded, a marked decrease of the systolic red blood cell velocity was noted, while the diastolic red blood cell velocity remained only minimally altered.

Stenosis also reduced the pressure in the venular system from 25 mm Hg to 12 mm Hg. The reduced pressure and flow in the capillary system was observed to have two effects: (1) reduction of perfused capillaries (called *decruitment*) and (2) increased permeability toward small molecules. This increased intercapillary distance adversely affects the diffusion distance for oxygen and nutrients in the ischemic tissue (Figure 5-8).

Reflow and Defective Reflow in Transient Ischemia

The clinical interest in reperfusion has been stimulated by the advances in clinical techniques that allow early opening of vessels by mechanical means or by thrombo-

Figure 5-8. The border zone of an infarct. The capillary bed was filled with Microfil (R), and the tissue was subsequently dehydrated and made transparent with oil of wintergreen. The vascular bed was religated prior to filling with Microfil (R).

lysis following infarction. Reflow has also gained the interest of the experimentalist because of the definition of viability that depends on the ability of cells to resume normality after reflow. The recent discovery of the oxygen and calcium paradox also has placed emphasis on the no reflow phenomenon. Another important event during reflow is hemorrhage into vascular beds.

Capillary Changes during Reperfusion

Following prolonged ischemia, the sudden influx of the full load of blood causes hemorrhage into the ischemic zone, an indicator of the loss of capillary integrity. In certain tissue regions, reflow becomes less and less possible because of or due to a combination of factors that include (1) a finite space, such as the calvarium for the brain or the renal capsule for the kidney; (2) interstitial edema combined with cell edema expanding the interstitial pressure on capillaries or raising the opening pressure of the vascular bed; and (3) local alteration of the capillary bed, such as endothelial blebs and microthrombi.[82] Although very few tissues fulfill the exact requirements for a total no reflow phenomenon, regional microscopic alteration can be shown in most instances.[83] A 90-minute regional ischemia in the heart causes a somewhat reduced reflow, but by no means the dramatic no reflow phenomenon that is seen in the kidney and brain. On a microscopic level, such defective capillary filling is limited to focal areas that may share both intravascular and extravascular components. A significant obliteration of reflow in the subendocardium and the midmyocardium was observed in the ATP-deficient contracture of the heart, a situation mainly seen in global ischemia of 90 minutes and more. The latter is due to the combined effects of ATP-deficit contracture, interstitial edema, and cellular edema (Figure 5-9).

Capillaries in Response to Ischemia

Kloner et al. described reperfusion of the myocardium following ischemia for 40 to 90 minutes as incomplete, and he termed the phenomenon the *no reflow phenomenon*.[82] This no reflow increase in the myocardium at risk as the periods of ischemia become longer.[83] Kloner's and subsequent observers reported significant change in the microvasculature, ranging from mechanical obstruction to active pathophysiologic reactions.

Complement Activation in Ischemia

Hill and Ward reported that ischemia in the heart released a tissue protease that cleaved the third component of the complement system.[84] Pretreatment with cobra

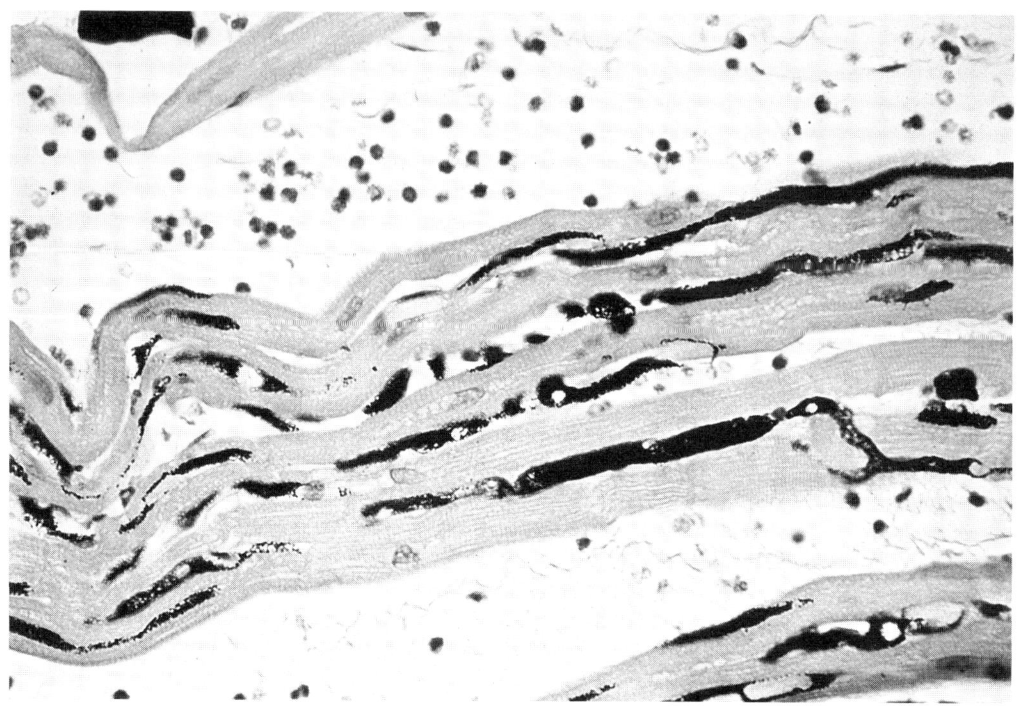

Figure 5-9. Border zone of an infarct. Note the Microfil (R) filled capillaries, the interstitial edema, and the inflammatory cells (90 minutes ischemia and 90 minutes reflow).

venom that depletes the C-3 portion of the complement did not result in liberation of C-3 after ischemia. It has been speculated that the liberation of C-3 by infarcts caused an increased chemotaxis and may be directly responsible, along with C-5, another component of complement, for the attraction of leukocytes stimulating the damaging effects on tissue by peroxidation, lysosomal enzymes, and endothelial injury.

Chemotaxis is even more damaging in reperfusion injury. Activated complement mediates injury to endothelial cells of other vascular systems such as the pulmonary system and may even be responsible for multiple-organ failure in hypoxia and shock. Nuytinck et al. tested the role of complement by injecting activated complement over a 4-hour interval into New Zealand white rabbits experiencing short periods of hypoxia.[85] Lung injury and sequestration of leukocytes in target organs, endothelial cell swelling, and interstitial edema were observed. The combination of hypoxia and complement activation was particularly damaging, causing changes associated with adult respiratory distress syndrome and with multiple-organ failure.

Corpuscular Blood Elements

The role of corpuscular elements of the blood may be threefold. It may cause (1) viscosity changes of the blood; (2) endothelial alteration and morphological unevenness of vascular surface; and (3) direct damage to the vascular wall and, beyond it, to the parenchyma. The first change is frequently associated with red blood cells. It has been determined that either changes in the cytoskeleton or the hemoglobin can lessen the deformability of red blood cells and therefore dramatically increase the viscosity in clinical situations of stasis or hypoperfusion. Furthermore, adhesiveness of red blood cells is associated with viscosity change. Mural platelet precipitates, both endothelialized and nonendothelialized, may cause disturbed regional flow. Granulocytes and macrophages have been associated with injury to the vascular structures and the adjacent parenchyma.

Red Blood Cells. Perfusion through the capillary system is determined by the driving pressure and by the resistance of the vascular system. Resistance is also influenced by the viscosity of the blood. Normally, the viscosity of the blood is close to the viscosity of plasma since the red blood cells are maximally pliable. Under ischemic conditions, the pliability of red blood cells may be reduced.[86] In addition, the red blood cells may aggregate in venules and small arteries. According to Tillmans and Kuebler, the effect of ischemia on the microcirculation is variable in the cat and rat, causing a heterogeneous pattern in respect to red blood cell distribution.[81] Some capillaries show an increase of red blood cell mass due to diffusion of plasma through the leaky capillary wall, and others show predominantly plasma perfusion.

Platelets. Myocardial platelet trapping after prolonged regional ischemia has been reported. After short periods of ischemia, such as 10 or 20 minutes, ^{51}Cr-labeled platelets accumulated in the ischemic tissue after reflow, with a peak of accumulation at 10 minutes of reflow. No data exist about platelet trapping in the partially perfused zone supplied by the collateral vessels. Therefore, it is uncertain what role the platelet trapping plays. Kloner et al. found multiple microthrombi in the microcirculation frequently associated with endothelial blebs. This is not commonly reported, but detection may well be a problem of sampling in electron microscopy.[82,87,88] Platelet aggregates have been linked to local vasoconstriction.[89,90]

Inflammatory Cells. Within minutes of vascular occlusion, leukocytes appear in slow flow capillaries. A number of messengers have been shown to be active, such as activated complement (mentioned previously), leukotriens, thromboxans, and phospholipids or their split products.[85,91-94] Leukocyte accumulation has been measured by Engler using [111]Indium during 3 hours of ischemia: leukocytes increased from 1.0×10^6 to 1.5×10^6 cells per g of tissue.[95] During a 5-minute reflow, leukocytes rose to 2.4×10^6 cells per g of tissue. It is proposed that the leukocytes have significant effects on tissue damage, causing edema, ventricular fibrillation, no reflow effect of 27% off the vascular bed, and restricted collateral flow. This effect of leukocytes could be reduced in the above-mentioned studies by agranulocytic blood and by adenosine, which acts as an inhibitor of leukocytes.[96]

Leukocyte adherence to endothelial cells appears to be related to the adherence-promoting leukocyte glycoprotein. Vedder et al. added monoclonal antibodies to the aherence-promoting leukocyte glycoprotein (CD-18) and reduced injury to organs in hemorrhagic shock and resuscitation in rabbits.[97] According to Tillmans and Kuebler, leukocytes may offer a far greater resistance to deforming forces.[81] This increased resistance increases the passage of time of leukocytes. The passage of leukocytes exceeds those of red blood cells in slow flowing capillaries. Thus, the presence of large numbers of leukocytes can plug capillaries. Leukocyte plugging may also limit reflow in an ischemic vascular bed, accordingly. Studies using inhibitors of leukocyte accumulation, such as ibuprofen, have demonstrated leukocyte accumulation reduced by 46% and a reduction of the final infarct size. A similar beneficial effect was also shown in studies where animals have been made leukocytopenic by antiserum against leukocytes before ischemia was reduced for long and short periods. Werns and Lucchesi used antiserum against leukocytes and inhibitors of leukocyte free radical production.[92] There was marked infarct size reduction in an infarction for 90 minutes, followed by reflow for 6 to 72 hours in the canine model. It is not entirely clear how this beneficial effect of leukocyte reduction is produced. Did a reduction in

leukocyte number also reduce the free radical injury of leukocytes on microcirculation and on myocytes? Was it mainly a coronary flow effect, or was it an effect on the coronary endothelial cells and microvasculature? Was it a reduction of lysosomal injury? A multiple-center report compared the effects of verapamil (a calcium blocking agent) and ibuprofen (an antiinflammatory agent) on infarct size reduction in the unconscious and conscious dog model.[98] When the data were controlled for the area at risk, collateral flow and rate pressure product evidence failed to show that ibuprofren reduced myocardial infarct size. Werns and Lucchesi linked the effect of accelerated infiltration of granulocytes in ischemia to the release of free radicals by the inflammatory cells (Figure 5-10).[92]

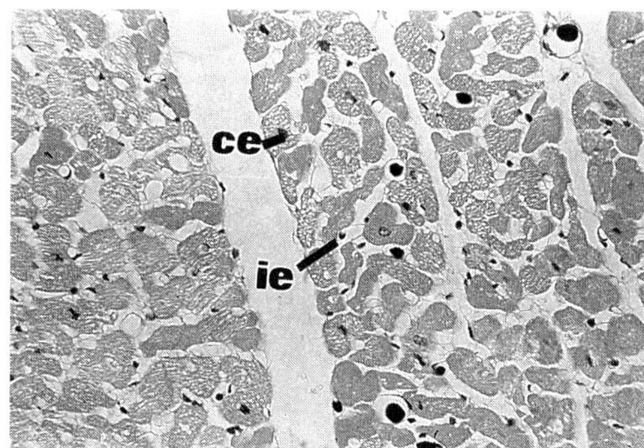

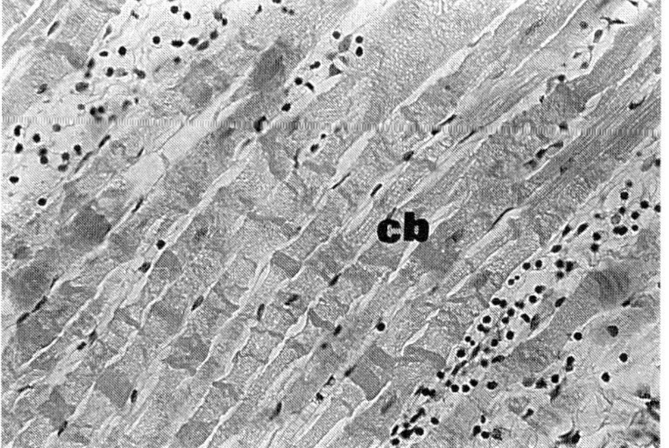

Figure 5-10. At (A) and beyond (B), the border zone of an infarct. The few Microfil filled vessels are seen. There is cell edema (ce) and interstitial edema (ie). Section B is taken from the NBT negative area. It shows contraction band necrosis and inflammatory cells.

The Endothelium of Capillaries. Tillmanns and Kuebler described the time course of changes in capillary endothelial cells after total regional ischemia.[81] The endothelial cells respond unevenly to the ischemic insult, some were considerably altered, while others remained intact. The ultrastructural changes range from loss of pinocytotic vesicles to cell swelling, as described by Armiger and Gavin and Kloner et al.[82,99] At 40 to 60 minutes, multiple endothelial blebs may be observed, and nuclear pyknosis is seen by light microscopy. Some of the less dramatic structural endothelial changes may have significant implication for the vascular integrity. They influence the rheological function of the capillaries. Separation of tight cell junctions allows leakage of fluid into the tissue. The formation of luminal blebs may form thrombotic nuclei and may obstruct the lumen between swollen endothelial cells. Many of these changes may be easily reversible if reflow is established. This has not been measured in a systematic way in the different vascular systems. It also has been shown that the hypoxic and anoxic endothelial cells are capable of releasing vasoconstrictor substances.[100-103] The compound isolated from porcine aorta is a 21-residue peptide with a long half-life of 40 to 60 minutes.

The question is, Why is ischemia damaging to vascular cells? Is anoxia itself damaging? Is the damage promoted by intravascular components such as leukocytes or thromboxans? Forman et al. described the central role played by endothelium in so-called reperfusion injury.[66] The hypoxia results in a rapid conversion from aerobic to anaerobic metabolism in the cells. High energy phosphates (ATP and ADP) are reduced, and protons and lactate accumulate. The consequence of acidosis inactivates Na^+/K^+-ATPase (see review on intermediate metabolism). Ferrari et al. postulated that the defenses are reduced during the early phase of endothelial cell injury.[104]

Repeated Ischemic Tissue Injury

A transient coronary occlusion of 15 minutes does not result in irreversible tissue injury but is connected with a depression of contractile function of the myocardium.[105] Similar intervals apply to other tissues and parenchymal functions. Since in the heart the depression of contractile function persists for days, the term *stunned myocardium* was coined. Originally, it was thought that defective energy metabolism is responsible for the prolonged functional depression, but recently, defective calcium metabolism has been strongly implicated. Repeated episodes of transient ischemia have been studied by a number of groups and it has been recently proposed that those intermittent short episodes of reperfusion are beneficial because they wash out metabolites and delay cell death.[2] But these episodes of intermittent ischemia cause considerable tissue alterations. Reimer et al. reported that 10 minutes of ischemia causes a 61% loss of ATP and a 41% loss of adenine nucleotides from the subendocardial zone.[106] Reperfusion may restore energy charge, but not the pool of adenine nucleotides; two, three, or even four additional 10-minute occlusions caused no additional adenine nucleotide loss, and less lactate. In contrast, 40 minutes of continuous occlusion caused an 87% loss of ATP and a 67% loss of adenine nucleotides from the same subendocardial region. Henrichs et al. even extended the periods of intermittent interruption of perfusion to 90 minutes, with similar results.[107]

New Vascular Growth

In ischemia, the patency of collateral channels is considered to be maximally open, and it is thought that no further pharmacological interference may influence collateral flow. This may be particularly true in the heart and kidney, but other tissues may experience similar phenomena on a different scale. However, it has been shown by

Schaper, that collateral flow in the dog heart does increase in the time interval from 6 to 24 hours after occlusion of an extramural coronary artery.[23,76] The observation of increased flow corresponds with evidence that the diameter of collateral vessels increases during the time interval.

Recent studies by Schaper and co-workers point to mechanisms that could promote vascular growth.[75] The authors observed the adherence of macrophages to the endothelium of collaterals shortly after coronary artery occlusion from 1 to 6 hours, with evidence of immigration of these macrophages into the intima of these vessels. Since macrophages execute a strong stimulus for vessels growth, it is thought that the mediator for cell division in the vessel wall and enlargement of the collateral lumen is the stimulated monocyte.

VASOSPASM

Recent patient data have shown the importance of abrupt progression of coronary occlusion in precipitation of ischemic events. Many cellular and humoral changes may contribute to this progression. Rupture of the plaque, bleeding into the plaque, ulceration, and thrombus formation are well-documented morbid anatomic causes connected with the precipitation of sudden ischemia.[108,109]

Among the many pathophysiological mechanisms, altered endothelial function seems to be a particularly likely candidate not only for major conducting and stenosed arteries, but also for the smaller intramural branches, because the endothelium acts to prevent vasoconstriction and platelet adherence. Although the protective function of the endothelium has been extensively documented in isolated normal tissue, much less is known about coronary endothelial action in the presence of atherosclerotic changes in the underlying intima. Such atherosclerosis is likely to cause anatomic disruption or

functional impairment of the endothelial layer. Even though endothelial regeneration may occur, it is possible that the restored layer lacks certain protective capabilities of the original endothelium, particularly in the presence of associated hyperlipidemia. These changes may explain the unique vasoconstrictor sensitivity exhibited by atherosclerotic coronary arteries.[110]

Clinical Background

Coronary occlusion due to vasospasm is now well accepted as a cause of myocardial ischemia. Ergonovine maleate, methacholine and more recently, histamine have been used to induce spasm for diagnostic purpose in patient with atypical angina.[111,112] Other investigators have studied the effect of various vasoactive substances *in vitro* and *in vivo* in human coronary arteries and have demonstrated hyperreactivity in severely stenotic vessels.[113-121] The difficulty in obtaining suitable autopsy material after short postmortem intervals has, however, severely hampered quantitative analysis. In these tests, the maximum response was obtained with U44069, a prostaglandin–endoperoxidase analog. This response was greater than the one reported with histamine. Morphological analysis showed that the atherosclerotic plaque within the vessel lumen did not qualitatively change contractile response, but severely atherosclerotic segments were supersensitive to histamine.[122] Kalsner and Richards observed that coronary arteries were hypersensitive to histamine and serotonin, and this reactivity paralleled tissue levels of histamine.[123]

Forman et al. observed an increased number of mast cells in atherosclerotic coronary arteries studied at autopsy.[124,125] These increased proportionately with the severity of narrowing in sudden-death patients. However, large numbers of mast cells were seen only in segments of the coronary artery in which spasm had been elicited and demonstrated clinically. The segment with spasm was only 70% narrowed in cross-sectional area by atherosclerotic plaque. Therefore, the release of some pharmacologic agent, probably histamine in this patient, may have been responsible for the conversion of chronic, stable atherosclerosis into an acutely progressive stenosis.

The Role of Hypercholesterolemia on Vascular Tone

Armstrong et al. have demonstrated the correlation of structural alteration and hemodynamic response in atherosclerotic and hyperlipidemic macaque monkeys.[126] Bush et al. showed the effect of alpha-2-adrenergic and serotonergic receptor antagonists on cyclic blood flow alteration in stenosed canine coronary artery.[127] Atherosclerotic vessels and coronary arteries have been shown to display "supersensitivity" to contractile stimuli by Henry and Yokoyama.[128] There is sufficient experimental evidence to support the hypothesis that hypercholesterolemia may augment vascular tone. Freiman et al. have identified an altered response of endothelium-dependent vascular relaxation to acetylcholine and thrombin in atherosclerotic primates.[65] Williamson et al. and Harrison et al. even demonstrated that improvement in vascular spasticity could be achieved and vascular relaxation response induced during the regression phase of atherosclerosis in primates.[129,130,131] Particularly clear evidence of atherosclerosis-related vasospasm in a similar model as selected for this study was first presented by Shimokawa et al. in atherosclerotic miniature swine.[132]

The question left open by these studies was, What is responsible for inducing vasospasm? Is the atherosclerosis itself responsible? Are associated structural changes in the vessel wall, such as increased mast cells, a factor? Is it the hyperlipidemia alone? Studies by Verbeuren et al. suggested that hypercholesterolemia alone may be sufficient to alter vascular tone and susceptibility for va-

sospasm.[133,134] Verbeuren et al. concluded that contractions by serotoninergic mechanisms tend to be augmented while those to alpha-adrenergic activation are decreased.[133]

Mast Cells in the Adventitia

Foreman et al. have shown that mast cells are significantly increased in the adventitia of atherosclerotic vessels as compared to normal vessels.[124] Maximum cellular response of mast cell proliferation was seen in 12 to 14 months in miniature pigs fed a high cholesterol diet. However, the maximum reactivity of coronary arteries to histamine and serotonin was demonstrated after shorter intervals of feeding, namely from 4 to 6 months. Heistad et al. have shown increased vasoconstrictor response to serotonin and norepinephrine in the large hind limb and atherosclerotic arteries compared to similar nonatherosclerotic vessels in either normal or hypercholesterolemia monkeys.[135] With regression of atherosclerosis. Harrison et al. have demonstrated improvement of receptor-mediated endothelial-dependent vascular relaxation.[130]

The question remains, What is the stimulus of vasoconstriction? and, what is the response system—the endothelial cells, the smooth muscle cells and/or the mast cells? Ludmer et al. called the observed vascular contractile response in atherosclerotic arteries a paradoxical response since it causes contraction following exposure to acetylcholine, an agent that normally should release EDRF.[136] This finding would suggest that the interface between blood vessel and blood is substantially controlled by the endothelium.

The Role of the Endothelium

While many mechanisms might explain the association of atherosclerosis and alteration in vascular tone, the suggestion that the endothelium plays a pivotal role is an intriguing one in light of the discovery by Furchgott et al. that endothelial cells may regulate vascular tone by releasing potent vasodilatory substance, EDRF.[67-69] The EDRF was extensively described in normal vessels by Furchgott et al. and many others.[67-69] Its release, tested by a bioassay of vascular relaxation, has become a standard test for endothelial function, the same function that is altered in hypercholesterolemia and severe atherosclerosis, as has recently been demonstrated.

Ganz et al. have shown that endothelial dysfunction occurs early in the course of coronary artery disease and can lead to inappropriate vasoconstriction.[114] The chemical nature of EDRF is still unknown. It has been suggested that EDRF released by acetylcholine may be a product of the oxidation of arachidonic acid or some other unsaturated fatty acid. Recently, evidence has been presented that the product released by endothelial cells is nitric oxide.[137] Some workers have reported that in certain smooth muscle cells there is a positive relationship between an increase of cGMP and relaxation.[138] The demonstration of the EDRF is based on an *in vitro* procedure. Is it only an *in vitro* phenomenon? We suggest that this is not the case in light of the report by Ganz et al., who found that vasoconstriction had already occurred in early stages of atherosclerosis in response to acetylcholine injected into human coronary arteries.[114]

Martin et al. have shown that hemoglobin and methylene blue each induce a dose-dependent inhibition of the endothelium independent of the relaxation (if any) produced by nitroglycerine, but neither affect the relaxation of isoproterenol.[139,140] The inhibitory effect of hemoglobin and methylene blue may be due to a blockade of guanylate cyclase. The rise in GMP content that accompanies relaxation induced by acetylcholine and glycerol trinitrate is abolished. Isoproterenol-induced relaxation causes no change in GMP content. We therefore propose to explore the role of GMP in atherosclerotic coronary arteries in miniature

swine and correlate these results with functional studies.[141,142]

The Mediators of Vascular Constriction in Atherosclerosis

Atherosclerotic vessels show supersensitivity toward contractile stimuli.[143] A proposed mediator of this constrictive response is serotonin. The mediation through vasa vasorum has been proposed by Heistad and Armstrong.[144] VanHoutte et al. discovered the paradox that products of coagulation, such as activated platelets, serotonin, thrombin, and adenine nucleotides, evoke endothelium-dependent relaxation.[145] Yet, in direct contact with the deendothelialized vessel wall, the same substances cause constriction. This finding points to the mediating function of the endothelium, which possibly reduces the ability of aggregating platelets to induce vasoconstriction. Therefore, in the absence of endothelium, platelet products such as 5-HT, adenine nucleotides, thromboxane A_2, and vasopressin produce vasoconstriction. This may also play a contributory role in atherosclerosis.[145-150]

CONCLUSION

The microvasculature is an intricate part of any organ. The response of the microvasculature determines greatly the effect of injury and the spread of tissue necrosis. Since the blood flowing through the vessels and the cells of the vessel wall is a reactant of injury, such as toxic injury and ischemia, the fate of parenchymal cells may be determined by the reaction of these cells. The cause of cell death can shift over the vascular structures.

Circulation is central to the discussion of cell death and acute cell injury since capillary flow and perfusion is one essential function defining the life of an organism. We understand that the circulatory system is one arm of the metabolic exchange system of tissues, complemented by the interstitial space, the cell compartments, and the lymphatic system. It is not surprising that the integrity of organs, cells, and tissues depends on constant flow, which means on intact circulation. What is surprising is the speed by which cells and tissues of highly active organs, such as the heart, disintegrate when circulation is interrupted. This disintegration has been called *autolysis* and was considered apart from life systems.

Recent advances in organ transplantation and tissue culture of human tissues, however, have cast new interest and attention to autolytic processes. It seems important to recognize that autolytic processes do not start when cell death has occurred, but much earlier. The balance and time gap between autolytic degradation and cell viability creates the window for organ preservation in organ transplantation and for growth of human tissue components and cells in culture. The regulation of the role of circulation in health and disease, and the ability of the circulatory system to recover and return to function after injury has been suffered, is basic to medical practice, such as surgery, angioplasty, treatment of shock, thrombosis, embolism, hemorrhage, trauma, heart attacks, and wound healing.

It may even be that physical characteristics of the perfusing blood are altered in ischemic and some toxic injuries. Perfusion, with altered fluids, acts as the inducer of multiple forms of injury. These injuries are seen following exposure to low osmotic pressures, anemic perfusion, perfusion during sickle cell crisis, perfusion of solutions low in calcium (the calcium paradox), and reflow creating or releasing free radicals.

The aspect of vascular regeneration and new growth of vessels has been omitted from this discussion since we are mainly concerned with acute cell injury and cell death. Instead, the contributions of the vascular system and of the blood system to viability were the main point of the discussion. We feel that learning about the flow—no reflow phenomenon after toxic and ischemic injury is equal in importance to learning about cellular mechanism that contribute to survival of parenchymal cells.

REFERENCES

1. Reimer KA, Long JB, Murry CE, Jennings RB. Three dimensional distribution of collateral flow within the anatomic area at risk after circumflex coronary artery occlusion in dogs. *Basic Res Cardiol.* 1987;82:473–485.
2. Murry CE, Jennings RB, Reimer KA. Preconditioning with ischemia: a delay of lethal cell injury in ischemia. *Circulation.* 1986;4:1124–1136.
3. Schaper W. Experimental infarcts and microcirculation. In: Hearse DJ, Yellon DM, eds. *Therapeutic Approaches to Myocardial Infarct Size Limitation.* New York: Raven Press; 1984:79.
4. Mergner WJ, Schaper J. Cellular and subcellular changes in myocardial infarction. In: *Cell Injury in Shock, Anoxia and Ischemia: Pathophysiology, Prevention and Treatment.* Trump BF, Cowley RA, eds. Baltimore: Williams & Wilkins; 1982:658–680.
5. Muller KD, Klein H, Schaper W. Changes in myocardial oxygen consumption 45 minutes after experimental coronary occlusion do not alter infarct size. *Cardiovasc Res.* 1980;14(12):710–18.
6. Olssen RA, Bugni WJ. Coronary circulation. In: Fozzard HA, Haber E, Jennings RB, Katz AM, Morgan HE, eds. *The Heart and Cardiovascular System: Scientific Foundation*, Vol. 2. New York: Raven Press;' 1986: p. 987.
7. Gottwick MG, Siebes M, Kirkeeide R, Schaper W, Hamodynamik von Koronarstenosen. *Z Kardiol.* 73(suppl.):1984;47–53.
8. Feigl EO, Buffington CW, Nathan HJ. Adrenergic vasoconstriction during myocardial underperfusion. *Circulation.* 1981;75(I)1–5.
9. Sobin SS, Tremer HM. Three dimensional organization of microvascular beds as related to function. In: Kaley G, Altura Bm, *Microcirculation, Vol. 1.* Baltimore: University Park Press; 1977: pp. 43–67.
10. Bagge U, Branemark PI, Romanus M. Vital microscopic observations of microvascular response to ischemia. *Scand J Plast Reconstr Surg.* 19(suppl):1982;10.
11. Horacek MJ, Earl AM, Gilmore JP. The renal microvasculature of the monkey: an anatomical investigation. *J Anat.* 1986;148:205–231.
12. Schraufnagel DE, Schmid A. Capillary structure in elastase induced emphysema. *Am J Pathol.* 1988;130:126–135.
13. Schraufnagel, DE. Microvascular corrosion casting of the lung: a state-of-the-art review. *Scanning Microsc.* 1987;1:1733–1747.
14. Porter GA, Bankston PW. Functional maturation of the capillary wall in the fetal and neonatal rat heart: permeability characteristics of developing myocardial capillaries. *Am J Anat.* 1987; 180:323–331.

15. Watanabe S. Pathophysiology of hypxaemic pulmonary vascular disease. *Bull Eur Physiopathol Respir.* 1987;23(suppl 11):207s–209s.
16. Minamikawa T, Miyake T, Takamatsu T, Fujita S. A new method of lectin histochemistry for the study of brain angiogenesis. *Histochemistry.* 1987;87:317–320.
17. Baez S. Microvascular terminology. In: Kaley G, Altura BM, ed. *Microcirculation* Vol. 1. Baltimore: University Park Press; 1977: pp. 23–35.
18. Chambers R, Zweifach BW. The topography and function of the mesenteric capillary microcirculation. *Am J Anat.* 1944;75:173–205.
19. Rhodin JAG. The ultrastructure of mammalian arterioles and precapillary sphincters. *J Ultrastruct Res.* 1967;18:181–223.
20. Bassingthwaighte JB, Ypintosi T, Harvey RB. Microvasculature of the dog left ventricle myocardium. *Microvasc Res.* 1974;7:229–249.
21. Ono T, Shimohara Y, Okada K, Irino S. Scanning electron microscopic studies on microvascular architecture of human coronary vessels by corrosion casts: normal and focal necrosis. *Scan Electron Microsc.* 1986;1:263–270.
22. Stoiko MA, Pendergrass RB, Mergner WJ. Aspects of the three dimensional structure of cardiac microcirculation in the dog. *J Mol Cell Cardiol.* 1983;15(suppl):40.
23. Schaper W. Heterogeneity in the coronary circulation. *J Cardiovasc Pharmacol.* 1985;7(suppl 3):S31–S35.
24. Maina JN. The morphology and morphometry of adult normal baboon lung (*Papio anubis*). *J Anat.* 1987;150:229–245.
25. Kvietys PR, Granger DN. Physiology and pathophysiology of the colonic circulation. *Clin Gastroenterol.* 1986;15:967–983.
26. Ypintsoi T, Bassingthwaighte JB. Circulatory transport of iodoantipyrine and water in the isolated dog heart. *Circ Res.* 1970;27:461–477.
27. Katz SA, Feigl EO. Little carbon dioxide diffusional shunting in coronary circulation. *Am J Physiol.* 1987;253:H 614–625.
28. Marcus ML. *The Coronary Circulation in Health and Disease.* New York: McGraw-Hill; 1983.
29. Sobol BJ, Wanlass SA, Joseph EB, Azarshahy I. Alteration of coronary blood flow in the dog by inhalation of 100 percent oxygen. *Circ Res.* 1962;11:797–802.
30. Feigl EO. Coronary physiology. *Physiol Rev.* 1983;63:1–205.
31. Molnar JI, Scott JB, Frohlich ED, Haddy FJ. Local effects of various anions and H^+ on dog limb and coronary vasculature. *Am J Physiol.* 1962;203:125–132.
32. Murray PA, Belloni FL, Sparks HV. The role of potassium in the metabolism control of coronary vasculature resistance of the dog. *Circ Res.* 1979;44:767–780.
33. Needleman P, Kulkarni P, Raz A. Coronary tone modulation: formation and actions of prostaglandins, endoperoxides and thromboxans. *Science.* 1977;195:409.
34. Liang CS, Lowenstein JM. Metabolic control of circulation: effects of acetate and pyruvate. *J Clin Invest.* 1978;62:1029–1038.
35. Berne RM. The role of adenosine in the regulation of coronary blood flow. *Circ Res.* 1980; 47:807–813.
36. Olafsson B, Forman MB, Puett DW, et al. Reduction of reperfusion injury in the canine preparation by intracoronary adenosine: importance of the endothelium and the noreflow phenomenon. *Circulation.* 1987;76:1135–1145.
37. Gerlach E, Deuticke B, Dreisbach RH. Der Nucleotidabbau im Herzmuskel in Sauerstoffmangel und seine mogliche Bedeutung fuer die Coronardurchblutung. *Naturwissenschaften.* 1963;50: 228–229.
38. Kroll K, Feigl EO. Adenosine is unimportant in controlling coronary blood flow in unstressed dog hearts. *Am J Physiol.* 1985;249:1176–1187.
39. Boatwright RB, Downey HF, Bashour FA, Crystal GJ. Transmural variation in autoregulation of coronary blood flow in hyperperfused canine myocardium. *Circ Res.* 1980;47:599.
40. Guyton RA, McClenathan JH, Newman GE, Michaelis LL. Significance of subendocardial S–T segment elevation caused by coronary stenosis in the dog. *Am J Cardiol.* 1977;40:373.
41. McRaven DR, Mark AL, Abboud Fm, Mayer HE. Response of coronary vessels to adrenergic stimuli. *J Clin Invest.* 1971;50:773.
42. Feigl EO. The paradox of adrenergic coronary vasoconstriction. *Circulation.* 1987;76:737–745.
43. Young MA, Vatner SF. Enhanced adrenergic constriction of iliac artery with removal of endothelium in conscious dog. *Am J Physiol.* 1986;250:H892.
44. Vatner SF, Pagani M, Mauders WT, Pasipoulardes AD. Alpha-adrenergic vasoconstriction and nitroglycerin vasodilatation of large coronary arteries in the conscious dog. *J Clin Invest.* 1980;65:5.
45. Vatner SF. Regulation of coronary artery resistance vessels and large coronary arteries. *Am J Cardiol.* 1985;56:16RE.
46. Woodman OL, Vatner SF. Cardiovascular responses to the stimulation of alpha-1 and alpha-2 adrenoreceptors in the conscious dog. *J Pharmacol Exp Ther.* 1986;237:86
47. Maturi MF, Greene R, Donohue B, et al. Functional consequences and intracoronary localization of alpha-adrenergic stimulation of the canine coronary circulation. *J Am Coll Cardiol.* 1986;8:885–893.
48. Huang AH, Feigl EO. Adrenergic vasoconstriction helps maintain uniform transmural blood flow distribution during exercise. *Circ Res.* 1988;62:286–298.
49. Cohen RA, Shepherd JT, VanHoutte PM. 5-Hydroxytryptamine can mediate endothelium-dependent relaxation of coronary arteries. *Am J Physiol.* 1983;245:H 1007.
50. Cohen RA, Shepherd JT, VanHoutte PM. Prejunctional and postjunctional actions of endogenous norepinephrine at the sympathetic neuroeffector junction in canine coronary arteries. *Circ. Res.* 1983;52:16.

51. Cox TM, Angus JA. Endothelium-dependent re-laxation of coronary arteries by noradrenalin and serotonin. *Nature.* 1983;305:627.
52. Davies PF, Ganz P, Diehl PS. Reversible micro-carrier-mediated junctional communication between endothelial and smooth muscle cell monolayers: an *in-vitro* model of vascular cell interaction. *Lab Invest.* 1985;53:710.
53. Ashton JH, Schmitz JM, Campbell WB, et al. Inhibition of cyclic flow variations in stenosed canine coronary arteries by thromboxane A2/prostaglandin H2 receptor antagonists. *Circ Res.* 1986;59:568.
54. Nathan HJ, Feigl EO. Adrenergic vasoconstriction lessens transmural steal during coronary hypoperfusion. *Am J Physiol.* 1986;250:H645-663.
55. Burton AC. On the physiological equilibrium of small blood vessels. *Am J Physiol.* 1951;164:319-329.
56. Guyton AC, Taylor AE, Granger HJ, Gibson WH. Regulation of interstitial fluid volume and pressure. *Adv Exp Med Biol.* 1972;33(0):111-18.
57. Guyton AC, Granger HJ, Taylor AE. Interstitial fluid pressure. *Physiol Rev.* 1971;51(3):527-563.
58. Guyton AC. Collaterals, blood flow and tissue nutrition. *N Engl J Med.* 1971;284(23):1323-1324.
59. Bush LR, Campbell WB, Kern K, et al. The effect of alpha2-adrenergic and serotonergic receptor antagonist on cyclic blood flow alterations in stenosed canine coronary arteries. *Circ Res.* 1984;55:642.
60. Sabiston DC, Gregg DE. Effect of cardiac contraction on coronary blood flow. *Circulation.* 1957;15:14-20.
61. Duling BR, Hogan RD, Langille BL, et al. Vasomotor control: functional hyperemia and beyond. *Fed Proc.* 1987;46:251.
62. Hintzc TH, Vatner SF. Reactive dilatation of large coronary arteries in the conscious dog. *Circ Res.* 1984;54:50.
63. Simionescu M, Simionescu N. Isolation and characterization of endothelial cells from the heart microvasculature. *Microvasc Res.* 1978;16:426-452.
64. Gerlach E, Ness S, Becker BF. The vascular endothelium: a survey of some newly evolving biochemical and physiological features. *Bas Res Cardiol* 1985;80:459-474.
65. Forman MB, Puett DW, Virmani R. Endothelial and myocardial injury during ischemia and reperfusion: pathogenesis and therapeutic implications. *J Am Coll Cardiol.* 1989;13(2):450-459.
66. Furchgott RF. Role of endothelium in response of vascular smooth muscle. *Circ Res.* 1983;53:557.
67. Furchgott RF, Martin W. Interaction of endothelial cells and smooth muscle cells of arteries. *Chest.* 1985;88(suppl):210S.
68. Furchgott RF, Zawadski JV. The obligatory role of endothelial cells in relaxation of arterial smooth muscle by acetylcholine. *Nature.* 1980;288:373.
69. Pohl U, Busse R. Endothelium-derived relaxant factor inhibits effects of nitro compounds in isolated arteries. Am J Physiol. 1987;252:H307.
70. Freiman PC, Mitchell GG, Heistad DD, Armstrong ML, Harrison DG. Atherosclerosis impairs endothelium-dependent vascular relaxation to acetylcholine and thrombin in primates. *Circ Res.* 1986;58:783.
71. Schaper W, Nienaber C, Gottwick M. The importance of the collateral circulation for myocardial survival. *Acta Med Scand.* 651(suppl.): 1981; 29-35.
72. Bloor CM. Functional significance of coronary collateral circulation. *Am J Pathol.* 1974;76:561.
73. Schaper W. Collateral anatomy and blood flow: its potential role in sudden coronary death. *Ann NY Acad Sci.* 1982;382:69-75.
74. Hearse DJ, Yellon DM. Why are we still in doubt about infarct size limitation? The experimentalist's viewpoint. In: Hearse DJ, Yellon DM, eds. *Therapeutic Approaches to Myocardial Infarct Size Limitation.* New York: Raven Press; 1984: p. 17.
75. Schaper J, Konig R, Franz D, Schaper W. The endothelial surface of growing coronary collateral arteries. *Virch Arch.* 1976;370:193.
76. Nienhaber C, Gottwick M, Winkler B, Schaper W. The relationship between the perfusion deficit, infarct size and time after experimental coronary artery occlusion. *Basic Res Cardiol.* 1983; 78:210-226.
77. Sage MD, Gavin JB. Microvascular function at the margins of early experimental myocardial infarcts in isolated rabbit hearts. *Heart Vessels.* 1986;2(2):81-86.
78. Hooper D, Mergner WJ. Studies on the border zone of regional myocardial infarction. Unpublished observation.
79. Mergner GW, Zakaria M, Mergner WJ, et al. The effect of systemic pressure on infarct size reduction under halothene anesthesia. *J Cardiovasc Anest.* 1988;2:194-203.
80. Braunwald E. Reduction of myocardial infarct size. *N Engl J Med.* 1974;291:525.
81. Tillmans H, Kuebler W. What happens in the microcirculation? In: Hearse DJ, Yellon DM, eds. *Therapeutic Approaches to Myocardial Infarct Size Limitation.* New York: Raven Press; 1984.
82. Kloner RA, Rude RE, Carlson N, Maroko PR, DeBoer LWV, Braunwald E. Ultrastructural evidence of microvascular damage and myocardial cell injury after coronary occlusion: which comes first? *Circulation.* 1980;62:945-952.
83. Ambrosio G, Becker LC, Hutchins GM, Weisman HF, Weisfeldt ML. Reduction in experimental infarct size by recombinant human superoxide dismutase: insights into the pathophysiology of reperfusion injury. *Circulation* 1986;74:1424-33.
84. Hill JH, Ward PA. C3 leukotactic factors produced by a tissue protease. *J Exp Med.* 1969;30(3):505-518.
85. Nuyintinck JK, Goris RJ, Weerts JG, Schilling PH, Stekhoven JH. Acute generalized microvascular injury by activated complement and hypoxia: the basis of adult respiratory distress syndrome and multiple organ failure. *Br J Exp Pathol.* 1986;67:537-548.
86. Dormandy J, Ernst E, Bennett D, Erythrocyto-

cyte deformability in the pathophysiology of the microcirculation. *Ric Clin Lab.* 1981;11(suppl 1):35.

87. Neri Seneri GG. Pathophysiological aspect of platelet aggregation in relation to blood flow rheology in microcirculation. *Ric Clin Lab.* 1981;11(suppl 1):39.

88. Laurindo FRM, Ezra D, Czaja J, Feuerstein G, Goldstein RE. Mechanisms augmenting flow resistance of thrombotic coronary artery stenosis in pigs. *Circulation.* 1986;74(suppl 2):245.

89. Houston DS, Shepherd JT, VanHoutte PM. Aggregating human platelets cause direct contraction and endothelium-dependent relaxation of isolated canine coronary arteries: role of serotonin, thromboxane A2 and adenine nucleotides. *J Clin Invest.* 1986;78:539.

90. Feuerstein G, Boyd LM, Ezra D, Goldstein RE. Effect of platelet-activating factor on coronary circulation in domestic pig. *Am J Physiol.* 1984;246:H466.

91. Lucchesi RB, Romson JL, Jolly SR. Do leukocytes influence infarct size? In: Hearse DJ, Yellon DM, eds. *Therapeutic Approaches to Myocardial Infarct Size Limitations.* New York: Raven Press; 1984:219.

92. Werns SW, Lucchesi BR. Leukocytes, oxygen radicals and myocardial injury due to ischemia and reperfusion. *Free Radic Biol Med.* 1988; 4:31–37.

93. Goldstein RE, Feuerstein G, Letts G, Ramwell RW, Ezra D. In: Bailey M, ed. *Proceedings of the Washington Symposium on Prostaglandins and Leukotrienes.* 1984.

94. Weiss SJ. Oxygen, ischemia and inflammation. *Acta Physiol Scand.* 1986;548(suppl):9.

95. Engler R. Consequences of activation and adenosine-mediated inhibition of granulocytes during myocardial ischemia. *Fed Proc.* 1987;46:2407–2412.

96. Vigorito C, Poto S, Picotti GB, Trigiani M, Marone G. Effect of activation of the H1 receptor on coronary hemodynamics in man. *Circulation.* 1986;73:1175.

97. Vedder NB, Winn RK, Rice CL, Chie EY, Arfors KE, Harlan JM. A monoclonal antibody to the adherence-promoting leukocyte glycoprotein, C18, reduces organ injury and improves survival from hemorrhagic shock and resuscitation in rabbits. *J Clin Invest.* 1988;81:939–944.

98. Reimer KA, Jennings RB, Cobb FR, et al. Animal models for protecting ischemic myocardium, results of the NHLBI Cooperative study: Comparison of unconscius and conscious dog models. *Circ Res.* 1985;56:651–665.

99. Arminger LC, Gavin JB. Changes in the microvasculation of ischemic and infarcted myocardium. *Lab Invest.* 1975;33:51–56.

100. Rubanyi GM, Lorenz RR, VanHoutte PM. Bioassay of endothelium-derived relaxing factor(s); inactivation by catecholamines. *Am J Physiol.* 1986;249:H95.

101. Rubanyi GM, Romero JC, VanHoutte PM. Flow-induced release of endothelium derived relaxing factor. *Am J Physiol.* 1986;250:H1145.

102. Rubanyi GM, VanHoutte PM. Oxygen-derived free radicals, endothelium and responsiveness of vascular smooth muscle. *Am J Physiol.* 1986; 250:H815.

103. Rubanyi GM, Vanoutte PM. Hypoxia releases a vasoconstrictor substance from the canine vascular endothelium. *J Physiol.* 1988;364:411–415.

104. Ferrari R, Ceconi C, Curello S, et al. Oxygen-mediated myocardial damage during ischemia and reperfusion: role of cellular defenses against oxygen toxicity. *J Mol Cell Cardiol.* 1985;17:937–945.

105. Ito BR, Tate H, Kobayashi M, Schaper W. Reversibily injured, post ischemic canine myocardium retains normal contractile reserve. *Circ Res.* 1987;61:834–846.

106. Reimer KA, Murry CE, Yamasawa I, Hill ML, Jennings RB. Four brief periods of myocardial ischemia cause no cumulative ATP loss or necrosis. *Am J Physiol.* 1987;251:H1306–1315.

107. Henrichs KJ, Matsucka H, Schaper J. Influence of repetitive coronary occlusions on myocardial adenine nucleotides, high energy phosphate and ultrastructure. *Basic Res Cardiol.* 1987;82:557–565.

108. Buja LM, Willerson JT. Clinicopathological correlates of acute ischemic heart disease syndromes. *Am J Cardiol.* 1981;47:343.

109. Buja LM, Willerson JT. The role of coronary artery lesions in ischemic heart disease: insights from recent clinicopathologic, coronary arteriographic, and experimental studies. *Human Pathol.* 1987;18:451.

110. Heistad DD, Armstrong ML, Amundsen S. Blood flow through vasa vasorum in arteries and veins: effect of luminal Po_2. *Am J Physiol.* 1986;250:H434.

111. Maseri A, Severi S, DeNes M, et al. Variant angina, one aspect of a continuous spectrum of vasospastic myocardial ischemia: pathogenetic mechanisms, estimated incidence and clinical and coronary arteriographic findings in 138 patients. *Am J Cardiol* 1978;42:1019.

112. McAlpin RN. Contribution of dynamic vascular wall thickening to luminal narrowing during coronary artery constriction. *Circulation.* 1980; 61:204.

113. Babour HG. The constricting influence of adrenalin upon the human coronary arteries. *J Exp Med.* 1912;15:404.

114. Ganz P, Davies PF, Leopold JA, Gimbrone MA, Alexander RW. Short and longterm interaction of endothelium and vascular smooth muscle in coculture: effects on cyclic GMP production. *Proc Natl Acad Sci.* 1986;83:3552.

115. Ganz P, Alexander RW. New insights into the cellular mechanism of vasospasm. *Am J Cardiol.* 1985;56:11E.

116. Ganz P, Abben R, Friedman PL, Garnic JD, Barry WH, Levin DC. Usefulness of trans-stenotic coronary pressure gradient measurements during diagnostic catheterization. *Am J Cardiol.* 1985;55:910.

117. Anderson R, Holmbert S, Sredmyr N, et al. Adrenergic alpha and beta receptors in coronary ves-

sels in man: an *in vitro* study. *Acta Med Scand.* 1972;191:241.

118. Martin W, Villani GM, Jothianandan D, Furchgott RF. Selective blockade of endothelium-dependent and glycerol dinitrate-induced relaxation by hemoglobin and methylene blue in the rabbit aorta. *J Pharmacol Exp Ther.* 1985;232:708.

119. Rooke TW, Rimele TJ, VanHoutte PM. Contraction of canine coronary artery in calcium free solution. *Am J Physiol.* 1984;247:H259.

120. Miller VM, VanHoutte PM. Endothelium-dependent contraction to arachidonic acid are mediated by products of cyclo-oxygenase. *Am J Physiol.* 1985;248:H432.

121. Ross G, Stinson E, Schroeder J, et al. Spontaneous phasic activity of isolated human coronary arteries. *Cardiovasc Res.* 1980;14:613.

122. Ginsberg R, Briston MR, Davis K, et al. Quantitative pharmacologic responses of normal and atherosclerotic isolated human coronary arteries. *Circulation.* 1984;69:430.

123. Kalsner S, Richards R. Coronary arteries of cardiac patients are hyperactive and contain stores of amines: a mechanism for coronary spasm. *Science.* 1984;206:1435.

124. Forman MB, Bates JA, Robertson D, et al. Increased adventitial mast cells in a patient with coronary spasm. *N Engl J Med.* 1985;313:1138.

125. Forman MB, Robertson RM. Current treatment of coronary spasm: advantages and disadvantages of beta blockers. In: Conti CR, ed. *Coronary Artery Spasm: Pathophysiology, Diagnosis and Treatment.* New York: Marcel Dekker; 1986.

126. Armstrong ML, Heistad DD, Marcus ML, Megan MB, Piegors DJ. Structural and hemodynamic responses of peripheral arteries of macaque monkeys to atherogenic diet. *Arteriosclerosis.* 1986;5:336.

127. Bush LR, Campbell WB, Kern K, Tilton GD, et al. The effects of alpha 2-adrenergic and serotonergic receptor antagonists on cyclic blood flow alterations in stenosed canine coronary arteries. *Circ Res* 1984;55:642–52.

128. Henry PD, Yokoyama M. Supersensitivity of atherosclerotic rabbit aorta to ergonovine: mediation by serotonergic mechanism. *J Clin Invest.* 1980;66:306.

129. Williams JK, Armstrong ML, Heistad DD. Regression of vasa vasorum in coronary arteries during treatment of atherosclerosis. *Circulation.* 1986;74(suppl II):335.

130. Harrison DC, Freiman PC, Armstrong ML, Donald D, Heistad DD. Improvement of receptor mediated endothelium dependent vascular relaxation following regression of atherosclerosis. *Circulation.* 1986;74:II286. (abstract)

131. Mustard JF, Murphy EA, Roswell HC, Downie HG. Platelets and atherosclerosis. *J Atheroscl Res.* 1964;4:1–28.

132. Shimokawa H, Tomoike H, Nabeyama S, et al. Coronary artery spasm induced in atherosclerotic miniature swine. *Science.* 1983;221:560.

133. Verbeuren TJ, Jordaens FH, Zonnekeyn LL, VanHove CE, Coene MC, Herman AG. Effect of hypercholesterolemia on vascular reactivity in rabbit, I: endothelium-dependent and endothelium-independent contraction and relaxation in isolated arteries of control and hypercholesterolemic rabbits. *Circ Res.* 1986;58:552.

134. Verbeuren TJ, Coene MC, Jordaens FH, Can Hove CE, Zonnekeyn LL, Herman AG. Effect of hypercholesterolemia on vascular reactivity in the rabbit, II: influence of treatment with dipyridamole on endothelium-dependent and endothelium-independent responses in isolated aortas of control and hypercholesterolemic rabbits. *Circ Res.* 1986;59:496.

135. Heistad DD, Armstrong ML, Marcus ML, Piegors DJ, Mark AL. Potentiation of vasoconstrictor responses to serotonin in the limb of atherosclerotic monkeys. *J Hypertension.* 1986;4 (suppl):S17.

136. Ludmer PL, Selwyn AP, Shook TL, et al. Paradoxical vasoconstriction induced by acetylcholine in atherosclerotic coronary arteries. *Engl J Med.* 1986;315:1046.

137. Palmer RM, Ferrige AG, Moncade S. Nitric oxide release accounts for biological activity of endothelium-derived relaxing factor. *Nature.* 1987;327:524–526.

138. Katsuki S, Murad F. Regulation of adenosine cyclic 3'5' monophosphate and guanosine cyclic 3'5'-monophosphate levels and contractility in bovine tracheal smooth muscle. *Mol Pharmacol.* 1977;13:330.

139. Martin W, Furchgott RF, Villani GM, Jothianandan D. Depression of contractile responses in rat aorta by spontaneously released endothelium-derived relaxing factor. *J Pharmacol Exp Ther.* 1986;337:529.

140. Martin W, Furchgott RF, Villani GM, Jothianandan D. Phosphodiesterase inhibitors induce endothelium-dependent relaxation of rat and rabbit aorta by potentiating the effect of spontaneously released endothelium-derived relaxing factor. *J Pharmacol Exp Ther.* 1986;237:539.

141. Goldberg ND, Graff G, Haddon MF, Stephen JH, Glaser DB, Moser ME. Redox modulation of splenic cell soluble guanylate cyclase activity: activation by hydrophilic and hydrophobic oxidants represented by ascorbic acid and dehydro-ascorbic acid, fatty acid hydroxides and prostaglandin endoperoxides. *Adv Cyclic Nucleotides Res.* 1978;9:101.

142. Murad F, Arnold WP, Mittal CK, Braugher JM. Properties and regulation of guanyl cyclase and some proposed function for cyclic GMP. *Adv Cyc Nucleotide Res.* 1979;11:175.

143. Henry PD, Yokoyama M. Supersensitivity of atherosclerotic rabbit aorta to ergonovine: mediation by serotonergic mechanism. *J Clin Invest.* 1980;66:306.

144. Heistad DD, Armstrong ML. Bloodflow through vasa vasorum in coronary arteries in atherosclerotic monkeys. *Arteriosclerosis.* 1986;6:326.

145. VanHoutte PM, Cohen RA. Effects of acetylcholine on the coronary artery. *Fed Proc.* 1984;43:2878.

146. Vanhoutte PM, Miller VM. Heterogeneity of endothelium-dependent responses in mammalian

blood vessels. *J Cardiovasc Pharmacol.* 1985; 7:S12.

147. VanHoutte PM, Lueschert TF. Serotonin and the blood vessel wall. *J Hypertension.* 1986;4:S29.

148. VanHoutte PM. Endothelium and vascular reactivity. *J Mal Vasc.* 1986;11:213.

149. VanHoutte PM. Could the absence or malfunction of vascular endothelium precipitate the occurrence of vasospasm. *J Mol Cell Cardiol.* 1986; 18:679.

150. VanHoutte PM. Endothelium and the control of vascular tissue. *News in Physiological Science.* 1987;2:18.

PART II

Mechanisms Propagating Cell Injury

6

Pathobiology of Lysosomes

Qian-Chun Yu
Louis Marzella

INTRODUCTION

The lysosome is an intracellular compartment responsible for the degradation of extracellular (through heterophagy) as well as intracellular (through autophagy) materials. These degradative activities play a role in several cellular metabolic pathways, in the normal turnover of biological molecules, in host defense against pathogens, and in removal of necrotic material from the blood stream and from tissues. A schematic representation of different types of lysosomes is shown in Figure 6-1; the lysosomal pathways for the uptake and degradation of intracellular and extracellular materials are also indicated.

Following the discovery in the early 1950s of the lysosomes, the enzymatic composition of this organelle was characterized and the substrate specificity was determined.[1,2] Several of the lysosomal enzymes have been purified. Procedures for the isolation of different types of intact lysosomes have been worked out.[3] The biogenesis of lysosomes and their functional relationship with other organelles has been studied.[4] The synthesis of lysosomal enzymes as well as the processing of polypeptide and carbohydrate moieties and intracellular translocation of the enzymes have been elucidated.[5-7] Several classes of lysosomal enzyme inhibitors have been characterized. Mechanisms of lysosomal acidification and of regulation of lysosomal degradation of cellular components through autophagy and heterophagy have been described.[8-20] The endocytosis of receptor–ligand complexes and the various pathways followed by these complexes to the lysosomes and to other cellular compartments have been extensively studied.[21] In addition to participating in several physiologic functions, the lysosomes are involved in the pathophysiology of several types of acute and chronic cell injury. Several disorders of lysosomal function (both congenital and acquired) have been characterized.[22-24]

It is the intention of the authors to summarize some of the recent progress in the understanding of the functions of lysosomes as well as of alterations in lysosomal functions induced by cell injury. For this purpose, the review will deal with these topics: the lysosome as regulatory compartment and characteristics and functions of lysosomal membranes and of lysosomal hydrolases in normal as well as pathologic conditions.

THE LYSOSOMES AS A REGULATORY COMPARTMENT

The lysosomes are acidic vacuoles of heterogeneous shape, size, and contents containing hydrolytic enzymes that digest proteins, lipids, carbohydrates, and nucleic acids derived from inside and outside the cell (see Figure 6-1). Heterogeneity of hy-

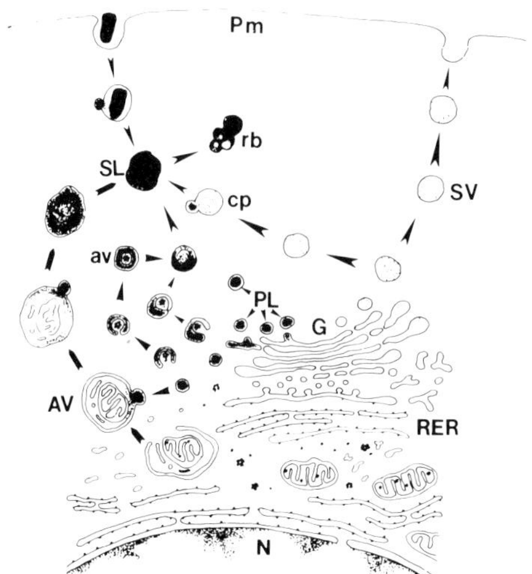

Figure 6-1. Schematic overview of lysosomal degradation pathways. The pathways for segregation of extracellular as well as cytoplasmic, cytosolic, and secretory materials are shown to converge to the secondary lysosomes (SL). SL degrade the segregated materials and reenter a new digestive cycle. Undegradable materials may accumulate in the lysosomes. These lysosomes are called *residual bodies* (rb). *Macro*autophagy: Cytoplasm is segregated into autophagic vacuoles (AV) formed by smooth cytomembranes. Primary (PL) and secondary (SL) lysosomes merge with AVs and provide hydrolytic enzymes for degradation of the segregated materials. *Micro*autophagy: Cytosolic components are segregated into lysosomes (av) by direct uptake via an endocyticlike process. Crinophagy: Secretory vacuoles (SV) are capable of merging with lysosomes and undergoing degradation inside crinophagic vacuoles (cp). Heterophagy: Segregation of extracellular materials occurs by formation of heterophagosomes at the plasma membrane (Pm), followed by degradation of phagocytized materials. Lysosomal enzymes are synthesized in the nuclear (N) envelope and rough endoplasmic reticulum (RER). Processing of the lysosomal enzymes occurs during translocation through the smooth ER and the Golgi (G) apparatus. The newly synthesized lysosomal enzymes are packaged into small vesicles (primary lysosomes, PL), which bud off from the trans region of the Golgi apparatus.

drolase levels exists among lysosomes within the same cell and among related cell types of an organ.[25] Lysosomal degradation serves several regulatory functions, namely: normal turnover of cell organelles and of extracellular components, regulation of secretory processes, amino acid and glucose homeostasis, physiologic or pathologic tissue growth, differentiation or atrophy. Figure 6-2 illustrates the ultrastructure of autophagic and heterophagic lysosomal degradation.

The lysosomes are the major site for intracellular degradation and turnover of cytosol and of cell organelles through autophagic mechanisms.[17,18,26] In response to conditions such as starvation and amino acid deprivation, the lysosomes increase the uptake and degradation of cell organelles to maintain normal levels of amino acids.[27-30] Several hormones can modulate lysosomal degradation. For example, in hypoinsulinemia and hyperglucagonemia induced by streptozotocin, the lysosomal system expands. Brekke and co-workers found that the increase in number and in volume density of lysosomes and autophagic vacuoles caused by streptozotocin-induced diabetes reverts to control values following pancreatic transplantation.[31] Selective autophagy of organelles has been reported and may represent a stage in the programmed differentiation of precursor cells into more mature or terminal phenotypes. For example, Heynen et al. have reported that autophagy of mitochondria in erythrocyte precursor cells in rat bone marrow starts at the time of nuclear extrusion.[32]

Autophagy therefore may be involved in the differentiation of these cells. The involvement of autophagic and heterophagic

Figure 6-2. Ultrastructural appearances of typical autophagic and heterophagic vacuoles. (A) *Macro*autophagic vacuoles (av) in the hepatocyte of a mouse liver after the administration of taxol and vinblastine. Stages of the evolution of the autophagic vacuoles are illustrated. av1: newly formed autophagosomes containing ultrastructurally undegraded organelles. av2: ultrastructural signs of degradation are evident. In late stages (av2), only barely discernible membrane remnants are seen. (× 16,720) *Inset*: *Micro*autophagic vacuoles (arrowheads) in a rat hepatocyte. (× 22,000) (B) Crinophagy of zymogen granules in hamster pancreatic acinar cells maintained in explant culture. Merging of zymogen granules (arrow) with secondary lysosomes and autophagic uptake of the granules (arrow) are illustrated (× 14,960). (C) Heterophagic vacuole (arrow) in a dog liver Kupffer cell (Kc) contains partially degraded remnants of a leukocyte. Two contiguous hepatocytes (h) are seen. (× 14,960)

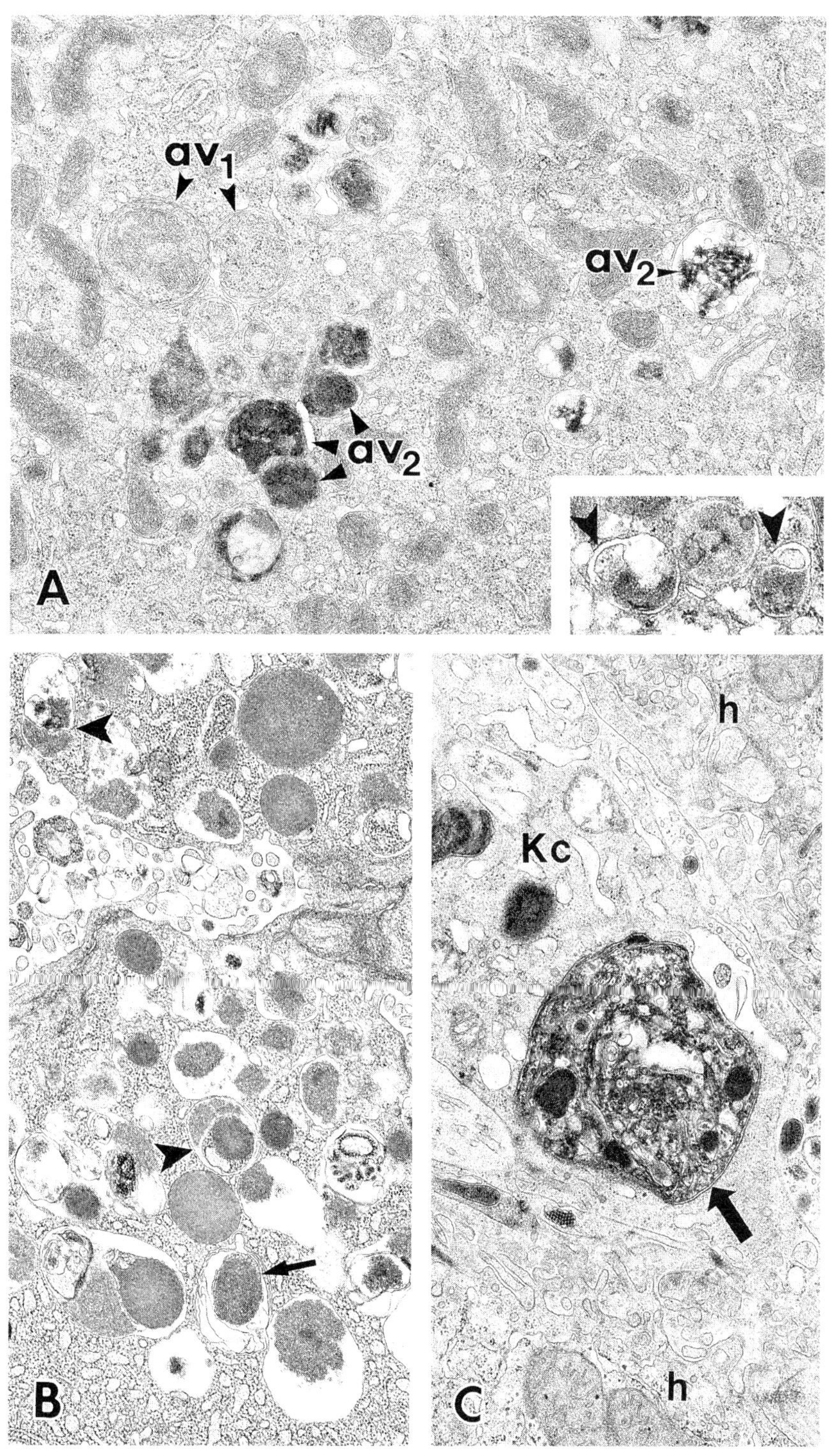

lysosomal degradation in the physiologic involution of organs is a well-known phenomenon.[33] The regulation of lysosomal numbers and size is accomplished in part by the invagination of lysosomal surface membrane. Membrane vesicles with trapped cytosolic components bud into the lysosomal matrix, where they are degraded. This phenomenon contributes to the turnover of cytosol and of lysosomal membranes, and has been designated *microautophagy*.[34] The phenomenon of self-regulation of lysosome numbers by microautophagy is particularly prominent in cells such as macrophages. Cultured mouse macrophages exposed to microtubule-active drugs manifest an ultrastructurally apparent process of "lysosomophagy." Clusters of small lysosomes in these cells appear to be engulfed by large lysosomes and degraded.[35]

The lysosomes participate in the intracellular degradation of secretory proteins. They thus play a role in the regulation of secretion. An example is the lysosomal degradation of thyroid colloid protein and the release of thyroid hormone.[36-40] In the anterior pituitary gland of the migratory Canada goose, a seasonal increase in the number of lysosomes with a corresponding decline in the number of secretory granules has been described. These changes seem to be associated with the regulation of secreted pituitary hormones.[41] In both the exocrine and endocrine pancreas, the lysosomes are able to merge with and degrade the secretory granules through crinophagy. This process may regulate postsynthetically the levels of pancreatic enzymes and insulin in this organ.[42-44]

An example of the biological significance of the ability of lysosomes to modify biological molecules and affect target cells was described by Thiele and Lipsky.[45] These workers found that neutrophil or macrophage lysosomes can convert the lysosomotropic compound L-leucine-methyl ester into a dipeptide methyl ester. This dipeptide shows preferential toxicity to human natural killer cells. Another example of the influence of lysosomal function on target cells is provided by the observation that a neutrophil lysosomal fraction can stimulate DNA synthesis and cell growth of cultured rat aortic smooth muscle cells and endothelial cells.[46] The lysosomes of macrophages are also known to be responsible for the proteolytic processing of foreign antigens. The transformed antigens are subsequently transferred to the surface membrane of the macrophage. The presentation of this antigen by the macrophage to lymphocytes is an important phenomenon in immune modulation.

In addition to the functions just described, the lysosomes are also known to participate in the regulation of numerous biochemical pathways. One of the most significant features of this type of regulation is that it is dependent not only on the activity of hydrolytic enzymes but also on some specific functions of lysosomal membranes. Two examples of this type of regulatory function are vitamin B12 and cholesterol metabolism.

Studies by Youngdahl-Turner and coworkers have demonstrated that lysosomes play an important role in vitamin B12 metabolism.[47,48] Normally cobalamin binds to a transport protein to form a protein–cobalamin complex. This complex binds to specific receptors on the cell surface and is then endocytosed by the cell. The endocytosed complex is transported to the lysosomes, where the binding protein is degraded and the free vitamin B12 is released and transferred through lysosomal membranes to the cytosol.[47,48] Rosenblatt and coworkers studied vitamin B12 release from fibroblast lysosomes of a patient who suffered from an inborn error of vitamin B12 metabolism.[49] A lysosomal fraction from the patient's fibroblasts accumulated cobalamin at levels higher than those of control cells at the same growth stage, and much higher than those of mutant fibroblasts, which are unable to accumulate cobalamin. In addition, most of the cobalamin that accumulated in the patient's lysosomes was found to be free cobalamin. The patient's lysosomes were thus defective in the transfer of intralysosomal free vitamin B12 into the cytosol.

The uptake and degradation of circulating low-density lipoprotein (LDL) particles regulates the biosynthesis of cholesterol. The LDL particles enter the cell through receptor-mediated endocytosis. Once inside the cell, the endocytic vesicles containing LDL fuse with lysosomes in which LDL is hydrolysed, and cholesterol is released into the cytosol.[50] Any defects in the endocytosis or lysosomal degradation of the LDL molecules will lead to impairment of the normal feedback regulation of cholesterol biosynthesis and induce cell injury. Defective internalization of LDL receptor–ligand complexes as well as abnormal distribution of LDL receptors on the plasma membrane have been described in cancer cells.[51] Normally, cells control the uptake of ligands by regulating the numbers of specific receptors present on the plasma membrane. Loss of intracellular potassium or hyperosmotic stress, such as may occur in cell injury, inhibits receptor-mediated endocytosis.[52,53]

CHARACTERISTICS OF LYSOSOMAL MEMBRANES

The Lysosomal Membrane as an Intracellular Barrier

The initial biochemical observation of the latency (partially hidden enzyme activity) manifested by a liver phosphatase eventually led de Duve to formulate the lysosome concept. According to this concept, the lysosomal membrane functions as a barrier separating the lysosomal hydrolytic interior from the remainder of the cytoplasm.[1,54] Electron microscopic studies readily confirmed the presence of membrane-bounded, electron-dense organelles in the same liver subcellular fractions analyzed by de Duve.

The lysosomal membrane possesses unique properties. The inner face of the membrane bilayer is constantly exposed to an acidic, highly degradative environment, while the outer face is in contact with the cytosol and other cytoplasmic organelles. By electron microscopy, most lysosomes show a clear, electron lucent halo along the inner face of the membrane (Figure 6-2, inset). This unstained halo has been postulated to be due to abundant sugar residues of lysosomal membrane glycoproteins. These glycoproteins are thought to render the inner face of the membrane highly resistant to the degradative enzymes. The cytoplasmic face of the lysosomal membrane, however, is susceptible to hydrolytic enzymes. This susceptibility has been shown by experiments in which isolated cellular organelles are injected into the blood stream of rats; the injected membranes, including isolated lysosomes, are taken up and readily degraded by Kupffer cells.[14–16] A proton pump located in the lysosomal membrane provides an acidic environment for the hydrolytic enzymes.[9,55] It is known that transport systems for amino acids, sugars, nucleotides, and other molecules are present in lysosomal membranes.[56–60] Evidence in support of the presence of ionic transport systems is also beginning to appear.[61] Disorders involving the permeation of vitamin B12 (discussed earlier) and of cystine out of the lysosome have been described. In the disease called cystinosis, the lysosomes accumulate high levels of free cystine derived from the normal degradation of proteins. It has been proposed that the major defect of cystinosis is the inability of the lysosomal membrane to transport free cystine from the lysosomal matrix into the cytosol.[62,63] This transport is thought to be an ATP-dependent exchange diffusion process.[64,65]

The capacity of the lysosomal enzymes to extensively degrade biological molecules provides a rationale for the evolution of cellular membrane compartments in which the lysosomal enzymes are segregated within a membrane barrier. The lysosomal membrane is normally permeable only to small (approximately 200 daltons) molecules. The entrance of other intracellular and extracellular material into the lysosomal compartment is a regulated process. *In vitro*, various factors such as pH, temperature, or osmotic stress decrease the stability of lysosomal membranes. *In vivo*, various pathologic insults may lead to the release of lysosomal enzymes either into the cytosol or into the extracellular space.[66-68] Lysosomes isolated from pathologically altered tissues are known to be more fragile.[69] Several fat-soluble vitamins and some hormones have been found to decrease the stability of lysosomal membranes, whereas cholesterol and vitamin E increase the stability of the membranes.[70] It has been proposed that lysosomal membranes may be damaged by lipid peroxidation induced by free radicals.[71,72] Lysosomes are known to accumulate heavy metal ions such as lead, copper, and iron. These metals may participate in redox reactions that generate free radicals and induce peroxidation of lysosomal membrane lipids.[73-77] Hydroxyl radical scavengers, such as dimethyl sulfoxide (DMSO), increase the lysosomal stability as well as the survival of cultured macrophages.[78] The viability of other cultured cells, such as rat hepatocytes, is also enhanced by DMSO.[79]

Factors that enhance the stability of lysosomal membranes have also been identified. The effect of ATP on isolated lysosomes has been evaluated in cells such as canine cardiocytes and rat hepatocytes.[80-88] ATP enhances the stability of isolated lysosomes incubated at different temperatures. This protection can be reversed by the addition of the H^+ ionophore m-chlorocarbonyl-cyanide phenylhydrazone (CCCP). It seems that maintenance of lysosomal membrane stability is related to the normal functioning of an ATP-driven electrogenic proton pump.[80]

An interesting positive correlation between prostaglandin I_2 levels and lysosomal membrane stability has been described.[69,84,85] This finding was investigated by Berberian and co-workers in cell mixtures from atherosclerotic aortas.[86,87] These authors established a correlation between increased levels of prostaglandin I_2 and enhanced stability of the lysosomal membrane. The enhanced stability is either a direct effect of prostaglandin I_2 or a result of the intralysosomal accumulation in these cells of free cholesterol, which may represent the initial stage of foam cell formation.[86,87] Following cell death and rupture of the cell surface membrane, intact lysosomal vesicles may be released to the extracellular space in tissues such as vascular walls. These lysosomal vesicles are identifiable by the presence of characteristic lysosomal "marker" enzymes.[88]

Lysosomal Membrane Turnover

Invaginations and flaplike extensions of lysosomal membranes as well as intralysosomal vesicles have been observed at the ultrastructural level in various types of cells and in isolated lysosomes. These membrane alterations may represent stages in the direct uptake of cytosolic material by the lysosomes. These phenomena have been named *microautophagy, lysosomal wrapping mechanism,* and *lysosomophagy.*

In contrast to macroautophagy, which involves the formation of autophagosomes by segregation of cytoplasm by cisternal membranes and the fusion of autophagosomes with lysosomes, microautophagy is postulated to be a means of direct uptake of cellular substances by lysosomes (see Figure 6-2). The lysosomal membrane invaginations and resultant intralysosomal vesicles contain predominantly cytosolic materials such as glycogen and ribosomes. Microautophagy is thought to be related to basal degradation of cellular proteins.[1,17,20,30,34,89-92] Ultrastructural analyses also suggest that the lysosomal membrane can wall-off relatively large amounts of cytoplasmic materials for degradation by means of pseudopodialike projections into cytoplasm. This phenomenon has been named *lysosomal wrapping mechanism* and is thought to be energy dependent.[93-96] The actual mechanism responsible for these conformational changes of lysosomal membranes is not known. To date there is no evidence of interaction between the materials being taken up and the proteins on the cytoplasmic side of the lysosomal membrane.

Recognition and Fusion of Lysosomal Organelles: Relationships to Cytoskeleton

The intracellular degradation of organelles by autophagy requires the segregation of cytoplasm in membrane-bound autophagic vacuoles, and fusion between autophagic vacuoles and preexisting lysosomes. This process is analogous to the endocytosis or phagocytosis of extracellular material; as a result of this segregation and fusion process, degradation by lysosomal hydrolytic enzymes always occurs within membrane-bound vacuoles.[12-15,26,97-100] In the case of autophagy, the segregation process appears to be nonselective in the sense that different types of cell organelles can be segregated in the same autophagic vacuole; the segregation does not appear to select pathologically altered organelles; and the composition of the segregated material reflects, in many cases, the relative abundance of the material in the cell.[101-104] In the case of phagocytosis and of endocytosis, recognition and fusion between phagosomes and endosomes with lysosomes occurs. This fusion process can be demonstrated within 4 to 10 minutes after the internalization of the vacuoles endocytosed through a receptor-mediated pathway in rat hepatocytes.[15,105-108].

Following endocytosis, the glycoproteins rapidly appear in a prelysosomal compartment of smooth-surfaced tubules and vesicles, and are then ferried to the lysosomes, where degradation occurs. The receptors for this ligand are recycled several times before they are degraded. It is thus suggested that the sorting of ligands from receptors might have taken place at the level of endosomes. Two kinds of transitional vesicles may be formed for different intracellular destinations: vesicles containing free ligand fuse with the lysosomes for degradation, while the vesicles carrying receptors are recycled to the plasma membrane. Selective transfer of plasma membrane proteins to the lysosomes for degradation occurs via endocytic pathways.[109,110]

Specific recognition of endosomes by lysosomes can also be observed during the endocytosis of several other ligands including immunoglobulins, epidermal growth factor (EGF), transferrin, and LDL. In the case of endocytosis of monovalent antibodies, the Fc receptors are not cross-linked, and the endosomes do not deliver the immunoglobulin to the lysosomes. However, if the receptors are cross-linked by polyvalent immune complexes, the endosomes deliver the immunoglobulin to the lysosomes for degradation.[111]

It is an interesting phenomenon that EGF and transferrin follow diverging intracellular pathways after uptake.[99,112] Both transferrin and EGF enter cells through receptor-mediated endocytosis at the same region of surface membrane. After internalization into endosomes, the EGF–receptor complexes are directed to lysosomes and degraded.[98,112,113] On the other hand, the transferrin–receptor complex is recycled back to the plasma membrane. The coated pits of transreticular elements of the Golgi apparatus have been proposed as the site for the sorting mechanism. Dunn and colleagues have further demonstrated that the EGF-induced receptor loss can be prevented by a lysosomal protease inhibitor (leupeptin) or by the inhibition of lyso some–endosome fusion with low temperature.[112] Recently Ballard reported that EGF may increase the lysosomal degradation of cellular proteins.[114] This increase can be inhibited by agents such as leupeptin, vinblastine, and ammonia, but not by cycloheximide.

It is not clear, however, whether this increased degradation is due to a direct effect on the function of lysosomes and autophagosomes, or whether it is mediated indirectly. Phorbol esters can induce transient internalization of unoccupied EGF receptors, but the endocytosed receptors cannot be recognized and degraded by lysosomes.[115,116] Examples from endocytosis of other ligands, such as IgA and LDL, also indicate the existence of selective recognition between endosomes (receptosomes) and lysosomal membranes.[117-121]

In summary, the lysosomes, through recognition of autophagic as well as heterophagic vacuoles and vesicles, occupy a pivotal position in the degradation of extracellular and intracellular materials. We still do not know the molecular basis for lysosomal recognition and fusion phenomena, and whether different populations of lysosomes are involved. Lipid and protein moieties of membranes, ionic species, and charge effects are known to affect fusion of biological membranes. However, with the exception of specific fusogenic effect of viral glycoproteins, the mechanism of membrane fusion in general is not known.[122-125]

The intracellular distribution and the movement of lysosomes have also received increased attention. It has been observed that various cytoplasmic organelles move along cytoskeletal elements, particularly along microtubules.[126] The intracellular distribution and movement of lysosomes have been found to be dependent upon the integrity of microtubules.[127,128] Treatment of macrophages with phorbol esters causes the extension and radial organization of microtubules, and promotes the movement of secondary lysosomes from their perinuclear location to the peripheral cytoplasm.[129,130] This effect can be prevented by colchicine and podophyllotoxin, but not by cytochalasin B or D.

Studies on autophagy demonstrate that microtubule depolymerizers such as vinblastine may inhibit the fusion between lysosomes and newly formed autophagic vacuoles, and this may influence the lysosomal degradation process.[15,101,102,131-133] Taxol, a microtubule stabilizing agent, enhances the ultrastructurally apparent process of degradation of autophagic vacuoles, as well as protein degradation rates, in lysosomal fractions from the livers of mice treated with vinblastine.[134] These findings strongly suggest that some spatial relationship exists between lysosomes and microtubules. Further evidence supporting this relationship comes from a study by Collot et al.[135] By using double immunofluorescent labeling for both lysosomes and cytoskeletal elements, a constant co-distribution between lysosomes and microtubules was demonstrated. The authors suggested that this association reflects some type of linkage and that such a linkage may play an important role in the subcellular distribution and intracellular transport of lysosomes.

Recently, translocator proteins responsible for the anterograde and retrograde organelle movement along microtubules have been described.[136,137] Trypsin treatment of the organelles was found to abolish such movement along microtubules, even in the presence of translocators.[138] Another example of possible translocator protein is a 100-kd protein, identified by Rodionov et al. This protein has been shown to be associated with microtubules, intermediate filaments, and coated vesicles.[139] These authors suggest that this 100-kd protein may play a role in the regulation of intracellular organelle movement by cytoskeletal elements. It is not known if the translocators for these movements are also recognized by components of lysosomal membranes and induce the saltatory movement of this organelle along the microtubules. *In vitro*, the lysosomes can complex with microtubules even in the absence of microtubule-associated proteins or cytosolic components.[140] A lysosomal membrane protein that can bind directly to microtubules has been identified.[141]

Compared to our detailed knowledge about the lysosomal enzymes, little is known about the composition of the lysosomal membrane.[142-144] The lysosomal membrane has a polypeptide composition different from that of the plasma membrane. In addition, the lysosomal membrane ATPase is not sensitive to inhibitors of plasma membrane ATPase such as ouabain and suramin. By using specific antibodies against lysosomal membranes, Reggio and co-workers have identified an integral membrane protein of approximately 100,000 d, which shows cross-reaction with purified H^+/K^+-ATPase from parietal cells of gastric mucosa, but not with the ouabain-sensitive ATPase of plasma membrane.

Identification of two other lysosomal membrane glycoproteins has been achieved by Chen et al. with monoclonal antibodies.[146] These two integral membrane proteins form a complex and are designated lysosome-associated membrane proteins (LAMP–1 of 105–115 kd and LAMP–2 of 100–110 kd). Three lysosomal integral membrane glycoproteins have been identified by Lewis et al.[147] These proteins, with molecular weights of 80 kd, 100 kd, and 120 kd, are all highly glycosylated. One of these proteins (100 kd) cross-reacts with the protein described by Reggio et al., whereas the 120-kd protein is recognized only in lysosomes in which it co-localizes with β-glucuronidase.[144,148] These lysosomal glycoproteins have been designated lgp 120, lgp 100, and lgp 80. It is not clear whether there is any relationship between lgp 120 and LAMP–1. The functional role of these lysosomal membrane-specific proteins remains to be elucidated, but a role in lysosomal acidification and lysosomal membrane resistance to digestion has been suggested.

Regulation of Autophagic Vacuole Formation

Autophagic protein degradation begins with the segregation of cellular components into membrane-bounded vacuoles to form autophagosomes. The two most important questions with respect to the formation of autophagosomes are (1) What is the origin of the autophagosome limiting membranes?

In other words, which internal membranes initiate the segregation process? (2) What intracellular signals mediate the process of segregation in response to a variety of physiologic as well as pathologic stimuli?

Origin of Autophagosome Membrane

The current concepts of the biogenesis of autophagic vacuoles are based upon ultrastructural and biochemical evidence. These types of analyses suggest that the membrane of autophagosomes may be derived primarily from cytomembranes of the endoplasmic reticulum (ER) and Golgi complex type as well as from the lysosomes themselves. In addition, a classification of autophagic vacuoles into *micro*autophagic and *macro*autophagic vacuoles has been proposed.[20] This classification is based upon the size of the autophagic vacuoles as well as on the postulated membrane of origin of the vacuoles. It has also been proposed that the induction of *micro-* and *macro*autophagy may be regulated differently.

The present challenge is to characterize these lysosomal structures by biochemical and immunochemical approaches to determine their functional significance in cellular degradation processes. In the case of *micro*autophagy and lysosomal wrapping mechanisms, the lysosomal membrane and, in some cases, smooth cytoplasmic membranes are apparently responsible for the uptake and segregation of components derived primarily from the cytosol. In all other *macro*autophagic processes, larger portions of cytoplasm are segregated by cisternal-like double membranes. Two hypotheses regarding the biogenesis of the limiting membranes of *macro*autophagic vacuoles have been suggested; (1) The membranes are derived from *de novo* synthesis of a new membranous structure around the cytoplasmic area being segregated to form autophagosomes, or (2) preexisting membranes wrap around the organelles or cytoplasm to be segregated, and finally seal up to form an autophagosome.[1,11]

The early idea of *de novo* synthesis has virtually been abandoned, mainly as a result of two lines of evidence. First, an increase in number of autophagic vacuoles can be noted after only a very short time lag from the application of a stimulus.[105] Moreover, protein synthesis inhibitors such as puromycin may enhance, rather than block, autophagy.[97,149,150] Finally, the idea of *de novo* synthesis is difficult to reconcile with currently accepted models of membrane biogenesis. Therefore, the second hypothesis — that is, that segregation membrane is derived from preexisting organelle membranes — is generally favored and is more in tune with the ultrastructural experimental evidence.

The available evidence indicates that ER and Golgi-associated membranes partici-pate in the formation of autophagosomes. The ER was recognized as a source of autophagic vacuole membranes by Swift and Hruban, who presented ultrastructural evidence showing apparent features of ER that constitute the segregation membrane of newly formed autophagosomes.[11] In the studies of autophagy induced by vinblastine and glucagon, the ER origin of autophagic vacuole membrane has been strongly suggested by electron microscopic observations.[19,102,131,151,152]

Figure 6-3 shows the ultrastructure of newly formed autophagosomes in cultured rat hepatocytes. The presence of direct continuity between the rough ER cisternae and the limiting membranes of the autophagosomes is suggested by these micrographs. The participation of ER and Golgi appa-

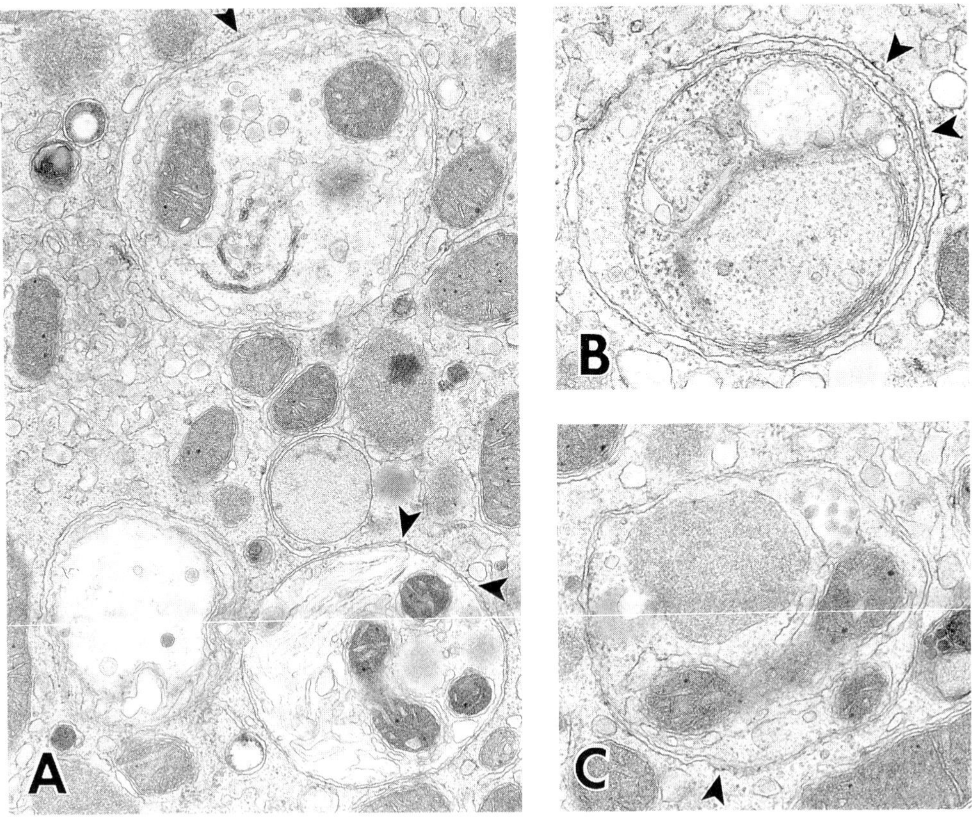

Figure 6-3. Origin of autophagosome membranes. Continuities between endoplasmic reticulum cisternae and limiting membranes (arrowheads) of autophagosomes are illustrated in cultured rat hepatocytes. (A) × 17,000; (B) × 24,000; (C) × 24,000

ratus in the formation of autophagic vacuoles is further suggested by unbuffered osmium staining techniques.[153] When the rat liver and hamster pancreas are stained by osmium impregnation, the autophagic vacuole and lysosomal membranes show dense osmium deposits (Figure 6-4).

The other sites of stain deposition in the hepatocytes are ER, nuclear envelope, and Golgi apparatus. The detailed electron microscopic study of hepatocytes by Novikoff and Shin showed by means of serial section techniques the existence of direct membrane continuities between autophagic vacuole and ER.[154] *In vitro* study of isolated autophagic vacuoles from rat liver shows the localization of a tightly bound ER membrane marker enzyme, G-6-Pase, to the early and late autophagic vacuole membranes.[155] The presence of ER marker enzymes in nascent autophagic vacuole membranes is not always clearly demonstrable. A rapid membrane modification or exclusion of specific proteins from nascent autophagic

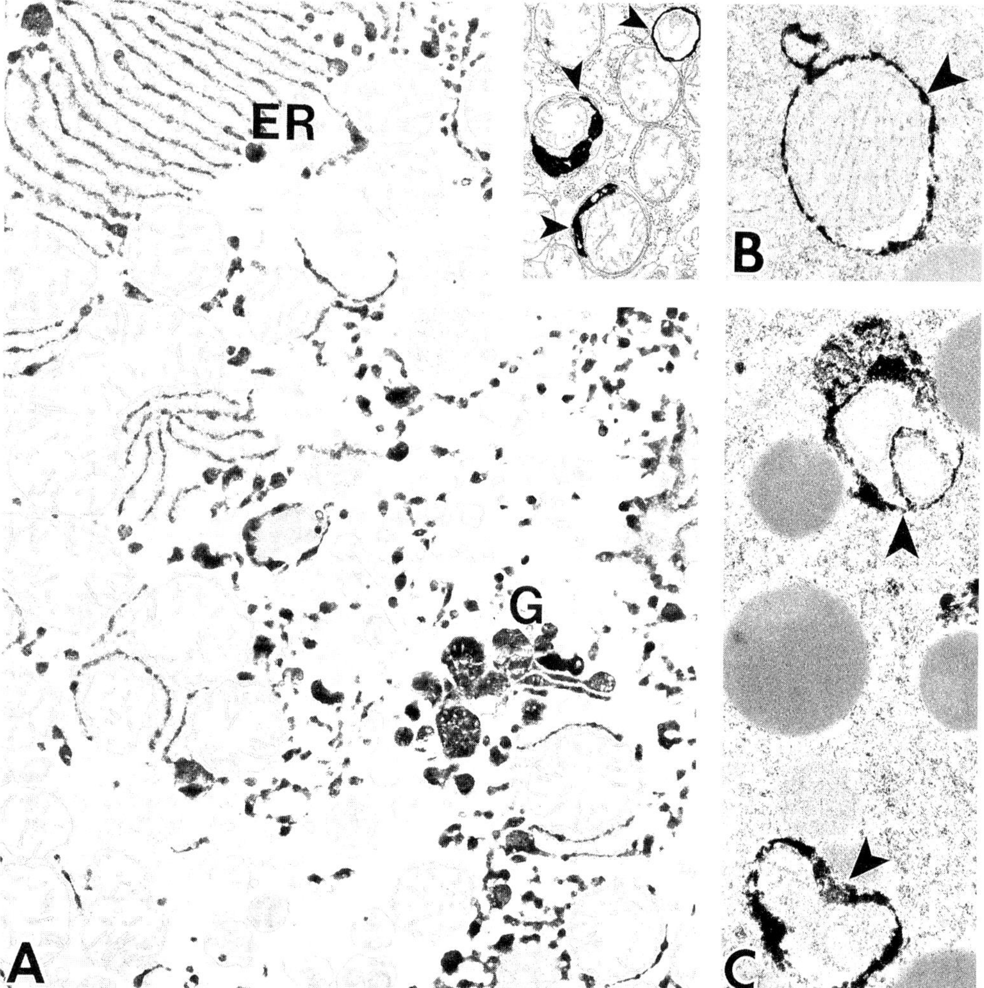

Figure 6-4. Origin of autophagosome membranes. (A) Osmium tetroxide stains endoplasmic reticulum (ER) and Golgi (G) apparatus of rat hepatocytes. (×14,000) *Inset*: Membranes of hepatocyte autophagic vacuoles (arrowheads) are also stained by osmium. (× 8,500) (BC) Osmium tetroxide staining of autophagic vacuole membrane (arrowheads) in cultured pancreatic acinar cells of hamster. (× 18,000)

vacuole membrane could account for the difficulty in demonstrating ER marker enzymes.[20] A positive correlation between increased numbers of autophagosomes and detachment of ribosomes from ER has further suggested a close relationship between autophagic vacuole membrane and ER.[97,150] Taken together, this evidence indicates that the ER is one of the sources for the limiting membrane of autophagic vacuoles.

Golgi apparatus and smooth ER vesicles are also thought to participate in the formation of autophagic vacuole membrane. Ultrastructural as well as cytochemical studies with the osmium impregnation technique suggest a relationship between the limiting membrane of autophagic vacuoles and cisternae and the small vesicles in the Golgi apparatus–smooth ER area.[156–158] Clusters of vesicles are present in this region and are thought to function as shuttles traveling between various cellular compartments. These vesicles are also seen in close proximity to autophagic vacuoles. It has been suggested that these vesicles may fuse with each other around certain parts of cytoplasm or organelles to form new autophagosomes. Some of these vesicles in close proximity to autophagic vacuoles may also represent fission vesicles that detach during the remodeling of the limiting membrane of autophagic vacuoles.

Intracellular Signals Inducing Autophagy

Several hypotheses have been postulated regarding the intracellular signals that link the intracellular and extracellular stimuli to the formation of autophagic vacuoles. The proposed intracellular signals include: decreased amino acid levels, detachment of ribosomes from ER, cAMP increase, and elevation of free calcium levels in the cytosol.

The first hypothesis is that a decreased level of intracellular free amino acids enhances autophagy. The consequent enhanced degradation of cell components thus maintains free amino acid levels for critical physiologic needs. This hypothesis is well supported by *in vivo* experiments and by experiments with perfused rat liver or isolated rat hepatocytes. In the latter models, factors such as amino acid composition and levels in the culture media can be easily manipulated.[29,30,89,159–168] These studies show clearly that amino acid deprivation and starvation increase autophagy in parallel with protein degradation, while refeeding or adding amino acids inhibit it. However, whether the decreased amino acid level causes autophagy directly or via another intracellular signal is not clear.

The postulated role of cAMP in the formation of autophagic vacuoles is based on the finding that increased levels of hormones such as glucagon cause increases in proteolysis and autophagy.[169] Cyclic AMP as well as dibutyryl cAMP have been shown to stimulate autophagic protein degradation in cultured hepatocytes.[170] While the exogenously administered cAMP taken up by cells is degraded rapidly, the dibutyryl cAMP taken up by cells is resistant to degradation and functions as an inhibitor of cAMP phosphodiesterase I, thus resulting in an increase of the endogenous cAMP.[171] Insulin inhibits the formation of autophagic vacuoles as well as protein degradation as a result of antagonism of glucagon effects.[29,105] It is interesting to note that 20-HE, an insect hormone that enhances autophagy in larval fat body cells, can markedly elevate cAMP content in those cells.[149,172] This effect can be blocked by cycloheximide, but not by puromycin; the latter drug actually increases the cellular level of cAMP.[149]

Other drugs used to increase the numbers of autophagic vacuoles, such as vinblastine, colchicine, nocodazole, and podophyllotoxin, have all been found to remarkably increase lymphocyte cAMP level in response to PGE1, and these effects can be dramatically inhibited by taxol.[173] All these drugs cause parallel increases in autophagosomes, levels of cytosolic cAMP, and free ribosomes. All of these effects can be inhibited by cycloheximide.

The association between the formation of

autophagic vacuoles and the detachment of ribosomes from the ER membrane has been shown.[97,149,150,174] Kovacs and co-workers studied the intracellular distribution of ribosomes after the administration of drugs that induce the accumulation of autophagic vacuoles in mouse pancreas. The drugs cause parallel increases in the levels of free ribosomes and in the numbers of autophagosomes. On the other hand, cycloheximide and emetine, which cause a decrease in the numbers of autophagosomes, do not induce a redistribution of the ribosomes from the ER to the cytosol. It was therefore proposed that drugs such as vinblastine, neutral red, dimethylnitrosamine, and puromycin, cause autophagosome formation through a stage of polyribosome detachment from ER membranes. According to this view, cycloheximide and emetine, both polyribosome stabilizers, inhibit autophagy mainly by anchoring ribosomes on the ER membranes. This hypothesis would explain why an accumulation of autophagosomes is seen in many pathologic conditions in which ER degranulation can be readily observed and is caused by sublethal cell injury.

That free calcium has a role in the formation of autophagic vacuoles has been suggested by several observations. Calcium plays an important regulatory role in many cellular metabolic pathways and functional activities, including protein turnover.[175,176] Calcium deregulation has been postulated to be one of the key factors in cell injury and death.[177-183] In injured cells, profound ultrastructural alterations of membranes occur, such as loss of polarity and fragmentation of ER cisternae and Golgi apparatus and blebbing of the plasma membrane. These membrane changes are associated with alterations in free calcium levels and cytoskeletal organization.[179,183]

Changes in intracellular free calcium are also associated with many physiologic membrane alterations. For example, phagocytosis is associated with calcium influx from the extracellular environment.[184] Intracellular calcium increase is also related to transport of membrane vesicles and to secretion. The elevation in calcium levels is responsible for the induction of secretory processes; the released calcium is derived from ER cisternae.[185]

The biochemical and ultrastructural alterations induced in the presence of elevated calcium levels in ER, Golgi apparatus, and plasma membranes share analogous features with the formation of autophagic vacuoles. Based on these analogies, the potential role of increased cytosolic free calcium in the process of autophagy, alone or in concert with cAMP, deserves consideration. Caution is called for in the development of mechanistic hypotheses based on the effects of agents that also induce a variety of other cellular reactions. For example, vanadate can activate adenylate cyclase and increase both cAMP and calcium levels in the cytosol. Yet vanadate inhibits rather than enhances lysosomal degradation of endogenous proteins.[186]

Lysosomal Protein Degradation through Autophagy

A balanced state between protein synthetic and degradative processes is necessary for the maintenance of normal cell functions. Alteration of this balanced state occurs rapidly in the cellular adaptation to local or systemic injury.[187] The mechanism of intracellular protein degradation is thought to depend in part on the life span of the proteins that are degraded.[29,188,189] A major part of degradation of short-lived abnormal as well as denatured proteins is carried out in the extralysosomal locations; while the degradation of long-lived proteins, and probably some short-lived proteins as well, is achieved by the lysosomal system through autophagy.[19,29,190]

Autophagic lysosomal proteolysis is a multistage process that involves several distinct but interrelated stages: segregation of material into autophagosomes, fusion between autophagosomes and lysosomes, and degradation of the segregated materials by hydrolytic enzymes within an acid environment.[1] Lysosomal protein degradation is sensitive to various physiologic regulators as well as pathologic perturbations.[170,191] Lysosomal protein degradation is also affected by neoplastic transformation. It has been proposed that in some neoplastic cells, lysosomal proteolysis is down-regulated to promote cellular growth.[192,193] For example, autophagic vacuoles and dense bodies isolated from neoplastic liver nodules manifest lower activities of proteinases and lower rates of protein degradation than normal liver.[194] Decreased response to physiologic inducers or blockers of autophagy are seen in neoplastic cells.[195,196]

Lysosomal protein degradation has been shown to be an active, energy-dependent process in various tissues and model systems.[197] The liver has been extensively studied because it is suitable for the preparation of cellular and subcellular fractions, and because lysosomal degradation can be readily modulated in this tissue. Agents that affect lysosomal degradation in rodent liver have been used *in vivo*,[102,105,132,134] in perfused liver,[89,165] in hepatocyte suspensions,[186,197-200] in hepatocyte monolayer culture,[167,168,170] and in hepatoma cells.[188]

Figure 6-5 shows the ultrastructural appearance of autophagic vacuoles isolated from rat liver following the administration of vinblastine. The autophagic vacuoles are isolated in a metrizamide gradient and are suitable for measurement of protein degradation.[19] Lysosomal degradation has also been studied in other tissues such as rat cardiocytes, rat skeletal muscle, brain, and in Ehrlich ascites tumor cells.[175,201-205] Figure 6-6 illustrates the typical appearance of autophagic vacuoles in various cell types.

Various agents have been employed to manipulate different stages of autophagy and to induce the accumulation of autophagic vacuoles.[97,160] Some of the commonly used agents include glucagon,[151,169] amino acid deprivation and starvation,[27] x-ray irradiation,[158,206,207] ischemia,[208-210] and

drugs such as colchicine and vinblastine[102,131,132,134,211,212] puromycin,[213] 20-hydroxyecdysone,[149] dimethylnitrosamine,[214] aflatoxin B$_1$,[215] and D-galactosamine.[216] These factors cause an increase in the numbers and size of autophagic vacuoles in the cells through different mechanisms.[20]

On the other hand, factors such as elevated levels of amino acids, insulin, cycloheximide, and 3-methyladenine have been found to decrease the numbers of autophagic vacuoles and to concomitantly inhibit lysosomal protein degradation.[29,165,199] The specific mechanisms responsible for such inhibitory effects are not known.

Because the fusion between autophagosomes and lysosomes is related to the normal function of cytoskeletal elements, particularly microtubules, microtuble disrupting drugs such as vinblastine, colchicine, and nocodazole have been used to affect this process. It has been shown that these agents may cause an accumulation of early stage autophagic vacuoles, many of which contain ultrastructurally identifiable cellular components.[97,101,102,134,212] It has been suggested that the increased number of autophagosomes is caused by toxic effects of these drugs as well as by microtubule depolymerization, which retards the fusion of autophagosomes with lysosomes. In order to understand this process, the effects of taxol, a microtubule stabilizing agent, on autophagic protein degradation in mouse liver were tested.[134]

Taxol is a plant alkaloid isolated from *Taxus brevifolia*, chemically characterized in the early 1970s and tested as an antineoplastic drug.[217] It is known that taxol can increase the rate and extent of microtubule assembly and stabilizes microtubules against several depolymerizing agents *in vitro*.[218-220] Taxol has been used to antagonize the effects of vinblastine, colchicine, and other related agents on lymphocyte functions.[173] Ultrastructural analyses of livers from mice treated with vinblastine alone show in the hepatocytes a large number of autophagic vacuoles, most of which contain identifiable

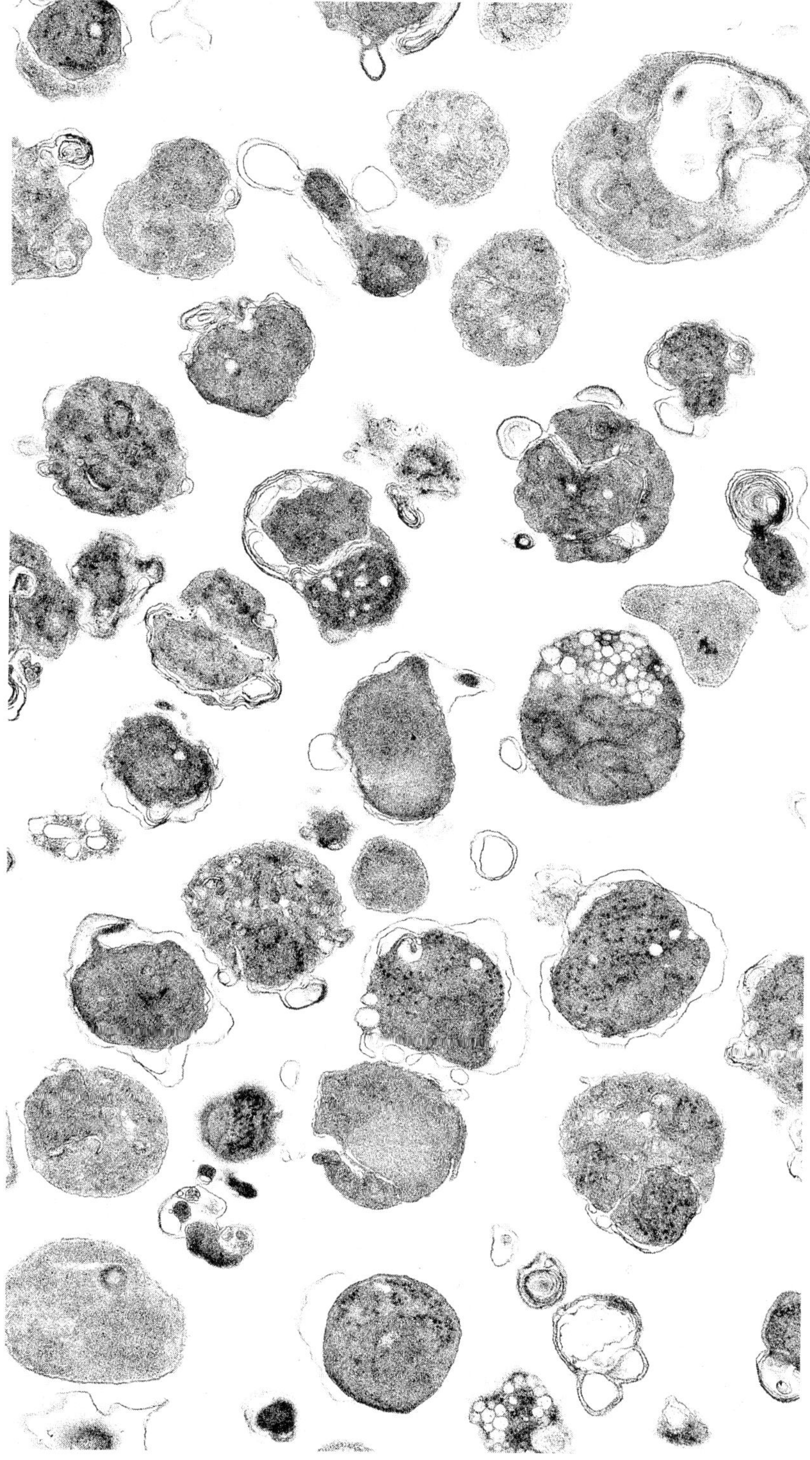

Figure 6-5. Ultrastructural appearance of isolated autophagic vacuoles. The vacuoles are from vinblastine-treated rat livers and were isolated in metrizamide density gradients. (× 22,750)

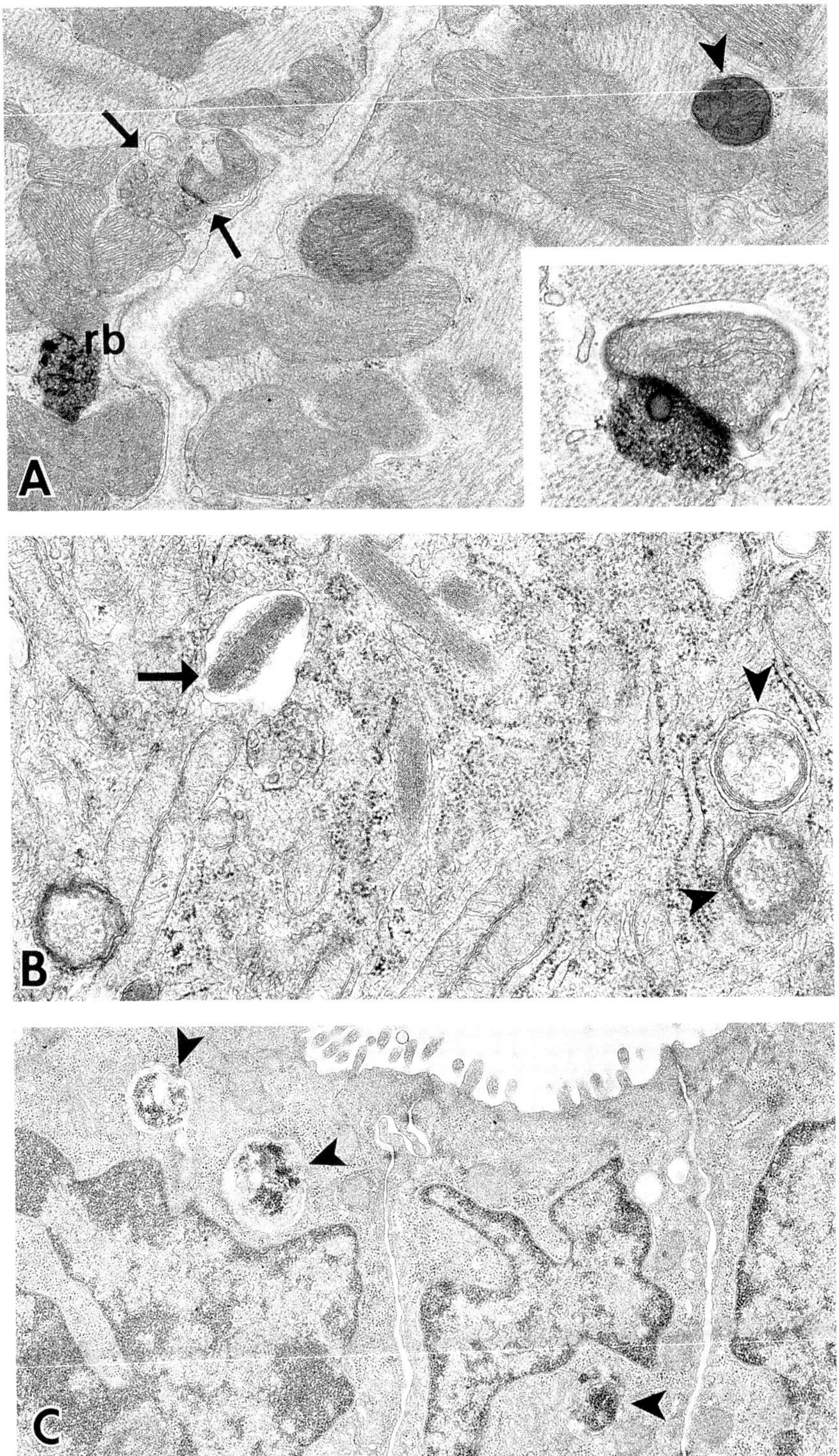

Figure 6-6. Ultrastructural appearance of typical autophagic vacuoles from various cell types. (A) Mouse cardiac myocytes. An autophagosome appears to be fusing with a secondary lysosome (arrows). A residual body (rb) and an autophagic vacuole containing an ultrastructurally degraded mitochondrion (arrowheads) are also seen. ($\times$ 13,050) *Inset*: Apparent merging between an autophagosome and a residual body. ($\times$ 29,450) (B) Human endothelial cell. Several autophagic vacuoles are present (arrowheads). One autophagic vacuole contains a Weibel-Palade body (arrow). ($\times$ 20,880) (C) Mouse bile ductular epithelium with several autophagic vacuoles (arrowheads). ($\times$ 10,440)

structures and are therefore considered to be in the early stages of degradation.[134] Animals treated with taxol alone do not show any specific changes in the lysosomal–autophagic compartment of hepatocytes. The hepatocytes of animals treated with taxol followed by vinblastine show many autophagic vacuoles, most of which are in the later stage of degradation, and contain electron-dense, unidentifiable material.

These observations indicate that taxol enhances the maturation of autophagic vacuoles induced by vinblastine. This enhancement is very likely achieved by enhancing the fusion process between autophagosomes and lysosomes. This notion is supported by the finding that taxol induces a parallel decline in the number of small dense bodies in the hepatocytes (Figure 6-7) and by the observation of apparent fusion between the

autophagic vacuoles and lysosomes. Measurement of protein degradation in a mitochondrial–lysosomal fraction shows that vinblastine alone causes some increase in the proteolytic rates, whereas animals treated with taxol followed by vinblastine show a greater increase in protein degradation rate in the mitochondrial–lysosomal fraction. It is likely that stabilization of microtubules by taxol enhances both the autophagosome–lysosome fusion and the lysosomal protein degradation.

In order to investigate the mechanism of lysosomal enzyme proteolysis, a variety of factors have been used in different experimental systems. Generally, these factors may be categorized into two major classes: (1) those that affect lysosome pH optimum and (2) those that alter lysosomal enzyme activities.

Ammonia, amines, and other weak bases are particularly effective in selectively inhibiting lysosomal degradation of long-lived proteins.[197,198] These molecules enter the acid lysosomal compartment, become protonated, and are trapped inside the lysosomes. As a result, these drugs elevate the lysosomal pH and inhibit protein degradation. Ammonia and other weak-base amines such as methylamine and propylamine have therefore been used to distinguish between lysosomal and nonlysosomal pathways of protein degradation.[198] However, the inhibition of lysosomal proteolysis by these agents is only partial, and cellular toxicity may also be induced.[92,221] Chloroquine, the antimalarial drug, can inhibit lysosomal protein degradation also, through alkalinization of lysosomal interior. Chloroquine enters the lysosomes, where it is protonated, and becomes impermeable to the lysosomal membrane due to charge effects.[222,223] After entering lysosomes, chloroquine is also bound with high affinity to lysosomal membranes; this also blocks the diffusion of this drug back to the cytosol.[224] Chloroquine also influences lysosomal enzyme activity by inhibiting the activity of cathepsins and by partially blocking recycling of lysosomal en-

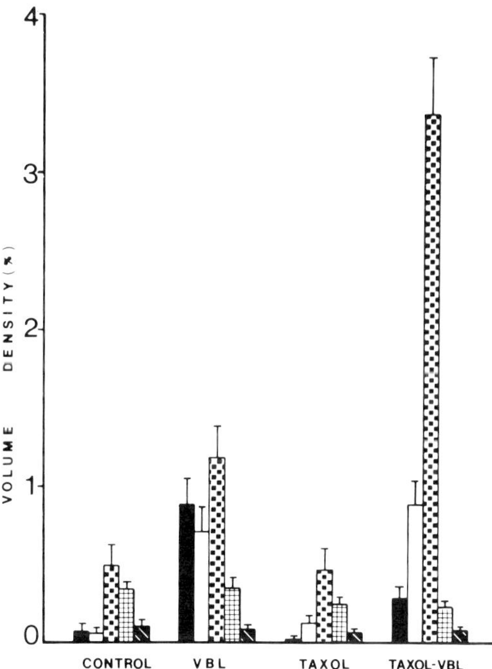

Figure 6-7. Influence of the microtubule active agents, vinblastine (VBL) and taxol, on the fractional volume of lysosomes in mouse hepatocytes. Solid black = autophagosomes; white = autophagolysosomes; black squares = large dense bodies; grid pattern = small dense bodies; diagonal stripe = microautophagy.

zyme receptors.[225] Interference with recycling in turn causes more lysosomal enzymes to be secreted to the outside of the cells instead of being segregated into the lysosome compartment; and this may also decrease lysosomal protein degradation.

The use of proteinase inhibitors has been a classic means of studying autophagic lysosomal degradation. The most frequently used proteinase inhibitor is leupeptin, an inhibitor of the lysosomal thiol proteinases cathepsins B, H, and L.[203] Leupeptin inhibits lysosomal protein degradation by approximately 80% in rat hepatocyte suspensions; lower inhibitory effects (20–30%) were reported by Neff and colleagues.[189,197,198,226] This inhibitory effect has been further confirmed by ultrastructural studies in tissues such as rat liver, isolated rat hepatocytes, mouse seminal vesicle cells, and rat brain.[204,226–229] In all these tissues, leupeptin causes an accumulation of autophagolysosomal structures. The lysosome and autophagic vacuole compartment in mouse seminal vesicle cells, for example, become 6 and 40 times larger, respectively, than in controls.[229] The autophagic vacuoles induced by these agents are irregular in shape and contain identifiable cytoplasmic components.[226,227] It is very likely that such accumulation of autophagic vacuoles and large irregular secondary lysosomes is due to impaired intralysosomal degradation as a result of leupeptin-induced proteinase inhibition. It is also suggested by Kovacs et al. that these enlarged lysosomes may have a reduced ability to fuse with other types of lysosomes.[226]

Other inhibitory agents, such as antipain (inhibits cathepsins A and B), chymostatin (inhibits cathepsins B and G), and vanadate, have also been used in the study of lysosomal protein degradation.[186,189,198] In all those studies, the lysosomal protein degradation was inhibited to various degrees when one of these agents was added.

The calcium dependence of several extralysosomal proteolytic enzymes is well known. There is also evidence that calcium levels influence lysosomal protein degradation. Cultured rat hepatocytes exposed to a defined calcium-free medium containing the calcium chelator EDTA (50 μmol) manifest a significant decrease in protein degradation.[166] This calcium effect has been further confirmed by Grinde who, by using chelators, calcium ionophore, and calcium channel blocker (verapamil), found that calcium is required for the lysosomal degradation of endogenous proteins in rat hepatocytes.[200] Since calcium has no effect on the degradation of endocytosed proteins, the author suggested that such calcium dependence is probably related to the formation of autophagic vacuoles or to the fusion of autophagic vacuoles with lysosomes, rather than to activation of lysosomal proteinases. Calcium stimulates protein degradation in rat skeletal muscles, and this stimulation is not mediated by the calcium-activated proteinases.[175]

The inhibitory effects of vanadate on protein degradation may also be related in part to modification of intracellular calcium levels.[186] Vanadate at a concentration of 10 mmol/L strongly inhibits endogenous protein degradation—a function that has been attributed to inhibition of lysosomal proteinases. Since vanadate inhibits calcium uptake by ER it is possible that alterations in calcium regulation may influence lysosomal protein degradation. Inhibition of the calcium-regulating protein calmodulin may also inhibit lysosomal degradation of endocytosed proteins such as EGF and LDL, with the resultant intralysosomal accumulation of those materials.[230–232]

A calmodulin inhibitor, thioridazine, causes many autophagosomes and altered lysosomes to appear in Hela cells. This suggests that calcium levels may directly or indirectly influence the formation and maturation of autophagosomes. Alternatively, since thioridazine also inhibits cathepsin B and acid phosphatase activity, it is possible that the accumulation of lysosomes may be due to the accumulation of undegraded materials.

Detailed ultrastructural studies correlated with measurements of proteolysis in lysosomal fractions are needed to further substantiate these calcium-related effects and to understand their mechanisms. The use of agents that affect intracellular calcium levels to explore the regulatory mechanisms of calcium in autophagic protein degradation is rendered more difficult by the induction of multiple cellular effects along with specific effects on lysosomes.

In conclusion, several intracellular regulatory functions performed by the lysosomes are accomplished through autophagic protein degradation. Many factors have been employed to manipulate this process at different stages; for example, formation of autophagosomes, fusion between autophagosomes and lysosomes, and intralysosomal degradation. With regard to the formation of autophagic vacuoles, it appears that different organelles have the potential for giving rise to the limiting membranes of autophagic vacuoles in different systems and experimental conditions. Endoplasmic reticulum, Golgi apparatus, lysosomes have all been implicated as the source of autophagic vacuole membranes. Four intracellular events have been proposed as the inducers of autophagy in response to physiologic or pathologic stimuli: decreased free amino acid levels, elevation of intracellular cAMP, increase in free cytosolic calcium, and detachment of polyribosomes from ER membrane. It is possible that the formation of autophagic vacuole may require more than one factor and that some of these signals may function sequentially or interactively.

CHARACTERISTICS OF LYSOSOMAL ENZYMES

Biogenesis of Lysosomal Enzymes

The lysosomes lack the capacity for the synthesis of proteins that is manifested by organelles such as ER and mitochondria.

Lysosomal membranes also lack mechanisms for the insertion or translocation of proteins across the lipid bilayer such as are displayed, for example, by mitochondrial membranes. Instead, the synthesis and translocation of newly synthesized lysosomal enzymes follows the secretory pathway common to proteins destined for export from the cell.[233] This pathway leads from the rough ER to the smooth ER, and finally to the Golgi apparatus.[3,234,235] In the Golgi apparatus, the lysosomal enzymes are targeted for intracellular "secretion" and packaged into small, coated vesicles (the primary lysosomes) that travel to and merge with other cellular lysosomal structures. The secretory proteins proper are packaged into secretory vacuoles that merge with the cell surface membrane and discharge their contents extracellularly. Many aspects of the biogenesis of the protein and carbohydrate chains of the lysosomal enzymes are known in detail. However, in comparison with what is known about secretory proteins, the information about lysosomal enzymes is more limited.

For the great majority of proteins, synthesis begins in the cytosol with the start of translation of messenger RNA by free ribosomes. In the case of all secretory and many membrane proteins, the growing protein chain becomes bound to the ER membrane by the following mechanism. A cytosolic protein complex, named the *signal recognition particle*, interrupts the growth of the protein chain in the cytosol. The signal recognition particle, with the associated protein chain and RNA complex, binds to a membrane protein (the docking protein) in the rough ER. This binding permits the insertion of the protein chain into the lumen of the ER. At this point, the signal recognition particle dissociates from the ER and the growth of the protein chain resumes.[236-238] The first 15 to 20 amino acids of the growing protein chain (at the NH2-terminus) are known as the "signal," or "leader," sequence. This leader sequence is responsible for the insertion of the growing protein chain into the ER. The leader sequence is usually removed from the growing protein chain by a protease enzyme, and carbohydrate residues are added onto the chain. Integral membrane proteins remain associated with the ER membrane and are transported to their destination in a membrane-associated form. On the other hand, most lysosomal enzymes and the secretory proteins are released into the lumen of the rough ER. The lysosomal enzymes move to the smooth ER via direct luminal continuities with the rough ER. Transport from the smooth ER to the *cis* region of the Golgi apparatus is via membrane vesicles. The Golgi apparatus is organized as parallel stacks of polarized cisternae. The stacks of cisternae have been distinguished in three regions with distinctive biochemical and cytochemical characteristics.[239,240]

Sequential modifications of the carbohydrate and protein chains of secretory proteins occur as the proteins transit across the stack of cisternae in a *cis* to *trans* direction. In the case of several of the lysosomal enzymes examined to date, a modification of the carbohydrate chain leads to binding of the enzyme to a 215-kd membrane receptor that recognizes a mannose 6-phosphate sugar and other conformational aspects of the enzyme molecule; a 46-kd protein able to bind phosphorylated high mannose oligosaccharide has also been identified.[242-244] The receptor-binding mechanism apparently sorts out the lysosomal enzymes from the secretory proteins and leads to the fission of vesicles containing lysosomal enzymes from the Golgi apparatus. It is thought that the specialized *trans* Golgi reticulum is the site where the lysosomal enzymes are packaged into coated vesicles.[245-250] These coated vesicles have a recycling function: They transport lysosomal enzymes to a prelysosomal compartment of acid pH where the enzymes dissociate from the mannose 6-phosphate receptor and go on to the lysosomes.[251-253]

The prelysosomal compartment through which this translocation occurs appears to be endocytic in nature.[254] Newly synthesized lysosomal membrane glycoproteins are also translocated to the lysosomes through the same intracellullar pathway.[253] Several important details of this overall scheme await further corroboration. For example, it appears that the protein coating seen ultrastructurally on vesicles associated with Golgi apparatus may not be analogous in some cases to the coating present on vesicles originating from the plasma membrane. Moreover, the transport of proteins through the secretory pathway may not be absolutely dependent on coated vesicles.[255,256] In addition, there are examples of correct transport of lysosomal enzymes to the lysosomes even in the absence of the carbohydrate residue (mannose 6-phosphate) thought to be responsible for sorting out the lysosomal enzymes from proteins destined for secretion.[257-261] Finally, the manner in which the lysosomal enzymes and other proteins are transported across the Golgi apparatus stacks and the precise manner in which lysosomes are formed are still not conclusively established.

In addition to the extensive modifications of the carbohydrate chains, the protein chains of lysosomal enzymes undergo several proteolytic modifications at different stages of their biogenesis.[262,263] In addition to the already-mentioned removal of the leader sequence within the ER, further proteolytic cleavage steps of the lysosomal enzymes occur in a post-Golgi prelysosomal compartment as well as in secondary lysosomes.[264-267]

Whereas carbohydrate modifications are necessary for intracellular sorting of the lysosomal enzymes, the significance of the modification of the protein chain is not known. It has been proposed that these proteolytic steps may be responsible for the activation of some lysosomal enzymes.[268,269] Alternatively, proteolytic cleavage of enzymes may uncover a signal necessary for the correct sorting of the enzymes or for the separation of the enzymes from the membrane receptors.[267]

Deficiencies of the biosynthetic enzymes responsible for the modification of the carbohydrate chains result in failure to transport the lysosomal enzymes to the lysosomes; the enzymes are instead secreted. Mucopolipidosis II and III are examples of autosomal recessive diseases with this type of defect.[270] Toxins that interfere with the synthesis of the carbohydrate chains (e.g., tunicamycin) or with the acidification of lysosomal and prelysosomal organelles (e.g., chloroquine) cause the pathologic secretion of lysosomal enzymes.[264,271] The secretion and reuptake of lysosomal enzymes are physiologic as well as pathologic phenomena manifested by many cells. For example, polymorphonuclear leukocytes and macrophages accumulate at sites of tissue injury and secrete their lysosomal contents in response to a variety of phagocytic stimuli.[272] The macrophages also possess receptors that function in the binding and endocytosis of lysosomal enzymes. The level of these receptors is low in macrophages activated against tumor cells and is elevated in macrophages treated with the steroid dexamethasone.[273,274] Therefore, lysosomal enzyme receptors on the surface of cells may regulate the extracellular levels of lysosomal enzymes in areas of tissue injury and profoundly affect the course of inflammatory reactions.

Regulation of Lysosomal Enzymes

With regard to control of proteolytic degradation, the action of a protease may be potentially controlled by several mechanisms, such as: synthesis and storage of proteases for release extracellularly in response to stimuli; presence of specific rate-limiting degradative steps that can be carried out by specific proteases only; modulation of protease activity by specific inhibitors or by activation of latent proteolytic activity through limited proteolysis. Particularly in cells secreting proteases in a constitutive manner, the control of degradation may be

achieved by stimulatory or inhibitory proteins that regulate the synthesis of the protease.[275-278] For example, in several types of tumor cells both synthesis and secretion of lysosomal enzymes such as cathepsin B and L can be induced by tumor promoters, some growth factors, and hormones.[279,280] With regard to lysosomal enzymes, endogenous compounds that directly inhibit enzyme activity have been described.[277,278] Proteins capable of stimulating lysosomal enzyme activity have also been identified, and genetic disorders in which specific activator proteins are lacking, have been described.[281] However, it is not known if these inhibitory or activating compounds serve any regulatory function. At present, there is also no evidence of regulation of lysosomal proteolysis through control of hydrolase synthesis. The degradative sequence is not known; however, it does not appear that rates of cleavage of substrate in the lysosomes are important in regulation of protein turnover.[282]

Lysosomal enzymes are separated from the cellular cytoplasm by a membrane. It appears that this membrane serves a key regulatory function. This membrane controls the merging of the enzyme-containing vacuoles with vacuoles that contain cellular or extracellular materials destined to be degraded. An acid pH optimum for the activity of the hydrolases is an additional protective and regulatory feature against uncontrolled degradation and cell injury induced by leakiness of lysosomal membranes. Because of the acid pH optimum of the lysosomal enzymes, the disruption of lysosomes results in decreased rather than increased rates of intracellular degradation.[283]

The disruption of lysosomes can be achieved by allowing the lysosomes to accumulate the methyl esters of certain amino acids such as leucine.[283] The methyl ester derivatives readily cross biological membranes. In the lysosomes, these derivatives are hydrolyzed to the less permeable free amino acids and become trapped. This approach of loading amino acid in the lysosomes has also been helpful in looking for the presence of amino acid transport systems in lysosomal membranes.[63] These types of experiments have provided evidence for a carrier system for the transport of cationic amino acids out of the lysosomes.[284] The observation that dipeptides traverse the lysosomal membrane more readily than amino acids suggests that systems for peptide transport may be present in the lysosomal membrane. Evidence for the existence of lysosomal membrane transport systems for sugars has been presented by Maguire et al.[285]

Functions of Lysosomal Enzymes

The lysosomes contain a variety of enzymes able to degrade nearly all biological molecules. The lysosomal enzymes are mostly free in the matrix of the lysosomes or are loosely attached to the membrane. A few enzyme such as acid phosphatase appear to be more tightly bound and are considered integral membrane proteins. The degradation of biosynthetic compounds by lysosomes is accomplished primarily by the action of enzymes (exoenzymes) that sequentially cleave off smaller molecules from the ends of molecular chains. These enzymes function in concert with some enzymes (endoenzymes) that split molecular chains into fragments by splitting off bonds in the interior of the chains. The degradation of proteins and oligopeptides is performed by endo- and exopeptidases acting synergistically. Free amino acids and dipeptides are the products of the hydrolysis.

Lysosomal enzymes also readily degrade nucleic acids to oligonucleotides and subsequently to free pyrimidine and purine nucleotides and inorganic phosphate. The degradation of carbohydrates and of carbohydrate chains of protein and lipid molecules also proceeds primarily by the action of exoenzymes (glycosidases) and yields monosaccharides. The degradation of neu-

tral lipids is accomplished by lipases and yields glycerol and fatty acids. These molecules are able to traverse the lysosomal membrane. Finally phospholipids, sphingolipids, and glycolipids are also degraded by lysosomal enzymes.

The lysosomal enzymes are active inside the acid lysosomal environment and, in physiologic conditions, participate in the normal degradation of intra- and extracellular materials. Several examples of specialized functions of lysosomal hydrolases exist. For example, invertebrates as well as non-mammalian vertebrates accumulate yolk proteins via receptor-mediated endocytosis into lysosomelike organelles. The yolk proteins are stored in this compartment after only a limited proteolytic modification. The yolk proteins are degraded to constituent amino acids at a later point during embryonic development, when the degradative enzymatic machinery is reconstituted.[286] The lysosomal enzymes are resistant to denaturation, and it is likely that in injured tissue they are released from dead cells and contribute to the autolytic digestion of necrotic tissue. The autolysis is favored by the pH drop that occurs in necrotic tissue.

The study of the latency of lysosomal enzyme activity by biochemical or, more recently, immunochemical assays has been a classic approach to study the development of lysosomal membrane fragility and the leakage of lysosomal enzymes during injury such as ischemia. The general consensus of opinion is that the fragility of the lysosomal membranes increases during cell injury.[287,288] However, it is not known if a causal relationship exists between intracellular leakage of enzymes from the lysosomes and cell death.

Lysosomal enzymes have the capacity to also degrade extracellular matrix components. As previously discussed, the secretion of lysosomal enzymes is a physiologic occurrence and can be further induced by pathologic conditions in cells such as neutrophils and macrophages. It has been hypothesized that the presence of charged groups and proton pumps present on the surface of these cells may create a microenvironment of slightly acidic pH that would favor the degradation of the extracellular matrix proteins by lysosomal enzymes secreted or associated with the cell surface membranes.[289]

Several inhibitors of lysosomal enzymes are known. Some are of microbiological or synthetic origin and have been primarily useful in elucidating the mechanisms of action of lysosomal enzymes. Several endogenous inhibitors, such as the cystatins, are present within cells; inhibitors are also present in extracellular fluids, where they function in the inactivation of lysosomal enzymes released in inflammatory reactions.[277,278,290] Several inhibitors of lysosomal function have toxicologic importance because their effects may contribute to the development of end organ damage. Heavy metal–protein complexes are known to be endocytosed and transported to the lysosomes. Degradation of the protein component occurs with the release of some of the metal to the cytosol. The metals also persist in the lysosomes, where they may impair the stability of the lysosomal membrane or the activity of the enzymes.[74,291,293]

The disease hemochromatosis exemplifies disorders that cause metal accumulation and tissue damage. Idiopathic hemochromatosis, an autosomal recessive disease, results from the hyperabsorption of iron from the intestinal epithelium. This leads to the gradual accumulation of iron in the lysosomes of the parenchymal cells.[294] Anemia characterized by ineffective erythropoiesis can also cause hyperabsorption of iron and accumulation of iron in lysosomes. This may be aggravated by further iron loading due to dependence on blood transfusions.

Amphiphilic drugs such as the aminoglycosides and chloroquine inhibit lysosomal phospholipases and induce lipid accumulation and cell injury.[295-297] Chloroquine, within 30 minutes of *in vivo* administration, causes an increase in the number of autophagic vacuoles and in the proteolytic rate of isolated lysosomes from livers.[298] Leu-

peptin also causes an increase in the number of autophagic vacuoles in liver. However, in the case of leupeptin, the proteolysis measured in isolated lysosomes is inhibited.[299] Accumulation of substrate in the lysosomes due to impaired degradation is primarily responsible for the accumulation of autophagic vacuoles induced by these drugs.

The opposite effects of the two drugs on the proteolytic activity of isolated lysosomes (increased by chloroquine, decreased by leupeptin) may be due to diffusion of chloroquine out of the isolated lysosomes, with consequent lessening of chloroquine-induced inhibition. It is also possible that the inhibition of degradation caused by agents that do not inactivate lysosomal enzymes may induce *in vivo* compensatory responses aimed at restoring proteolytic rates toward normal. The lysosomal storage disorders induced by amphiphilic drugs have been reviewed by Hruban.[300]

An increase in the levels of lysosomal enzymes is commonly seen in injured tissues. This increase is sometimes due to increased levels of lysosomal enzymes in the parenchymal cells.[301] In most cases, the enzyme elevation is due to the accumulation of phagocytic cells and other nonparenchymal cells rich in lysosomal enzymes at these sites.[302] For example, an elevation in the levels of hydrolases such as cathepsin D has been demonstrated in the central nervous system in multiple sclerosis and in experimental models of other demyelinating diseases. Hematogenous influx of phagocytes as well as proliferation of glial cells are for the most part responsible for these elevations in cathepsins.[303]

Cathepsin D can also degrade cartilage proteoglycan; extracellular localization of this and other lysosomal enzymes has been demonstrated, and it is thought to be responsible for physiologic bone and cartilage remodeling.[304] Elevated levels of extracellular cathepsin D and B are seen in joints affected by diseases such as rheumatoid arthritis. Enzyme secretion by cells such as phagocytes and cells lining the synovial surface of joints is thought to be responsible for these elevated lysosomal enzyme levels and consequent joint destruction.[305] Secretion of lysosomal enzymes by cells is induced by a cAMP-dependent mechanism that involves G proteins.[306]

Cathepsin B can also degrade collagen and is secreted in large amounts by some neoplastic cells. This phenomenon has led to the hypothesis that cathepsin B may facilitate the metastatic potential of neoplastic cells.[307-309] Tumor promoters such as phorbol esters are also known to induce secretion of lysosomal enzymes from inflammatory cells.[309] Mouse fibroblasts transformed by Kirsten virus secrete abundant quantities of a lysosomal cysteine protease (cathepsin L) in a precursor form. This enzyme is activated at an acid pH and can degrade various proteins, including extracellular matrix proteins such as collagen, laminin, and fibronectin.[310] Both increased synthesis (25-fold greater) and low affinity of the enzyme for the mannose 6-phosphate receptor account for the selective secretion of this enzyme.[280] The presence of increased numbers of lysosomes in some types of cancer cells has been used in experimental attempts to target cytotoxic compounds such as detergents to the cancer cells.[311]

The acid environment of the lysosomal compartment is taken advantage of, to try to increase the specificity of cytotoxic drugs used for therapy. The cytotoxic drugs are bound to protein carriers by linkages that are acid-sensitive. Therefore, the drugs are inactive at neutral pH and are only activated after being internalized by target cells and cleaved in the lysosomal acid millieu.[312]

Accumulation of Undegraded Material in the Lysosomes

The accumulation of undegraded material in the lysosomes has been demonstrated histochemically, ultrastructurally, and by biochemical studies of isolated lysosomes. This material is primarily a lipid known as *lipopigment* because of its yellowish to dark brown color and because, biochemically, it consists of oxidized and polymerized fatty acids. Lipopigment is subdivided into ceroid and lipofuscin. Ceroid pigment is found in macrophages and is the undegraded remnant of heterophagocytosed material.[313] Lipofuscin is thought to be the remnant of autophagocytosis. These lipids remnants may be resistant to further hydrolysis and therefore accumulate and polymerize in the lysosomes. Alternatively, as a result of cell injury, the lipid may have undergone peroxidative change and may have become resistant to degradation before uptake into the lysosomes.

In contrast to ceroid, as much as one-half the weight of lipofuscin consists of protein. Tissue necrosis leads to the accumulation of ceroid in the macrophages. Aging, as well as cell injury, causes an increase in lipofuscin-containing lysosomes.[314] Some variation in the structure and composition of lipofuscins occurs in different cell types and as a result of different pathologic conditions. It appears that the accumulation of lipopigment can be induced rapidly by injury.[204] The accumulation of lipopigment is prominent in the livers of patients who die from trauma and sepsis (Figure 6-8). Accumulation of undegraded material in the lysosomes can lead to regurgitation of the material from the lysosomes to the extracellular space.[315] Low molecular weight substances are regurgitated more easily, suggesting that a process of "reverse endocytosis" may be occurring.[316]

Lysosomes in Antimicrobial Defense

The lysosomes of phagocytic cells are important defense mechanisms against microbial agents. Oxidative mechanisms, cationic and other nonenzymatic proteins, are primarily responsible for the killing of microorganisms in the phagolysosomes of phagocytes. The neutral and acid proteases are thought to play a secondary role in killing and are primarily responsible for the degra-

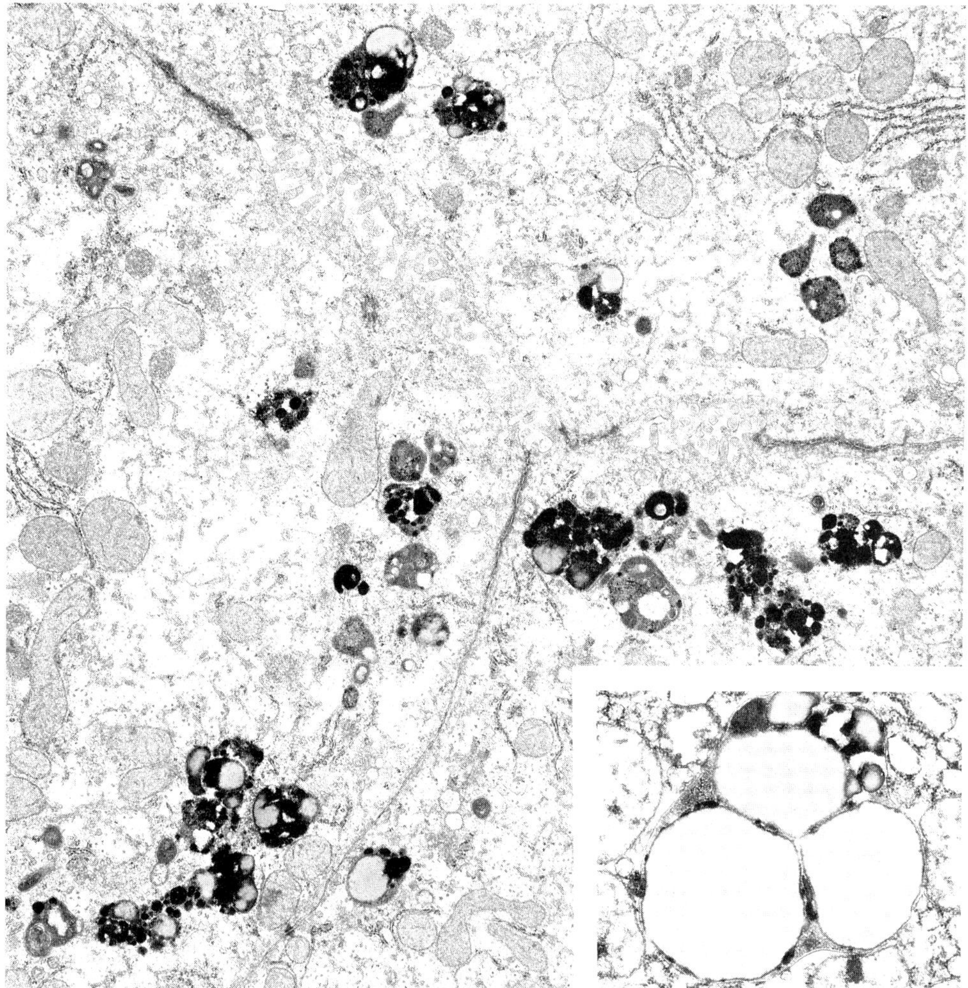

Figure 6-8.   Lipofuscin-containing lysosomes in human liver obtained from an immediate autopsy. (× 8,800; Inset: × 15,000)

dation of the pathogenic organisms. Microorganisms have evolved several interesting adaptations to escape lysosomal digestion, and may actually use lysosomal organelles to complete stages in their growth and development cycles. Viruses use an acidic prelysosomal compartment to enter cells.

The infection of cells by viruses begins with the random interaction of a virion and the cell surface membrane. If sufficient numbers of receptors on the cell surface bind to the viral envelope glycoproteins, the virion becomes anchored and progressively more receptors are recruited to the zone of attachment. The engulfment of the virion begins by infolding of the surface membrane to form a pouch (coated pit) that pinches off from the cell surface and gives rise to a coated vesicle. Coated vesicles lose the ultrastructurally visible coat, which corresponds to the protein clathrin, and move toward the cell interior to a vacuole called the *endosome.* Merging of the vesicles with the endosome occurs by membrane fusion. This process is known as *receptor-mediated endocytosis* and is the mechanism for entry into cells by enveloped viruses of various taxonomic groups such as orthomixo-, rabdo-, and togaviruses.

The fate of membrane receptor–ligand complexes and associated soluble complexes delivered to the endosomal compartment varies. Recycling to the plasma membrane or delivery to the lysosomes for degradation are the most common events. Transit to other compartments such as Golgi apparatus or to other plasma membrane domains also occurs, depending on the type of substance being endocytosed. Endocytosed viruses may escape lysosomal digestion because the viral envelope can fuse with the endosomal membrane and release the viral nucleoid into the cytosol. A low pH is the prerequisite for this fusion. Entry of the virus into the cytosol directly through the plasma membrane can be achieved experimentally *in vitro* by acidifying the extracellular medium. Conversely, the alkalinization of the endosomal–lysosomal compartment by the administration of weak bases inhibits the entry of viruses into the cell's cytosol.

These events have been studied with Semliki Forest, vesicular stomatitis, influenza, and retroviruses.[317-319] In addition to viruses, rickettsiae and some trypanosomes can escape from phagocytic vacuoles before fusion with lysosomes occurs; how the membrane breach occurs in these latter cases is not precisely known.[320,321] Several microorganisms such as algae, protozoans, and bacteria are able to prevent the fusion of phagosomes with the lysosomes or the acidification of lysosomes.[322-327] Some intracellular pathogens such as *Leishmania donovani* and *Coxiella burnetii* show a metabolic dependence on an acid environment.[328,329] Finally, an acidic environment is necessary for the entry into the cytosol of bacterial toxins such as diphtheria toxin and *Pseudomonas* toxin A. Cells with a defect in acidification of endosomes and lysosomes are resistant to viral infection and to damage by acidification-dependent bacterial and chemical toxins.[330-332]

ACKNOWLEDGMENTS

We wish to acknowledge Robert Pendergrass, Seung Chang, Perry Comegys, and Janet Fowler of the Electron Microscopy Laboratory. We also thank Wayne J. Ivusich for his editorial assistance.

REFERENCES

1. de Duve C, Wattiaux R. Functions of lysosomes. *Annu Rev Physiol.* 1966;28:435.
2. de Duve C. Lysosomes revisited. *Eur J Biochem.* 1983;137:391.
3. Marzella L, Glaumann H. Biogenesis, translocation and function of lysosomal enzymes. *Int Rev Exp Pathol.* 1983;25:239.
4. Novikoff PM, Novikoff AB, Quintana N, Hauw J. Golgi apparatus, GERL, and lysosomes of neurons in rat dorsal root ganglia, studied by thick sections and thin section cytochemistry. *J Cell Biol.* 1971;50:859.
5. Creek KE, Sly WS. The role of the phosphomannosyl receptor in the transport of acid hydrolases to lysosomes. In: Dingle JT, Dean R, Sly WS, eds. *Lysosomes in Biology and Pathology*, Vol. 7. New York: Elsevier; 1984:63.
6. Shepherd VL, Stahl PD. Macrophage receptors for lysosomal enzymes. In: Dingle JT, Dean R, Sly WS, eds. *Lysosomes in Biology and Pathology*, Vol. 7. New York: Elsevier; 1984:83.
7. Goldberg D, Gabel C, Kornfeld S. Processing of lysosomal enzyme oligosaccharide units. In: Dingle JT, Dean R, Sly WS, eds. *Lysosomes in Biology and Pathology*, Vol. 7. New York: Elsevier; 1984:45.
8. Dell'Antone P. Electrogenicity of the lysosomal proton pump. *FEBS Lett.* 1984;168:15.
9. Reeves JP. The mechanisms of lysosomal acidification. In: Dingle JT, Dean R, Sly WS, eds. *Lysosomes in Biology and Pathology*, Vol. 7. New York: Elsevier, 1984:175.
10. Ashford TP, Porter KR. Cytoplasmic components in hepatic cell lysosomes. *J Cell Biol.* 1962;12:198.
11. Swift H, Hruban Z. Focal degradation as a biological process. *Fed Proc.* 1964;23:1026.
12. Glaumann H, Berezesky I, Ericsson JLE, Trump BF. Lysosomal degradation of cell organelles, I: ultrastructural analysis of uptake and digestion of intravenously injected mitochondria by Kupffer cells. *Lab Invest.* 1975;33:239.
13. Glaumann H, Berezesky I, Ericsson JLE, Trump BF. Lysosomal degradation of cell organelles, II: ultrastructural analysis of uptake and digestion of intravenously injected microsomes and ribosomes by Kupffer cells. *Lab Invest.* 1975;33:252.
14. Glaumann H, Trump BF. Lysosomal degradation of cell organelles, III: uptake and disappearance in Kupffer cells of intravenously injected isotope-

labelled mitochondria and microsomes *in vivo* and *in vitro*. *Lab Invest*. 1975;33:262.

15. Glaumann H, Marzella L. Degradation of membrane components by Kupffer cell lysosomes. *Lab Invest*. 1981;45:479.

16. Henell F, Ericsson JLE, Glaumann H. Degradation of phagocytosed lysosomes by Kupffer cell lysosomes. *Lab Invest*. 1983;48:556.

17. Marzella L, Ahlberg J, Glaumann H. Autophagy, heterophagy, microautophagy and crinophagy as the means for intracellular degradation. *Virchows Arch (Cell Pathol)*. 1981;36:219.

18. Khairallah EA. Quantitative assessment of the contribution of autophagy to intracellular protein breakdown. *Biochem Soc Trans*. 1985;13:1012.

19. Marzella L, Ahlberg J, Glaumann H. Isolation of autophagic vacuoles from rat liver: morphologic and biochemical characterization. *J Cell Biol*. 1982;93:144.

20. Marzella L, Glaumann H. Autophagy, crinophagy and microautophagy as mechanisms of protein degradation. In: Glaumann H, Ballard FJ, eds. *Lysosomes, Their Role in Protein Degradation*. New York: Academic Press; 1987:319.

21. Pastan IH, Willingham MC. Receptor-induced endocytosis: coated pits, receptosomes, and the Golgi. *Trends Biochem Sci*. 1983;8:250.

22. Hers HG. Inborn lysosomal diseases. *Gastroenterology*. 1965;48:625.

23. Kolodny EH. Lysosomal storage disease. *N Engl J Med*. 1976;294:1217.

24. Glew RH, Basu A, Prence EM, Remaley AT. Biology of disease: lysosomal storage disease. *Lab Invest*. 1985;53:250.

25. Yokota S, Kato K. Heterogeneity of rat kidney lysosomes revealed by immunoelectron microscopic staining for cathepsins B and H. *Histochemistry*. 1988;89:499.

26. Glaumann H, Ericsson JLE, Marzella L. Mechanisms of intralysosomal degradation with special reference to autophagocytosis and heterophagocytosis of cell organelles. *Int Rev Cytol*. 1981; 73:149.

27. Pfeifer U. Cellular autophagy and cell atrophy in the rat liver during long-term starvation: a quantitative morphological study with regard to diurnal variations. *Virchows Arch (Cell Pathol)*. 1973;12:195.

28. Neely AN, Cox JR, Fortney JA, Schworer LM, Mortimore GE. Alterations of lysosomal size and density during rat liver perfusion: suppression by insulin and amino acids. *J Biol Chem*. 1977; 252:6948.

29. Mortimore GE. Mechanisms of cellular protein catabolism. *Nutri Rev*. 1982;40:1.

30. Mortimore GE, Hutson NJ, Surmacz CA. Quantitative correlation between proteolysis and macro- and microautophagy in mouse hepatocytes during starvation and refeeding. *Proc Natl Acad Sci USA*. 1983;80:2179.

31. Brekke IB, Danielsen H, Reith A. Normalization of hepatic lysosomal autophagy in streptozotocin diabetic rats after pancreatic transplantation. *Virchows Arch (Cell Pathol)*. 1983;43:189.

32. Heynen MJ, Tricot G, Verwilghen RL. Auto-phagy of mitochondria in rat bone marrow erythroid cells: relation to nuclear extrusion. *Cell Tissue Res*. 1985;239:235.

33. Henell F, Ericsson JLE, Glaumann H. An electron microscopic study of the post-partum involution of the rat uterus. *Virchows Arch (Cell Pathol)*. 1983;42:271.

34. Marzella L, Ahlberg J, Glaumann H. *In vitro* uptake of particles by lysosomes. *Exp Cell Res*. 1980;129:460.

35. Thyberg J, Blombren K, Hellgren D, Hedin U. Lysosomophagy in cultured macrophages treated with the anti-microtubular drug nocodazole. *Eur J Cell Biol*. 1982;27:279.

36. Novikoff AB, Essner E, Quintana N. Golgi apparatus and lysosomes. *Fed Proc*. 1964;23:1010.

37. Wollman SH, Spicer SS, Burstone MS. Localization of esterase and acid phosphatase in granules and colloid droplets in rat thyroid epithelium. *J Cell Biol*. 1964;21:191.

38. Wetzel BK, Spicer SS, Wollman SH. Changes in fine structure and acid phosphatase localization in rat thyroid cells following thyrotropin administration. *J Cell Biol*. 1965;25:593.

39. Williams JA, Wolff J. Possible role of microtubules in thyroid secretion. *Proc Natl Acad Sci USA*. 1970;67:1901.

40. Selmi S, Rousett B. Identification of two subpopulations of thyroid lysosomes: relation to the thyroglobulin proteolytic pathway. *Biochem J*. 1988;253:523.

441. John TM, George JC. Seasonal ultrastructural changes in the anterior pituitary of the migratory Canadian goose; 3: Autumn (Fall) migration. *Cytobios*. 1985;42:99.

42. Resau JH, Marzella L, Trump BF, Jones RT. Degradation of zymogen granules by lysosomes in cultured pancreatic explants. *Am J Pathol*. 1984;115:134.

43. Schnell AH, Borg LAH. Lysosomes and pancreatic islet function: glucose-dependent alterations of lysosomal morphology. *Cell Tissue Res*. 1985;239:537.

44. Halban PA, Mutkoski R, Dodson G, Orci L. Resistance of insulin crystal to lysosomal proteases: implications for pancreatic B-cell crinophagy. *Diabetologia*. 1987;30:348.

45. Thiele DL, Lipsky PE. Regulation of cellular function by products of lysosomal enzyme activity: elimination of human natural killer cells by a dipeptide methyl ester generated from L-leucine methyl ester by monocytes or polymorphonuclear leukocytes. *Proc Natl Acad Sci USA*. 1985;82:2468.

46. Sasahara M, Hazama F, Amano S, Yamada E. Effects of the lysosomal fraction of polymorphonuclear leukocytes on proliferation of cultured vascular cells. *Virchows Arch (Cell Pathol)*. 1985;49:121.

47. Youngdahl-Turner P, Rosenberg LE. Binding and uptake of transcobalamin II by human fibroblasts. *J Clin Invest*. 1978;61:133.

48. Youngdahl-Turner P, Mellman IS, Allen RH, Rosenberg LE. Protein-mediated vitamin uptake: adsorptive endocytosis of the transcobalamin II–

cobalamin complex by cultured human fibroblasts. *Exp Cell Res.* 1979;118:127.

49. Rosenblatt DS, Hosack A, Matiaszuk NV. Defect in vitamin B-12 release from lysosomes: newly described inborn error of vitamin B-12 metabolism. *Science.* 1985;228:1319.

50. Brown MS, Goldstein JL. A receptor-mediated pathway for cholesterol homeostasis. *Science.* 1986;232:34.

51. Gahl D, Simpson ER, Porter JC, Snyder JM. Defective internalization of low density lipoprotein in epidermoid cervical cancer cells. *J Cell Biol.* 1982;92:597.

52. Larkin JM, Brown MS, Goldstein JL, Anderson RGW. Depletion of intracellular potassium arrests coated pit formation and receptor-mdiated endocytosis in fibroblasts. *Cell.* 1983;33:273.

53. Daukas G, Zigmond SH. Inhibition of receptor-mediated but not fluid-phase endocytosis in polymorphonuclear leukocytes. *J Cell Biol.* 1985; 101:1673.

54. de Duve C, Pressman BC, Gianetto R, Wattiaux R, Applemans F. Tissue fractionation studies, 6: intracellular distribution patterns of enzymes in rat liver tissue. *Biochem J.* 1955;60:604.

55. Schneider DL. The proton pump ATPase of lysosomes and related organelles of the vacuolar apparatus. *Biochem Biophys Acta.* 1987;895:1.

56. Bernar J, Tietz F, Kohn LD, et al. Characteristics of a lysosomal membrane transport system for tyrosine and other neutral amino acids in rat thyroid cells. *J Biol Chem.* 1986;261:17107.

57. Renlund M, Kovanen PT, Raivio KO, Aula P, Gahmberg CG, Ehnholm C. Studies on the defect underlying the lysosomal storage of sialic acid in Salla disease. *J Clin Invest.* 1986;77:568.

58. Forster S, Lloyd JB. Solute translocation across the mammalian lysosomal membrane. *Biochim Biophys Acta.* 1988;947:465.

59. Tietze F, Kohn LD, Kohn AD, et al. Carrier-mediated transport of monoiodotyrosine out of thyroid cell lysosomes. *J Biol Chem.* 1989;264: 4762.

60. Pisoni RL, Thoene JG. Detection and characterization of a nucleoside transport system in human fibroblast lysosomes. *J Biol Chem.* 1989;264: 4850.

61. Moriyama Y. Potassium ion dependent proton efflux and depolarization from spleen lysosomes. *Biochem Biophys Res Commun.* 1988;156:211.

62. Gahl WA, Tietze F, Bashan N, Steinherz R, Schulman JD. Defective cystine exodus from isolated lysosome-rich fractions of cystinotic leucocytes. *J Biol Chem.* 1982;257:9570.

63. Gahl WA, Bashan N, Tietze F, Bernardini I, Schulman JD. Cysteine transport is defective in isolated leucocyte lysosomes from patients with cystinosis. *Science.* 1982;217:1263.

64. Kooistra T, Schulman JD, Lloyd JB. Cystinosis and lysosomal handling of cystine residues. In: Dingle JT, Dean R, Sly WS, eds. *Lysosomes in Biology and Pathology,* Vol. 7. New York: Elsevier; 1984:443.

65. Jonas AJ, Smith ML, Schneider JA. ATP-dependent lysosomal cystine efflux is defective in cystinosis. *J Biol Chem.* 1982;257:13185.

66. Sawant PL, Desai ID, Tappel AL. Factors affecting the lysosomal membrane and availability of enzymes. *Arch Biochem Biophys.* 1964;105:247.

67. Ericsson JLE, Brunk UT. Alterations in lysosomal membranes as related to disease process. In: Trump BF, Arstila AU, eds. *Pathobiology of Cell Membranes,* Vol. I. New York: Academic Press; 1975;217.

68. Hawkins HK. Reactions of lysosomes to cell injury. In: Trump BF, Arstila AU, eds. *Pathobiology of Cell Membranes,* Vol. II. New York: Academic Press; 1980:251.

69. Hieda N, Toki Y, Sugiyama S, Ito T, Satake T, Ozawa T. Prostaglandin I_2 analogue and propanolol prevent ischemia induced mitochondrial dysfunction through the stabilization of lysosomal membranes. *Cardiovasc Res.* 1988;22:219.

70. de Duve C, Wattiaux R, Wibo M. Effects of fat soluble compounds on lysosomes *in vitro. Biochem Pharmacol.* 1962;9:97.

71. Desai ID, Sawant PL, Tappel AL. Effects of peroxidative and radiation damage to lysosomes. *Biochim Biophys Acta.* 1964;86:277.

72. Fong KL, McCay PB, Poyer LJ. Evidence that peroxidation of lysosomal membrane is initiated by hydroxyl free radicals produced during flavin enzyme activity. *J Biol Chem.* 1973;248:7792.

73. Lindquist RR. Studies on the pathogenesis of hepatoenticular degeneration, 3: the effects of copper on rat liver lysosomes. *Am J Pathol.* 1968;53:903.

74. Brun A, Brunk U. Histochemical indications for lysosomal localization of heavy metals in normal rat brain and liver. *J Histochem Cytochem.* 1970;18:820.

75. Brunk U, Brun A. Histochemical evidence for lysosomal uptake of lead in tissue cultured fibroblasts. *Histochimie.* 1972;29:140.

76. Selden C, Owen M, Hopkins JMP, Peters TJ. Studies on the concentration and intracellular localization of iron proteins in liver biopsy specimens from patients with iron overload with special reference to their role in lysosomal disruption. *Br J Haematol.* 1980;44:593.

77. Abok K, Hirth T, Ericsson JLE, Brunk U. Effects of iron on the stability of macrophage lysosomes. *Virchows Arch (Cell Pathol).* 1983;43:85.

78. Abok K, Rundquist I, Forsberg B, Brunk U. Dimethyl-sulfoxide increases the survival and lysosomal stability of mouse peritoneal macrophages exposed to low-LET ionizing radiation and/or ionic iron in culture. *Virchows Arch (Cell Pathol).* 1984;46:307.

79. Isom HC, Secott T, Georgoff I, Woodworth C, Mummaw J. Maintenance of differentiated rat hepatocytes in primary culture. *Proc Natl Acad Sci USA.* 1985;82:3252.

80. Spanier AM, Dickens BF, Weglicki WB. Response of canine cardiocyte lysosomes to ATP. Am J Physiol. 1985;249;H20.

81. Ruth RC, Weglicki WB. Mechanisms of loss of latency of lysosomal enzymes: effects of incubation on the properties of lysosomal membranes. *Biochem J.* 1980;186:243.

82. Ruth RC, Weglicki WB. Effects of ATP on lyso-

somes: inhibition of the loss of latency caused by cooling. *Am J Physiol.* 1982;242:C192.

83. Ruth RC, Weglicki WB. Effects of ATP on lysosomes: protection against hyperosmolar KCl. *Am J Physiol.* 1983;245:C68.

84. Lefer AM, Ogletree ML, Smith JB, et al. Prostacyclin: a potentially valuable agent for preserving myocardial tissue in acute myocardial ischemia. *Science.* 1978;200:52.

85. Ogletree ML, Lefer AM. Prostaglandin-induced preservation of the ischemic myocardium. *Circ Res.* 1978;42:218.

86. Berberian PA, Jenison MW. Arterial prostaglandins and lysosomal function during atherogenesis, I: homogenate of diet-induced atherosclerotic aortas of rabbit. *Exp Mol Pathol.* 1985;43:22.

87. Berberian PA, Jenison MW, Roddick V. Arterial prostaglandins and lysosomal function during atherogenesis, II: isolated cells of diet-induced atherosclerotic aortas of rabbit. *Exp Mol Pathol.* 1985;43:36.

88. Seydewitz V, Staubesand J. Immunocytochemical demonstration of lysosomal matrix vesicles in the arterial wall of the rat. *Histochemistry.* 1988;88:463.

89. Mortimore GE, Ward WF. Internalization of cytoplasmic proteins by hepatic lysosomes in basal and deprivation induced proteolytic states. *J Biol Chem.* 1981;256:7659.

90. Dean RT. Concerning a possible mechanism for selective capture of cytoplasmic proteins by lysosomes. *Biochem Biophys Res Commun.* 1975; 67:604.

91. Ahlberg J, Marzella L, Glaumann H. Uptake and degradation of proteins by isolated rat liver lysosomes: suggestion of a microautophagic pathway of proteolysis. *Lab Invest.* 1982;47:523.

92. Ahlberg J, Glaumann H. Uptake-microautophagy and degradation of exogenous proteins by isolated rat liver lysosomes: effects of pH, ATP and inhibitors of proteolysis. *Exp Mol Pathol.* 1985;42:78.

93. Sakai M, Ogawa K. Energy-dependent lysosomal wrapping mechanism (LWM) during autophagolysosome formation. *Histochemistry.* 1982;76: 479.

94. Sakai M, Ogawa K. Relationship between lysosomal wrapping mechanism (LWM) and cytoskeletal elements during autophago-lysosome formation. *Acta Histochem Cytochem.* 1984;17:1.

95. Krstic R. Formation of autophagosomes in the pinealocytes of the rat and Mongolian gerbil: a lysosomal wrapping mechanism? *Cell Tissue Res.* 1985;241:677.

96. Sakai M, Araki N, Ogawa K. Lysosomal movements during heterophagy and autophagy with special reference to nematolysosome and wrapping lysosome. *J Electr Microsc Tech.* 1989; 12:101.

97. Kovacs J, Rez G. Autophagocytosis (Review). *Acta Biol Acad Sci Hung.* 1979;30:177.

98. Willingham MC, Pastan I. Transit of epidermal growth factor through coated pits of the Golgi system. *J Cell Biol.* 1982;94:207.

99. Willingham MC, Hanover JA, Dickson RB, Pastan I. Morphologic characterization of the pathway of transferrin, endocytosis and recycyling in human KB cells. *Proc Natl Acad Sci USA.* 1984;81:175.

100. Barbieri CL, Brown K, Rabinovitch M. Depletion of secondary lysosomes in mouse macrophages infected with *Leishmania mexicana amazonensis*: a cytochemical study. *Z. Parasitenkd.* 1985;71:159.

101. Hirsimaki P, Pilstrom L. Studies on vinblastine-induced autophagocytosis in mouse liver, III: a quantitative study. *Virchows Arch (Cell Pathol).* 1982;41:51.

102. Marzella L, Glaumann H. Increased degradation in rat liver induced by vinblastine, II: morphologic characterization. *Lab Invest* 1980;42:18.

103. Kominami E, Hashida S, Khairallah EA, Katunuma N. Sequestration of cytoplasmic enzymes in autophagic vacuole–lysosomal system induced by injection of leupeptin. *J Biol Chem.* 1983; 258:6093.

104. Rogers SW, Rechsteiner M. Degradation of structurally characterized proteins injected into HeLa cells: effects of intracellular location and the involvement of lysosomes. *J Biol Chem.* 1988;263:19843.

105. Pfeifer U. Inhibition by insulin of the formation of autophagic vacuoles in rat liver: a morphologic approach of the kinetics of intracellular degradation by autophagy. *J Cell Biol.* 1978;78:152.

106. Thyberg J, Hedin U, Stenseth K. Endocytic pathways and time sequence of lysosomal transfer of macromolecules in cultured mouse peritoneal macrophages. *Cell Tissue Res.* 1985;241:299.

107. Wall D, Hubbard AL. Receptor-mediated endocytosis of asialoglyco-proteins by rat liver hepatocytes: biochemical characterization of the endosomal compartment. *J Cell Biol.* 1985;101:2104.

108. Wall DA, Wilson G, Hubbard AL. The galactose-specific recognition system of mammalian liver: the route of ligand internalization in rat hepatocytes. *Cell.* 1980;21:79.

109. Haylett T, Thilo L. Limited and selective transfer of plasma membrane glycoproteins to membrane of secondary lysosomes. *J Cell Biol.* 1986; 103:1249.

110. Draye J-P, Courtoy PJ, Quintart J, Baudhuin D. A quantitative model of traffic between plasma membrane and secondary lysosomes: evaluation of inflow, lateral diffusion and degradation. *J Cell Biol.* 1988;170:405.

111. Mellman I. Membrane recycling during endocytosis. In: Dingle JT, Dean R, Sly WS, eds. *Lysosomes in Biology and Pathology*, Vol. 7. New York: Elsevier; 1984:201.

112. Dunn WA, Connolly TP, Hubbard AL. Receptor-mediated endocytosis of epidermal growth factor by rat hepatocytes: receptor pathway. *J Cell Biol.* 1986;102:24.

113. Beguinot L, Lyall RM, Willingham MC, Pastan I. Down-regulation of the epidermal growth factor receptor in KB cells is due to receptor internalization and subsequent degradation in lysosomes. *Proc Natl Acad Sci USA.* 1984;81:2384.

114. Ballard FJ. Regulation of protein breakdown by epidermal growth factor in A431 cells. *Exp Cell Res.* 1985;157:172.

115. Klausner RD, Harford J, Van Renswoude J. Rapid internalization of the transferrin receptor in K562 cells is triggered by ligand binding or treatment with a phorbol ester. *Proc Natl Acad Sci USA.* 1984;81:3005.

116. Beguinot L, Hanover JA, Ito S, Richert ND, Willingham MC, Pastan I. Phorbol esters induce transient internalization without degradation of unoccupied epidermal growth factor receptors. *Proc Natl Acad Sci USA.* 1985;82:2774.

117. Mullock BM, Hinton RH, Dobrota M, Peppard J, Orlans E. Endocytic vesicles in liver carry polymeric IgA from serum to bile. *Biochim Biophys Acta.* 1979;587:381.

118. Takahashi I, Nakane PK, Brown WR. Ultrastructural events in the translocation of polymeric IgA by rat hepatocytes. *J Immunol.* 1982;128:1181.

119. Schiff JM, Fisher MM, Underdown BJ. Receptor-mediated biliary transport of immunoglobulin A and asialoglyprotein: sorting and missorting of ligand revealed by two radiolabeling methods. *J Cell Biol.* 1984;98:79.

120. Finck MH, Reichen J, Vierling JM, Kloppel TM, Brown WR. Hepatic uptake and disposition of human polymeric IgA1 in perfused rat liver: evidence for incomplete biliary excretion and intrahepatic degradation. *Am J Physiol.* 1985;248: G450.

121. Mostov KE, Simister NE. Transcytosis. *Cell.* 1985; 43:389.

122. White J, Matlin K, Helenius A. Cell fusion by Semliki forest, influenza, and vesicular stomatitis viruses. *J Cell Biol.* 1981;89:674.

123. Lucy JA. Do hydrophobic sequences cleaved from cellular polypeptides induce membrane fusion reactions *in vivo? FEBS Lett.* 1984;166: 223.

124. Roos DS, Choppin PW. Biochemical studies on cell fusion, II: control of fusion response by lipid alteration. *J Cell Biol.* 1985;101:1591.

125. Knudsen KA. The calcium-dependent myoblast adhesion that precedes cell fusion is mediated by glycoproteins. *J Cell Biol.* 1985;101:891.

126. Schliwa M. Mechanisms of intracellular organelle transport. In: Shaw JW, ed. *Cell Muscle Motility,* Vol. 5. New York: Plenum; 1984:1.

127. Matteoni R, Kreis TE. Translocation and clustering of endosomes and lysosomes depends on microtubules. *J Cell Biol.* 1987;105:1253.

128. Mithieux G, Audebet C, Rousset B. Association of purified thyroid lysosomes to reconstituted microtubules. *Biochim Biophys Acta.* 1988;969:121.

129. Wang E, Goldman RD. Functions of cytoplasmic fibers in intracellular movements in BHK-21 cells. *J Cell Biol.* 1978;79:708.

130. Phaire-Washington L, Silverstein SC, Wang E. phorbol myristate acetate stimulates microtubule and 10-nm filament extension and lysosome redistribution in mouse macrophages. *J Cell Biol.* 1980;86:641.

131. Arstila AU, Nuuja IJM, Trump BF. Studies on cellular autophagocytosis: vinblastine-induced autophagy in the rat liver. *Exp Cell Res.* 1974; 87:249.

132. Marzella L, Glaumann H. Increased degradation in rat liver induced by vinblastine, I: biochemical characterization. *Lab Invest.* 1980;42:8.

133. Marzella L, Glaumann H. Inhibitory effects of vinblastine on protein degradation. *Virchows Arch (Cell Pathol).* 1980;34:111.

134. Yu QC, Marzella L. Modification of lysosomal protein degradation in mouse liver with taxol. *Am J Pathol.* 1986;122:553.

135. Collot M, Louvard D, Singer SJ. Lysosomes are associated with microtubules and not with intermediate filaments in cultured fibroblasts, *Proc Natl Acad Sci USA.* 1984;81:788.

136. Vale RD, Schnapp BJ, Reese TS, Sheetz MP. Organelle, bead and microtubule translocations provided by soluble factor from the squid giant axon. *Cell.* 1985;40:559.

137. Vale RD, Reese TS, Sheetz MP. Identification of a novel force generating protein (kinesin) involved in microtubule-based motility. *Cell.* 1985;41:39.

138. Vale RD, Schapp BJ, Mitchson T, Steuer E, Reese TS, Sheetz MP. Different axoplasmic proteins generate movement in opposite directions along microtubules *in vitro. Cell.* 1985;43:623.

139. Rodionov VI, Nadezhdina ES, Leonova EV, Vaisberg EA, Kuznetsov SA, Gelfand VI. Identification of a 100 kD protein associated with microtubules, intermediate filaments and coated vesicles in cultured cells. *Exp Cell Res.* 1985; 159:377.

140. Mithieux G, Rousett B. Regulation of the microtubule–lysosome interaction: activation by Mg^{2+} and inhibition by ATP. *Biochim Biophys Acta.* 1988;971:29.

141. Mithieux G, Rousett B. Identification of a lysosome membrane protein which could mediate ATP-dependent stable association of lysosomes to microtubules. *J Biol Chem.* 1989;264:4664.

142. Burnside J, Schneider DL. Characterization of rat liver lysosomes: composition, enzyme activities and turnover. *Biochem J.* 1982;204:525.

143. Ohsumi Y, Ishikawa T, Kato K. A rapid and simplified method for the preparation of lysosomal membranes from rat liver. *J Biochem (Tokyo).* 1983;93:547.

144. Reggio H, Bainton D, Harms E, Louvard D. Antibodies against lysosomal membranes reveal a 100,000-mol-wt protein that cross-reacts with purified H^+,K^+ ATPase from gastric mucosa. *J Cell Biol.* 1984;99:1511.

145. Chen JW, Murphy TL, Willingham MC, Pastan I, August T. Identification of two lysosomal membrane glycoproteins. *J Cell Biol.* 1985; 101:85.

146. Mane SM, Marzella L, Bainton DF, et al. Purification and characterization of human lysosomal membrane glycoproteins. *Arch Biochem Biophys.* 1989;268:360.

147. Lewis V, Green SA, Marsh M, Vihko P, Helenius A, Mellman I. Glycoproteins of the lysosomal membrane. *J Cell Biol.* 1985;100:1839.

148. Rodman JS, Seidman L, Farquhar MG. The

membrane composition of coated pits, microvilli, endosomes, and lysosomes is distinctive in the rat kidney proximal tubule cell. *J Cell Biol.* 1986; 102:77.

149. Komuves LG, Sass M, Kovacs J. Autophagocytosis in the larval midgut cells of *Pieris brassicae* during metamorphosis: induction by 20-hydroxyecdysone and the effects of puromycin and cycloheximide. *Cell Tissue Res.* 1985;240:215.

150. Rez G, Kiss A, Bucsek MJ, Kovacs J. Attachment of ribosomes to endoplasmic membranes in mouse pancreas: degranulation *in vivo* caused by the inducers of autophagocytosis neutral red, vinblastine, purdmycin, and cadmium ions, and prevention by cycloheximide. *Chem Biol Interact.* 1976;13:77.

151. Arstila AU, Trump BF. Studies on cellular autophagocytosis: the formation of autophagic vacuoles in the liver after glucagon administration. *Am J Pathol.* 1968;53:687.

152. Hirsimaki Y, Arstila AU, Trump BF. Autophagocytosis: *in vitro* induction by microtubule poisons. *Exp Cell Res.* 1975;92:11.

153. Marzella L, Yu QC, Mergner WJ, Trump BF. Unbuffered osmium staining of cell organelles: alterations induced by cell injury. *Virchows Arch (Cell Pathol).* 1984;45:273.

154. Novikoff AB, Shin WY. Endoplasmic reticulum and autophagy in rat hepatocytes. *Proc Natl Acad Sci USA.* 1978;75:5039.

155. Gray RH, Sokol M, Brabec R, Brabec MJ. Characterization of chloroquine-induced autophagic vacuoles isolated from rat liver. *Exp Mol Pathol.* 1981;34:72.

156. Locke M, Sykes AK. The role of the Golgi complex in the isolation and digestion of organelles. *Tissue and Cell.* 1975;7:143.

157. Locke M, Collins JV. Organelle turnover. In: Trump BF, Arstila AU, eds. *Pathobiology of Cell Membranes,* Vol. II. New York: Academic Press; 1980;223.

158. Hamberg H. Cellular autophagocytosis induced by X-irradiation and vinblastine: on the origin of the sequestering membranes. *Acta Path Microbiol Scand.* 1983;91(Sect A):317.

159. Henell F, Glaumann H. Participation of lysosomes in basal proteolysis in perfused rat liver. *Exp Cell Res.* 1985;158:257.

160. Seglen PO, Gordon PB, Tolleshaug H, Hoyrik H. Autophagy and protein degradation in isolated rat hepatocytes. *Biochem Soc Trans.* 1985; 13:1007.

161. Poso AR, Wert JJ, Mortimore GE. Multifunctional control by amino acids of deprivation-induced proteolysis in liver: role of leucine, *J Biol Chem.* 1982;257:12114.

162. Seglen PO, Gordon PB, Poli A. Amino acid inhibition of the autophagic/lysosomal pathway of protein degradation in isolated rat hepatocytes. *Biochim Biophys Acta.* 1980;630:103.

163. Seglen PO. Regulation of autophagic protein degradation in isolated liver cells. In: Glaumann H, Ballard FJ, eds. In: *Lysosomes, Their Role in Protein Degradation.* New York: Academic Press; 1986.

164. Kovacs AL, Grinde B, Seglen PO. Inhibition of autophagic vacuole formation and protein degradation by amino acids in isolated hepatocytes. *Exp Cell Res* 1981;133:431.

165. Schworer CM, Schiffer KA, Mortimore GE. Quantitative relationship between autophagy and proteolysis during graded amino acid deprivation in perfused rat liver. *J Biol Chem.* 1981;256:7652.

166. Hopgood MF, Clark MG, Ballard FJ. Inhibition of protein degradation in isolated rat hepatocytes. *Biochem J.* 1977;164:399.

167. Hopgood MF, Clark MG, Ballard FJ. Protein degradation in hepatocyte monolayers: effects of glucagon, adenosine 3′5′-cyclic monophosphate or insulin, *Biochem J.* 1980;186:71.

168. Hopgood MF, Clark MG, Ballard FJ. Stimulation by glucocorticoids of protein degradation in hepatocyte monolayers. *Biochem J.* 1981;196:33.

169. Schworer CM, Mortimore GE. Glucagon-induced autophagy and proteolysis in rat liver: mediation by selective deprivation of intracellular amino acids. *Proc Natl Acad Sci USA.* 1979; 76:3169.

170. Yu QC, Marzella L. Response of autophagic protein degradation to physiologic and pathologic stimuli in rat hepatocyte monolayers. *Lab Invest.* 1988;58:643.

171. Hsie AW, Kawashima K, O'Neill JP, Schroder CH. Possible role of adenosine cyclic 3′:5′-monophosphate phosphodiesterase in the morphological transformation of Chinese hamster ovary cells mediated by N6,02-dibutyryl adenosine cyclic 3′:5′-monophosphate. *J Biol Chem.* 1975;250: 984.

172. Sass M, Csikos GY, Komuves L, Kovacs J. Cyclic AMP in the fat body of *Mamestra brassicae* during the last instar and its possible involvement in the cellular autophagocytosis induced by 20-hydroxyecdysone. *Gen Comp Endocrinol.* 1983;50: 116.

173. Wolberg G, Stopford CR, Zeimmerman TP. Antagonism by taxol of effects of microtubule-disrupting agents on lymphocyte cAMP metabolism and cell function. *Proc Natl Acad Sci USA.* 1984;81:3496.

174. Kovacs J, Rez G, Kiss A. Vinblastine-induced autophagocytosis and its prevention by cycloheximide and emetine in mouse pancreatic acinar cells *in vivo. Cytobiologie.* 1975;11:309.

175. Baracos V, Greenberg RE, Goldberg AL. Influence of calcium and other divalent cations on protein turnover in rat skeletal muscle. *Am J Physiol.* 1986;250:E702.

176. Silver G, Etlinger JD. Regulation of myofibrillar accumulation in chick muscle cultures: evidence for the involvement of calcium and lysosomes in non-uniform turnover of contractile proteins. *J Cell Biol.* 1985;101:2383.

177. Farber JL, El-Mofty SK. The biochemical pathology of liver cell necrosis. *Am J Pathol.* 1975; 81:237.

178. Trump BF, McDowell EM, Arstila AU. Cellular reaction to injury. In: Hill RB, La Via MF, eds. *Principles of Pathobiology.* New York: Oxford University Press; 1980:20.

179. Trump BF, Berezesky IK, Phelps PC. Sodium and calcium regulation and the role of the cytoskeleton in the pathogenesis of disease: a review and hypothesis. *Scan Electr Microsc.* 1981;2:435.

180. Trump BF, Berezesky IK. Role of sodium and calcium regulation in toxic cell injury. In: Mitchell JR, Horning MG, eds. *Drug Metabolism and Drug Toxicity.* New York: Raven Press; 1984:261.

181. Laiho KU, Berezesky IK, Trump BF. The role of calcium in cell injury. *Survey Synth Pathol Res.* 1983;2:170.

182. Farber JL. The role of calcium in cell death. *Life Sci.* 1981;29:1289.

183. Jewell SA, Bellomo G, Thor H, Orrenius S, Smith M. Bleb formation in hepatocytes during drug metabolism is caused by disturbances in thiol and calcium ion homeostasis. *Science.* 1982;217:1257.

184. Sawyer DW, Sullivan JA, Mandell GL. Intracellular free calcium localization in neutrophils during phagocytosis. *Science.* 1985;230:663.

185. Somlyo AP. Cellular site of calcium regulation. Nature. 1984;309:516.

186. Seglen PO, Gordon PB. Vanadate inhibits protein degradation in isolated rat hepatocyts. *J Biol Chem.* 1981;256:7699.

187. Marzella L, Trump BF. Cell injury and its meaning in shock and resuscitation. In: Siegel J, ed. *Trauma: Emergency Surgery and Critical Care.* New York: Churchill Livingston; 1987:35.

188. Knowles SE, Ballard FJ. Selective control of the degradation of normal and aberrant proteins in Reuber H35 hepatoma cells. *Biochem J.* 1976;156:609.

189. Neff NT, Demartino GN, Goldberg AL. The effect of protease inhibitors and decreased temperature on the degradation of different classes of proteins in cultured hepatocytes. *J Cell Physiol.* 1979;101:439.

190. Ahlberg J, Berkenstam A, Henell F, Glaumann H. Degradation of short and long lived proteins in isolated rat liver lysosomes: effects of pH, temperature and proteolytic inhibitors. *J Biol Chem.* 1985;260:5847.

191. Yu Q-C, Lipsky M, Trump BF, Marzella L. Response of human hepatocyte lysosomes to postmortem anoxia. *Hum Pathol.* 1988;19:1174.

192. Polet H, Swager J. Effects of serum and conditioned medium on protein degradation, migration of nonhistone proteins to the nucleus and DNA synthesis in transformed cells. *J Cell Physiol.* 1987;130:436.

193. Tayek JA, Blackburn GL, Bistrian BR. Alterations in whole body, muscle, liver and tumor tissue protein synthesis and degradation in Novikoff hepatoma and Yoshida sarcoma tumor growth studied *in vivo. Cancer Res.* 1988; 48:1554.

194. Ahlberg J, Yucel T, Ericsson L, Glaumann H. Characterization of the proteolytic compartment in rat hepatocyte nodules. *Virch Arch B Cell Pathol.* 1987;53:79.

195. Yucel T, Ahlberg J, Glaumann H. Overall proteolysis in perfused and subfractionated chemically induced malignant hepatoma of rat: effects of amino acids. *Exp Mol Pathol.* 1989; 50:38.

196. Lee HK, Myers RAM, Marzella L. Stimulation of autophagic protein degradation by nutrient deprivation in a differentiated murine teratocarcinoma (F9, 12-1a) cell line. *Exp Mol Pathol.* 1989; 50:139.

197. Plomp PJAM, Gordon PB, Meijer AJ, Hovrik H, Seglen PO. Energy dependence of different steps in the autophagic–lysosomal pathway. *J Biol Chem.* 1989;264;6699.

198. Grinde B, Seglen PO. Differential effects of proteinase inhibitors and amines on the lysosomal and non-lysosomal pathways of protein degradation in isolated rat hepatocytes. *Biochim Biophys Acta.* 1980;632:73.

199. Seglen PO, Gordon PB. 3-Methyladenine: specific inhibitor of autophagic/lysosomal protein degradation in isolated rat hepatocytes. *Proc Natl Acad Sci USA.* 1982;79:1889.

200. Grinde B. Role of calcium for protein turnover in isolated rat hepatocytes. *Biochem J.* 1983; 216:529.

201. Chua BHL, Long WM, Lautensack N, Lins JA, Morgan HE. Effects of diabetes on cardiac lysosomes and protein degradation. *Am J Physiol.* 1983;245:C91.

202. Long WM, Chua BHL, Lautensack N, Morgan H. Effects of amino acid methyl esters on cardiac lysosomes and protein degradation. *Am J Physiol.* 1983;245(Cell. Physiol. 14):C101.

203. Libby P, Goldberg AL. Leupeptin, a protease inhibitor, decreases protein degradation in normal and diseased muscles. *Science.* 1978;199;534.

204. Ivy GO, Schottler F, Wenzel J, Baudry M, Lynch G. Inhibitors of lysosomal enzymes: accumulation of lipofuscin-like bodies in the brain. *Science.* 1984;226:985.

205. Hirsimaki Y, Hirsimaki P. Vinblastine-induced autophagocytosis: the effects of disorganization of microfilaments by cytochalasin B. *Exp Mol Pathol.* 1984;40:61.

206. Hamberg H, Brunk U, Ericsson JLE, Jung B. Cytoplasmic effects of x-irradiation on cultured cells in a non-dividing stage, 2: alterations in lysosomes, plasma membrane, Golgi apparatus and related structures. *Acta Path Microbiol Scand.* 1977;85(Sect A):625.

207. Hamberg H, Edman P. Induced autophagocytosis in macrophages: origin of the segregating membranes. *Acta Path Microbiol Scand.* 1983;91(Sect A):91/1.

208. Cole S, Matter A, Karnovsky MJ. Autophagic vacuoles in experimental atrophy. *Exp Mol Pathol.* 1971;14:158.

209. Marzella L, Glaumann H. Effects of *in vivo* liver ischemia on microsomes and lysosomes. *Virchows Arch (Cell Pathol).* 1984;45:97.

210. Salminen A, Vihko V. Autophagic response to strenuous exercise in mouse skeletal muscle fibers. *Virchows Arch (Cell Pathol).* 1984;45:97.

211. Terry RD, Wisniewski H, Johnson AB. Studies on the formation of autophagic vacuoles in neurons treated with spindle inhibitors (colchi-

cine and vinblastine). *J Neuropath Exp Neurol.* 1970;29:142.

212. Yu QC, Trump BF, Marzella L. Modification of autophagic proteolysis with taxol. *J Cell Biol.* 1984;99:369a.

213. Longnecker DS. Modification of puromycin-induced changes in pancreatic acinar cells by cycloheximide pre-treatment in rats. *Lab Invest.* 1972;26:459.

214. Hendy R, Grasso P. Autophagy in acute liver damage produced in the rat by dimethylnitrosamine. *Chem Biol Interact.* 1972;5:401.

215. Rao MS, Svoboda DJ, Reddy JK. The ultrastructural effects of aflatoxin B_1 in the rat pancreas. *Virchows Arch (Cell Pathol).* 1975;18:119.

216. Rumpelt HJ, Albring M, Thoenes W. Prevention of D-galactosamine induced hepatocellular autophagocytosis by cycloheximide. *Virchows Arch (Cell Pathol).* 1974;16:195.

217. Wani MC, Taylor HL, Wall M, Cozzan P, McPhail AT. Plant antitumor agents, VI: the isolation and structure of taxol, a novel antileukemic and antitumor agent, from *Taxus brevifolia. J Am Chem Sci.* 1971;93:2325.

218. Schiff PB, Fant J, Horwitz SB. Promotion of microtubule assembly *in vitro* by taxol. *Nature.* 1979;277:665.

219. Schiff PB, Horwitz SB. Taxol stabilizes microtubule in mouse fibroblast cells. *Proc Natl Acad Sci USA.* 1980;77:1561.

220. Parness J, Horwitz SB. Taxol binds to polymerized tubulin *in vitro. J Cell Biol.* 1981;91:479.

221. Kaplan G, Keogh EA. Analysis of the effect of amines on inhibition of receptor-mediated and fluid-phase pinocytosis in rabbit alveolar macrophages. *Cell.* 1981;24:925.

222. Ohkuma S, Poole B. Cytoplasmic vacuolation of mouse peritoneal macrophages and the uptake into lysosomes of weakly basic substances. *J Cell Biol.* 1981;90:656.

223. Poole B, Ohkuma S. Effects of weak bases on the intralysosomal pH in mouse peritoneal macrophages. *J Cell Biol.* 1981;90:665.

224. Colombo MI, Bertini F. *In vitro* interaction between mouse liver lysosomes and chloroquine. *Biol Cell.* 1985;54:73.

225. Gonzales-Noriega A, Grubb JH, Talkad R, Sly WS. Chloroquine inhibits lysosomal enzyme pinocytosis and enhances lysosomal enzyme secretion by impairing receptor recycling. *J Cell Biol.* 1980;85:839.

226. Kovacs AL, Reith A, Seglen PO. Accumulation of autophagosomes after inhibition of hepatic protein degradation by vinblastine, leupeptin or a lysosomotropic amine. *Exp Cell Res.* 1982;137:191.

227. Furuno K, Ishikawa T, Kato K. Appearance of auto-lysosomes in rat liver after leupeptin treatment. *J Biol Chem.* 1982;91:1485.

283. Glaumann H, Ahlberg J, Berkenstam A, Falk M, Henell F. Protein degradation in the lysosomes. *Biochem Soc Trans.* 1985;13:1010.

229. Kovacs J. Morphometric study of the effects of leupeptin, vinblastine, estron acetate and cycloheximide on the autophagic vacuole–lysosomal

230. Van Berkel TJC, Nagelkerke JF, Kruijt JK. The effects of Ca^{++} and trifluoperazine on the processing of human acetylated low density lipoprotein by non-parenchymal liver cells. *FEBS Lett.* 1981;132:61.

231. Akiyama S, Tomita K, Kuwano M. The effects of calcium antagonists on the proteolytic degradation of low-density lipoprotein in HeLa cells. *Exp Cell Res.* 1985;158:192.

232. Kuratomi Y, Akiyama S, Ono M, et al. Thioridazine enhances lysosomal accumulation of epidermal growth factor and toxicity of conjugates of epidermal growth factor with *Pseudomonas* exotoxin. *Exp Cell Res.* 1986;162:436.

233. Nishimura Y, Kato K. Intracellular transport and processing of lysosomal cathepsin B. *Biochem Biophys Res Comm.* 1987;148:254.

234. Sabatini D, Kreibich G, Morimoto T, Adesnik M. Mechanisms for the incorporation of proteins in membranes and organelles. *J Cell Biol.* 1982;92:1.

235. Wickner WT, Lodish HF. Multiple mechanisms of protein insertion into and across membranes. *Science.* 1985;230:400.

236. Walter P, Blobel L. Translocation of proteins across the endoplasmic reticulum, II: signal recognition protein (SRP) mediates the selective binding to microsomal membranes of *in vitro*–assembled polysomes synthesizing secretory protein. *J Cell Biol.* 1981;91:551.

237. Walter P, Blobel L. Translocation of proteins across the endoplasmic reticulum, III: signal recognition protein (SRP) causes signal sequence dependent and site-specific arrest of chain elongation that is released by microsomal membranes. *J Cell Biol.* 1981;91:557.

238. Meyer DI, Krause E, Dobberstein B. Secretory protein translocation across membranes—the role of the 'docking protein.' *Nature.* 1982;297:647.

239. Goldfischer S. The internal reticular apparatus of Camillo Golgi: a complex, heterogeneous organelle, enriched in acid, neutral and alkaline phosphatases, and involved in glycosylation, secretion, membrane flow, lysosome formation, and intracellular digestion. *J Histochem Cytochem.* 1982;30:717.

240. Dunphy WG, Rothman JE. Compartmental organization of the Golgi stack. *Cell.* 1985;42:13.

241. Sly WS, Fischer HD. The phosphomannosyl recognition system for intracellular and intercellular transport of lysosomal enzymes. *J Cell Biochem.* 1982;18:67.

242. Kornfeld R, Kornfeld S. Assembly of asparagine-linked oligo-saccharides. *Ann Rev Biochem.* 1985;54:31.

243. Hoflack B, Kornfeld S. Lysosomal enzyme binding to mouse P388D1 macrophage membranes lacking the 215-kDa mannose 6-phosphate receptor: evidence for the existence of a second mannose 6-phosphate receptor. *Proc Natl Acad Sci USA.* 1985;82:4428.

244. Stein M, Zijderhand-Bleekemolen JE, Geuze HJ,

Hasilik A, von Figura K. MR 46,000 mannnose 6-phosphate specific receptor: its role in targeting of lysosomal enzymes. *EMBO J.* 1987;6:2677.

245. Nichols BA. Uptake and digestion of horseradish peroxidase in rabbit alveolar macrophages: formation of a pathway connecting the lysosomes to the cell surface. *Lab Invest.* 1982;47:235.

246. Campbell CH, Rome HL. Coated vesicles from rat liver and calf brain contain lysosomal enzymes bound to mannose 6-phosphate receptors. *J Biol Chem.* 1983;258:13347.

247. Geuze HJ, Slot JW, Strous GJAM, Hasilik A, von Figura K. Ultrastructural localization of the mannose 6-phosphate receptor in rat liver. *J Cell Biol.* 1984;98:2047.

248. Brown WJ, Farquhar WG. Accumulation of coated vesicles bearing mannose 6-phosphate receptors for lysosomal enzymes in the Golgi region of I-cell fibroblasts. *Proc Natl Acad Sci USA.* 1984;81:5135.

249. Doine AI, Oliver C, Hand AR. The Golgi apparatus and GERL during postnatal differentiation of rat parotid acinar cells: an electron microscopic cytochemical study. *J Histochem. Cytochem.* 1984;32:477.

250. Schultze-Lohoff E, Hasilik A, von Figura K. Cathepsin D precursors in clathrin-coated organelles from human fibroblasts. *J Cell Biol.* 1985;101:824.

251. Pfeffer SR. The endosomal concentration of a mannose 6-phosphate receptor is unchanged in the absence of ligand synthesis. *J Cell Biol.* 1987;105:229.

252. Griffiths G, Hoflack B, Simons K, Mellman I, Kornfeld S. The mannose 6-phosphate receptor and the biogenesis of lysosomes. *Cell.* 1988;52:324.

253. Geuze HJ, Stoorvogel W, Strous GJ, Slot JW, Bleekemolen JE, Mellman I. Sorting of mannose 6-phosphate receptors and lysosomal membrane proteins in endocytic vesicles. *J Cell Biol.* 1988;107:2491.

254. Brown WG, Goodhouse J, Farquhar MG. Mannose-6-phosphate receptors for lysosomal enzymes cycle between the Golgi complex and endosomes. *J Cell Biol.* 1986;103:1235.

255. Griffiths G, Pfeiffer S, Simons K, Matlin K. Exit of newly synthesized membrane proteins from the trans cisterna of the Golgi complex to the plasma membrane. *J Cell Biol.* 1985;101:949.

256. Payne GS, Schekman R. A test of clathrin function in protein secretion and cell growth. *Science.* 1985;230:1005.

257. Schwaiger H, Hasilik A, von Figura K, Wiernken A, Tanner W. Carbohydrate-free carboxypeptidase Y is transferred into the lysosome-like yeast vacuole. *Biochem Biophys Res Comm.* 1982;104:950.

258. Owada M, Newfeld EF. Is there a mechanism for introducing acid hydrolyases into liver lysosomes that is independent of mannose 6-phosphate recognition? Evidence from I-cell disease. *Biochem Biophys Res Comm.* 1982;105:814.

259. Waheed A, Gottschalk S, Hille A, et al. Human lysosome acid phosphatase is transported as a transmembrane protein to lysosomes in transfected baby hamster kidney cells. *EMBO J.* 1988;7:2351.

260. Tsuji A, Omura K, Suzuki Y. I-cell disease: evidence for a mannose 6-phosphate independent pathway for translocation of lysosomal enzymes in lymphoblastoid cells. *Clin Chimica Acta.* 1988; 176:115.

261. Aerts JMFG, Schram AW, Strijland A, et al. Glucocerebroside, a lysosomal enzyme that does not undergo oligosaccharide phosphorylation. *Biochim Biophys Acta.* 1988;964:303.

262. Portnoy DA, Erickson AH, Kochan J, Ravetch JV, Unkeless JC. Cloning and characterization of a mouse cysteine proteinase. *J Biol Chem.* 1986;263:14697.

263. Nishimura Y, Furuno K, Kato K. Biosynthesis and processing of lysosomal cathepsin L in primary cultures of rat hepatocytes. *Arch Biochem Biophys.* 1988;263:107.

264. Rosenfeld MG, Kreibich G, Popov D, Kato K, Sabatini DD. Biosynthesis of lysosomal hydrolases: their synthesis in bound polysomes in determining their subcellullar distribution. *J Cell Biol.* 1982;93:135.

265. Erickson AH, Blobel G. Carboxyl-terminal proteolytic processing during biosynthesis of the lysosomal enzymes beta-glucuronidase and cathepsin D. *Biochemistry.* 1983;22:5201.

266. Gieselmann V, Hasilik A, von Figura K. Processing of human cathepsin D in lysosomes *in vitro.* J Biol Chem. 1985;260:3215.

267. Richardson JM, Woychik NA, Ebert DL, Dimond RL, Cardelli JA. Inhibition of early but not late proteolytic processing events leads to missorting and oversecretion of precursor forms of lysosomal enzymes in *Dictyostelium discoideum. J Cell Biol.* 1988;107:2097.

268. Puizdar V, Turk V. Cathepsinogen D: characterization and activation to cathepsin D and inhibitory peptides. *FEBS Lett.* 1981;132:299.

269. Mort JS, Leduc M, Recklies AD. A latent thiol protease from ascitic fluid of patients with neoplasia. *Biochim Biophys Acta.* 1981;662:173.

270. Varki AP, Reitman ML, Kornfeld S. Identification of a variant of mucolipidosis III (pseudo Hurler polydystrophy): a catalytically active N-acetyl-glucosaminyl-phosphotransferase that fails to phosphorylate lysosomal enzymes. *Proc Natl Acad Sci USA.* 1981;78:7773.

271. Brown JA, Novak EK, Swank RT. Effects of ammonia on processing and secretion of precursor and mature lysosomal enzyme from macrophages of normal and pale ear mice: evidence for two distinct pathways. *J Cell Biol.* 1985;100:1894.

272. Steinman RM, Mellman IS, Muller WA, Cohn ZA. Endocytosis and the recycling of plasma membrane. *J Cell Biol.* 1983;96:1.

273. Imber MJ, Pizzo SV, Johnson WJ, Adams D. Selective diminution of the binding of mannose by murine macrophages in the late stage of activation. *J Biol Chem.* 1982;257:5129.

274. Shepherd VL, Konish MG, Stahl P. Dexamethasone increases expression of mannose receptors and decreases extracellular lysosomal enzyme ac-

cumulation in macrophages. *J Biol Chem.* 1985;260:160.

275. Johnson-Wint B, Gross J. Regulation of connective tissue collagenase production: stimulators from adult and fetal epidermal cells. *J Cell Biol.* 1984;98:90.

276. Brinckerhoff CE, Benoit MC, Culp WJ. Autoregulation of collagenase production by a protein synthesized and secreted by synovial fibroblasts: cellular mechanism for control of collagen degradation. *Proc Natl Acad Sci USA.* 1985;82:1916.

277. Pontremoli S, Melloni E, Salamino F, Sparatore B, Michetti M, Horecher BL. Endogenous inhibitors of lysosomal proteinases. *Proc Natl Acad Sci USA.* 1983;80:1261.

278. Kirshke H, Barrett AJ. Chemistry of lysosomal proteases. In: Glaumann H, Ballard FJ, eds. *Lysosomes: Their Role in Protein Degradation.* New York: Academic Press, 1987:193.

279. Vignon F, Capony F, Chambon M, Freiss G, Garcia M, Rochefort H. Autocrine growth stimulation of the MCF 7 breast cancer cells by the estrogen-regulated 52 K protein. *Endocrinology.* 1986;118:1537.

280. Dong J, Prence EM, Sahagian GG. Mechanism for selective secretion of a lysosomal protease by transformed mouse fibroblasts. *J Biol Chem.* 1989;264:7377.

281. Inui K, Emmett M, Wenger DA. Immunological evidence for deficiency in an activator protein for sulfatide sulfatase in a variant form of metachromatic leukodystrophy. *Proc Natl Acad Sci USA.* 1983;80:3074.

282. Buktenica S, Frankfater A. Effect of subunit size and conformation on the rate of lysosomal degradation of extracellular proteins in cultured mouse peritoneal macrophages. *J Biol Chem.* 1987;262:11611.

283. Reeves JP, Decker RS, Crie JS, Wildenthal K. Intracellular disruption of rat heart lysosomes by leucine methyl ester: effects on protein degradation. *Proc Natl Acad Sci USA.* 1981;78:4426.

284. Pisoni RL, Thoene JG, Christensen HN. Detection and characterization of carrier-mediated cationic amino acid transport in lysosomes of normal and cystinotic human fibroblasts' role in therapeutic cysteine removal? *J Biol Chem.* 1985;260:4791.

285. Maguire GA, Docherty K, Hales CN. Sugar transport in rat liver lysosomes: direct demonstration by using labelled sugars. *Biochem J.* 1983;212:211.

286. Wall DA, Meleka I. An unusual lysosome compartment involved in vitellogenin endocytosis by Xenopus oocytes. *J Cell Biol.* 1985;110:1651.

287. Decker RS, Wildenthal K. Role of lysosomes and latent hydrolytic enzymes in ischemic damage and repair of the heart. In: Wildenthal K, ed. *Degradative Processes in Heart and Skeletal Muscle.* Amsterdam: Elsevier; 1980:390.

288. Wattiaux R, Wattiaux-DeConinck S. Effects of ischemia on lysosomes. *Intern Rev Pathol.* 1984; 26:85.

289. Baron R, Neff L, Louvard D, Courtoy PJ. Cell-mediated extracellular acidification and bone absorption: evidence for a low pH in resorbing lacunae and localization of a 100-kD lysosomal membrane protein at the osteoclast ruffled border. *J Cell Biol.* 1985;85:2210.

290. Kunze H, Bohn E, Loffler B-M. Inhibitors of liver lysosomal acid phospholipase A₁. *Eur J Biochem.* 1988;177:591.

291. Madsen KM, Christensen EI. Effects of mercury on lysosomal protein digestion in the kidney proximal tubule. *Lab Invest.* 1978;38:165.

292. Verbueken AH, Van de Vyver FL, Van Grieken RE, et al. Ultrastructural localization of aluminum in patients with dialysis-associated osteomalacia. *Clin Chem.* 1984;30:763.

293. Stein G, Laske V, Muller A, Braunlich H, Linb W, Fleck C. Aluminum induced damage of the lysosomes in the liver, spleen and kidneys of rats. *J Appl Toxicol.* 1987;7:253.

294. Cleton MI, de Bruijn WC, van Blokland WTM, Marx JJM, Roelofs JM, Rademakers LHPM. Iron content and acid phosphatase activity in hepatic parenchymal lysosomes of patients with hemochromatosis before and after phlebotomy treatment. *Ultrastrc Pathol.* 1988;12:161.

295. Hostetler KY, Reasor M, Yazahi PJ. Chloroquine-induced phospholipid fatty liver: measurement of drug and lipid concentrations in rat liver lysosomes. *J Biol Chem.* 1985;260:215.

296. Tulkens P, Van Hoof F. Comparative toxicity of aminoglycoside antibiotics towards lysosomes in a cell culture model. *Toxicology.* 1980;17:195.

297. Mingeot-Leclercq M-P, Laurent G, Tulkens PM. Biochemical mechanism of aminoglycoside-induced inhibition of phosphatidylcholine hydrolysis by lysosomal phospholipases. *Biochem Pharmacol.* 1988;37:591.

298. Glaumann H, Ahlberg J, Berkenstam A, Henell F. Rapid isolation of rat liver secondary lysosomes–autophagic vacuoles following chloroquine administration. *Exp Cell Res.* 1986;163:151.

299. Henell F, Glaumann H. Effects of leupeptin on the autophagic vacuole system of rat hepatocytes: correlation between ultrastructure and degradation. *Lab Invest.* 1984;51:46.

300. Hruban Z. Pulmonary and generalized lysosomal storage induced by amphiphilic drugs. *Env Health Perspec.* 1984;55:53.

301. Sasahara M, Hazama F, Amano S. et al. Effect of hypertension on lysosomal enzyme activities in aortic endothelial cells. *Atherosclerosis.* 1988; 70:53.

302. Parmacek MS, Decker ML, Lesch M, Samarel AM, Decker RS. Lysosomal changes during thyroxine-induced left ventricular hypertrophy in rabbits. *Am J Physiol.* 1986;251:C737.

303. Hirsch HE. Proteinases and demyelination. *J Histochem Cytochem.* 1981;29:425.

304. Baron R, Neff L, Brown W, Courtoy PJ, Louvard D, Farquhar MG. Polarized secretion of lysosomal enzymes: co-distribution of cation-independent mannose-6-phosphate receptors and lysosomal enzymes along the osteoclast exocytic pathway. *J Cell Biol.* 1988;106:1863.

305. Poole AR, Mort JS. Biochemical and immuno-

logical studies of lysosomal and related proteinases in health and disease. *J Histochem Cytochem.* 1981;29:495.

306. Warren L. Stimulated secretion of lysosomal enzymes by cells in culture. *J Biol Chem.* 1989; 264:8835.

307. Pietras RJ, Szego CM, Roberts JA, Seeler BJ. Lysosomal cathepsin B–like activity: mobilization in prereplicative and neoplastic epithelial cells. *J Histochem Cytochem.* 1981;29:440.

308. Sloane BF, Dunn JR, Honn KV. Lysosomal cathepsin B: correlation with metastatic potential. *Science.* 1981;212:1151.

309. Kelly BA, Carchman RA. The relationship between lysosomal enzyme release and protein phosphorylation in human monocytes stimulated by phorbol esters and opsonized zymosan. *J Biol Chem.* 1987;262:17404.

310. Gal S, Gottesman MM. The major excreted protein of transformed fibroblasts is an activable acid-protease. *J Biol Chem.* 1986;261:1760.

311. Miller DK, Griffiths E, Lenard J, Firestone RA. Cell killing by lysosomotropic detergents. *J Cell Biol.* 1983;97:1841.

312. Diener E, Diner UE, Sinha A, Xie S, Vergidis R. Specific immune suppression by immunotoxins containing daunomycin. *Science.* 1986;231:148.

313. Gedigk P, Totovic V. Lysosomes and lipopigments. In: Trump BF, Laufer A, Jones RT, eds. In: *Cellular Pathobiology of Human Disease.* New York: Gustav Fisher; 1983:205.

314. Travis DF, Travis A. Ultrastructural changes in the left ventricular rat myocardial cells with age. *J Ultrastruc Res.* 1972;39:124.

315. Buktenica S, Olenick SJ, Solgia R, Frankfater A. Degradation and regurgitation of extracellular proteins by cultured mouse peritoneal macrophages and baby hamster kidney fibroblasts: kinetic evidence that the transfer of proteins to lysosomes is not irreversible. *J Biol Chem.* 1987; 262:9469.

316. Buckmaster MJ, LoBraico D, Ferris AL. Retention of pinocytosed solute by CHO cell lysosomes correlates with molecular weight. *Cell Biol Intern Reports.* 1987;11:501.

317. Marsh M, Helenius A. Adsorptive endocytosis of Semliki–Forest virus. *J Mol Biol.* 1980;142: 439.

318. Anderson KB, Nexo BA, Entry of murine retrovirus into mouse fibroblasts. *Virology.* 1983;125: 85.

319. Marsh M. The entry of enveloped viruses into cells by endocytosis. *Biochem J.* 1984;218:1.

320. Nogueira N, Gordon S, Cohn ZA. *Trypanosoma cruzi*: modification of macrophage function during infection. *J Exp Med.* 1978;146:157.

321. Marsh M, Wellsteed J, Kern H, Harms E, Helenius A. Monensin inhibits Semliki Forest virus penetration into cultured cells. *Proc Natl Acad Sci USA.* 1982;79:5297.

322. Karakashian SJ, Rudzinska MA. Inhibition of lysosomal fusion with symbiont-containing vacuoles in *Paramecium bursaria. Exp Cell Res.* 1981; 131:387.

323. Horwitz MA. The Legionnaire's disease bacterium (*Legionella pneumophila*) inhibits phagosome–lysosome fusion in human monocytes. *J Exp Med.* 1983;158:2108.

324. Horwitz MA, Maxfield FA. *Legionella pneumophila* inhibits acidification of its phagosome in human monocytes. *J Cell Biol.* 1984;99:1936.

325. Black CM, Paliesheskey M, Beaman BL, Donovan RM, Goldstein E. Acidification of phagosomes in murine macrophages: blockage by *Nocardia asteroides. J Infect Dis.* 1986;154:952.

326. Frehel C, Rastogi N. *Myocbacterium leprae* surface components intervene in the early phagosome–lysosome fusion inhibition event. *Infect Immun.* 1987;55:2916.

327. Chicurel M, Garcia E, Goodsaid F. Modulation of lysosomal pH by *Mycobacterium tuberculosis*–derived proteins. *Infect Immun.* 1988;56:479.

328. Hackstadt T, Williams JC. pH dependence of the *Coxiella burnetii* glutamate transport system. *J Bacteriol.* 1983;154:598.

329. Mukkada AJ, Meade JC, Glaser TA, Bonventre PF. Enhanced metabolism of *Leishmania donovani* at acid pH: an adaptation for intracellular growth. *Science.* 1985;229:1099.

330. Merion M, Schlesinger P, Brooks RM, Mohering JM, Mohering TJ, Sly WS. Defective acidification of endosomes in Chinese hamster ovary cell mutants "cross-resistant" to toxins and viruses. *Proc Natl Acad Sci USA.* 1983;80:5315.

331. Timchak LM, Kruse F, Marnell MH, Draper RK: A thermosensitive lesion in a Chinese hamster cell mutant causing differential effects on the acidification of endosomes and lysosomes. *J Biol Chem.* 1986;261:14154.

332. Cain CC, Murphy RF. A chloroquine-resistant Swiss 3T3 cell line with a defect in late endocytic acidification. *J Cell Biol.* 1988;106:269.

7

Biochemical Mechanisms of Oxidative Cell Injury in Isolated Hepatocytes

Margo M. Moore
Pierluigi Nicotera
Sten Orrenius

IN RECENT YEARS, THE FIELD OF TOXI-cology has been increasingly concerned with the relevance of activated metabolic intermediates, especially in the form of free radical compounds, to the cell injury and death produced by a variety of xenobiotics. Our laboratory has primarily employed the isolated hepatocyte model to study these biochemical alterations, with an emphasis on determining the critical event(s) that irreversibly damage cell function.[1] In particular, the role of the glutathione system in preventing hepatotoxicity has been a primary focus of our research. In this chapter, we shall attempt to provide a general overview of the mechanisms that are currently thought to be important in causing oxidative cell injury and discuss the involvement of the glutathione system using examples from our own and others' work in this field. The broad scope of this report precludes an exhaustive overview of the literature, and further information regarding the individual subjects can be obtained from many detailed reviews.

REACTIVE INTERMEDIATES AND THE PRODUCTION OF OXIDATIVE STRESS

Metabolism of Lipophilic Compounds

The oxidative metabolism of endogenous and exogenous lipid-soluble compounds is primarily performed in the liver by the cytochrome P_{450} monooxygenase system present in the endoplasmic reticulum (ER) (for reviews see refs. 2 and 3). Briefly, after binding of the substrate to the hemoprotein, reducing equivalents are transferred from the cofactor, NADPH, to the heme iron of the cytochrome P_{450} by NADPH-cytochrome P_{450} reductase. Molecular oxygen is then bound to the reduced heme iron, and the second electron from NADPH is transferred to O_2, producing reduced "active oxygen." One atom of the bound O_2 is combined with $2H^+$ to form H_2O, while the other is inserted into the bound substrate molecule. This addition of oxygen to lipophilic substrates increases their hydrophilic-

179

ity, thus rendering them more suitable for excretion (so-called Phase I reactions). Water solubility can be further enhanced by subsequent conjugation with sulfate, glucuronate, or glutathione (Phase II reactions).

The products of cytochrome P_{450} metabolism are not always less toxic than the parent compound. For example, the one-electron transfer processes during monooxygenation can produce highly reactive free radical intermediates (i.e., metabolites possessing an unpaired electron) that can interact with intracellular components. If the reactive intermediates can regenerate the parent compound by interaction with intracellular electron acceptors (e.g., O_2), a continuous process of oxidation and reduction will occur, resulting in the progressive oxidation of cellular molecules—an event known as *redox cycling. Oxidative stress* is a term that has been used to describe this process in which the generation of oxygen radicals and the oxidation of critical chemical species, for example, membrane lipids and sulfhydryl groups of proteins, can lead to functional impairment and, ultimately, cell death.

Although the microsomal cytochrome P_{450} monooxygenase is the most well-characterized system, other enzymes within the ER and mitochondria can also produce reactive intermediates during the metabolism of lipid-soluble compounds. For example, results from our laboratory and others have shown that one electron reduction of a quinone, menadione (2-methyl-1,4-naphthoquinone), occurs with a variety of flavoprotein reductases, including microsomal NADPH-cytochrome P_{450} reductase and mitochondrial NADH-ubiquinone oxidoreductase.[4-6] NADH-cytochrome b_5 reductase, a microsomal enzyme that can catalyze the oxidative desaturation of fatty acids, can also participate in the one-electron reduction of certain drug substrates.[7,8] In addition, peroxidases, such as prostaglandin synthase and horseradish peroxidase, have also been shown to generate reactive intermediates during the oxidation of various xenobiotics.[9-11] This reaction will be quanti-

tatively more important in tissues containing a high concentration of these enzymes, such as the kidney.[12]

In summary, the cell possesses a variety of oxidative enzymes that can generate electrophilic intermediates. The proportion of the parent molecule that is converted to a potentially toxic metabolite is dependent on a variety of conditions, including the type of substrate and its concentration within the cell. The availability of the conjugating species, sulfate, glucuronate and, as will be discussed in a later section, glutathione, are also critical determinants of the final intracellular concentration of the "activated" intermediates.

Active O_2 Species

As stated earlier, compounds possessing an unpaired electron generally have a high chemical reactivity and can directly initiate structural damage by binding to cell components. In addition, the reactive intermediates may also reduce O_2, to yield superoxide radical, O_2^-, which itself can undergo further reduction to potentially toxic products (Figure 7-1). The reaction of free radical metabolites with molecular oxygen is favored for at least two reasons. First, O_2 is present within the cytosol and ER at concentrations comparable to those found in the extracellular space (≈ 20 mm Hg).[13] (Note that a gradient does exist, however, from the extracellular space to the inner mitochondrial membrane, under conditions of low P_{O_2}.[13]) In addition, O_2 will preferentially partition into the lipid bilayer, thereby creating a constantly replenished supply available to the products of membrane-bound oxidative enzyme systems. The feature of O_2 that determines its reactivity is its electronic configuration. Molecular oxygen is a diradical because the two unpaired electrons in the $*\pi$ antibonding orbitals have parallel spins; since electrons in atomic or molecular orbitals must have opposing spins (the Pauli exclusion principle), electrons

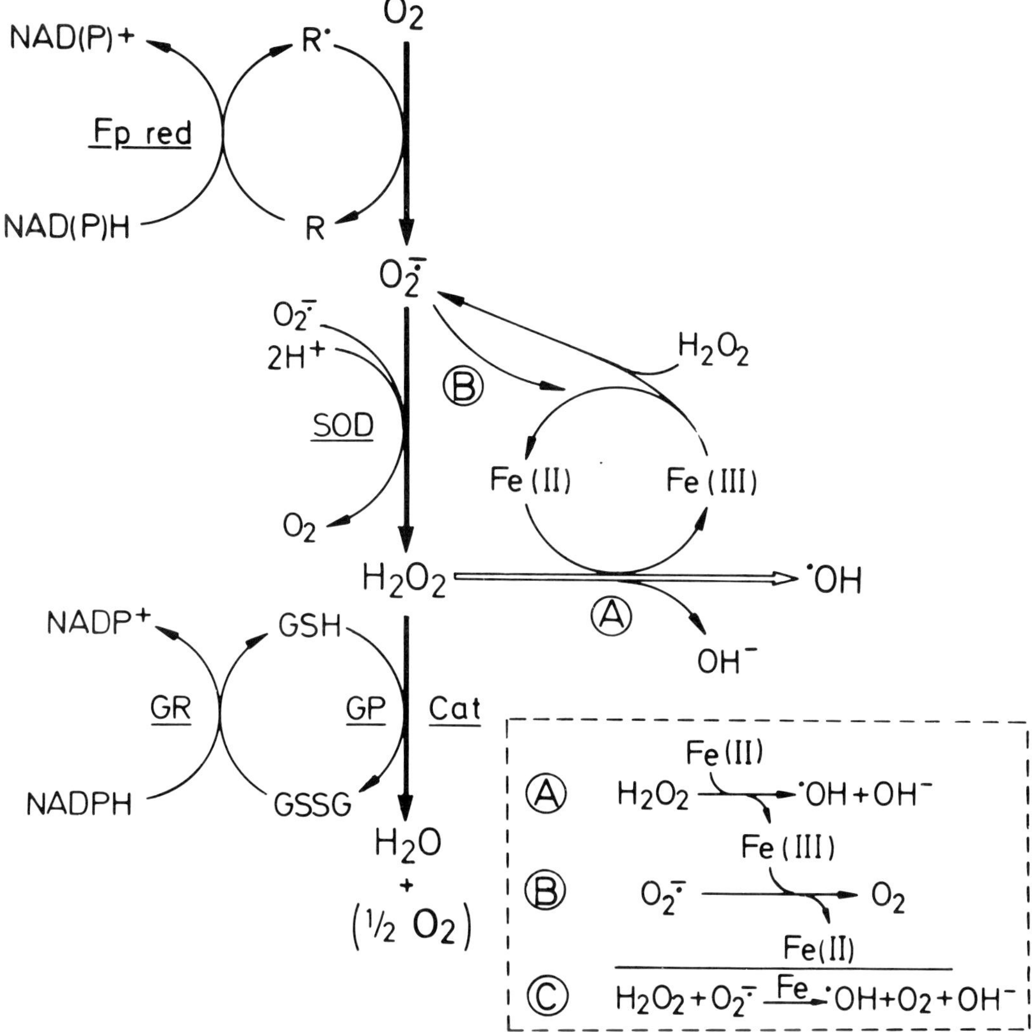

Figure 7-1. The production of reactive oxygen species by electrophilic intermediates. A = Fenton reaction; B = O_2^- reduction of Fe (III); C = iron-catalyzed $\cdot OH$ production; Fp = flavoprotein reductases; SOD = superoxide dismutase; GP = glutathione peroxidase; GR = glutathione reductase; Cat = catalase; R = xenobiotic.

must be transferred to oxygen singly. Consequently, oxygen can readily accept the unpaired electron from a radical intermediate yielding O_2^- and the parent molecule, thereby setting up conditions for redox cycling.

Under physiological conditions, O_2^- spontaneously dismutates to form hydrogen peroxide, H_2O_2. Both O_2^- and H_2O_2 are generated during normal cell metabolism, and

mechanisms exist for their removal. The superoxide dismutases, present in both the cytosol (CuZnSOD) and the mitochondria (MnSOD), can accelerate the spontaneous process by a factor of approximately 10^{10}, according to the calculations of Fridovich.[14] The H_2O_2 produced by dismutation can then be metabolized either by catalase or the selenoprotein glutathione peroxidase. Catalase, present in peroxisomes, can metabolize only H_2O_2, whereas glutathione peroxidase can also metabolize organic hydroperoxides (including lipid hydroperoxides) (for review, see ref. 15). Glutathione peroxidase can reduce hydroperoxides at the expense of glutathione (GSH), forming glutathione disulfide (GSSG), which can be re-reduced by the enzyme glutathione reductase, which consumes NADPH as a cofactor.

The relative importance of catalase *versus* glutathione peroxidase in the catabolism of H_2O_2 has been investigated in our laboratory, and the results have demonstrated that glutathione peroxidase is more important in the metabolism of endogenously generated H_2O_2 at low rates of H_2O_2 production. As the H_2O_2 level rises, the action of catalase becomes more important.[16] H_2O_2 can also participate in a transition metal–catalyzed (especially Cu(I) or Fe(II)) conversion to the hydroxyl radical, ^-OH, via a process that is known as the Fenton reaction:

$$Fe^{2+} + H_2O_2 \rightarrow Fe^{3+} + \cdot OH + OH^-$$

The additive effects of superoxide dismutase and catalase administration in reducing oxygen-mediated toxicity has led Fridovich to postulate that a direct interaction of H_2O_2 and O_2^- could also generate $\cdot OH$, a reaction first described by Haber and Weiss in 1934. More recently, the mechanism of $\cdot OH$ production has been modified to include catalysis by transition metals, as the direct rate of interaction of these two oxygen species is thought to be negligible under physiological conditions (see Figure 7-1).[17]

The relative toxicities of O_2^-, H_2O_2, and $\cdot OH$ have been primarily investigated by using "specific" scavengers for a particular radical or protection by enzymes such as superoxide dismutase and/or catalase. There is evidence that all three species are able to produce some toxicity, although the effects of H_2O_2 are thought to be primarily due to the generation of hydroxyl radical.[15] For example, DiGiuseppi and Fridovich have demonstrated that SOD prevented O_2-dependent quinone toxicity in a bacterial cell culture.[18] Since the hydroxyl radical scavenger, dimethylsulfoxide, was unable to protect the cells, they concluded that $\cdot OH$ was not involved and O_2^- was directly responsible for the toxic effects. This conclusion is based on the assumption that any H_2O_2 formed did not itself contribute significantly to the toxicity, an assumption that is borne out by experimental evidence. The inability of H_2O_2 *per se* to inhibit mitochondrial respiratory control or initiate lipid peroxidation has reinforced the idea that it is the iron (II)-catalyzed $\cdot OH$ formation that makes the production of H_2O_2 deleterious to the cell.[19,20] The pathways of active oxygen generation and removal are depicted in Figure 7-1. Note that there are several other possible reactions among the radical species that have not been included for the sake of simplicity.

Although $\cdot OH$ has been implicated as the ultimate toxic oxygen species in a number of experimental systems because of its ability to initiate lipid peroxidation, there is evidence that another highly reactive oxygen species, singlet oxygen (designated 1_{O_2} or $O_2{}^1\Delta g$), may also be able to contribute to oxidative cell injury. Exposing a biological pigment to light in the presence of O_2 can cause a reversal of one of the spins of the unpaired electrons of molecular oxygen, thereby generating 1_{O_2}. Singlet oxygen has also been shown to be produced from cytochrome P_{450} metabolism of ROOH to ROH by chemiluminescence in the presence of microsomes and oxene donors, and indirect evidence has demonstrated that 1_{O_2} can have toxic effects.[21] Lynch and Fridovich have observed protection using the 1_{O_2} scavenger, histidine, against the autoinactivation of xanthine oxidase during acetaldehyde metabolism.[22]

In certain experimental systems, it has been difficult to attribute the toxicity of an agent to a particular reactive O_2 species. Preventing toxicity by superoxide dismutase or catalase does not necessarily imply that the toxicity is a direct result of either O_2^- or H_2O_2 but, as stated previously, may be due to other active oxygen species produced from their interaction. Furthermore, the scavengers employed are only relatively specific for one chemical species, and caution should be exercised in interpreting the results of studies employing these agents.

In summary, electrophilic products of xenobiotic metabolism are able to initiate cellular damage by several mechanisms: by direct arylation or alkylation of cellular macromolecules by the reactive intermediates themselves; by redox cycling with O_2 to produce O_2^-, which can, in turn, form other cytotoxic oxygen species; and by direct oxidation of cellular components by the active intermediate. The relative importance of these processes in the hepatotoxic effects of many xenobiotics has been the subject of much debate in this field. The importance of the glutathione system in protecting against the oxidative stress produced by all of these processes supports the critical role of the glutathione status in the development of cytotoxicity.

GLUTATHIONE AND DRUG METABOLISM

Mechanisms of GSH Depletion during Oxidative Stress

The concentration of glutathione in mammalian cells is in the millimolar range (0.5 – 10 mmol), and it consists primarily of GSH, although a small fraction exists in the form of mixed disulfides (mostly GSS-protein) and, to a lesser extent, GSSG (for review of glutathione synthesis and metabolism, see ref. 23). A prominent feature of oxidative metabolism is the depletion of soluble and protein-bound thiols. Figure 7-2 summarizes the various pathways of glutathione depletion induced by the production of reactive metabolites.

Electrophilic intermediates can be conjugated with GSH to form the S-conjugate directly or in reactions catalyzed by glutathione S-transferases—a family of enzymes with broad substrate specificity.[24] Conjugation with GSH is considered a detoxication

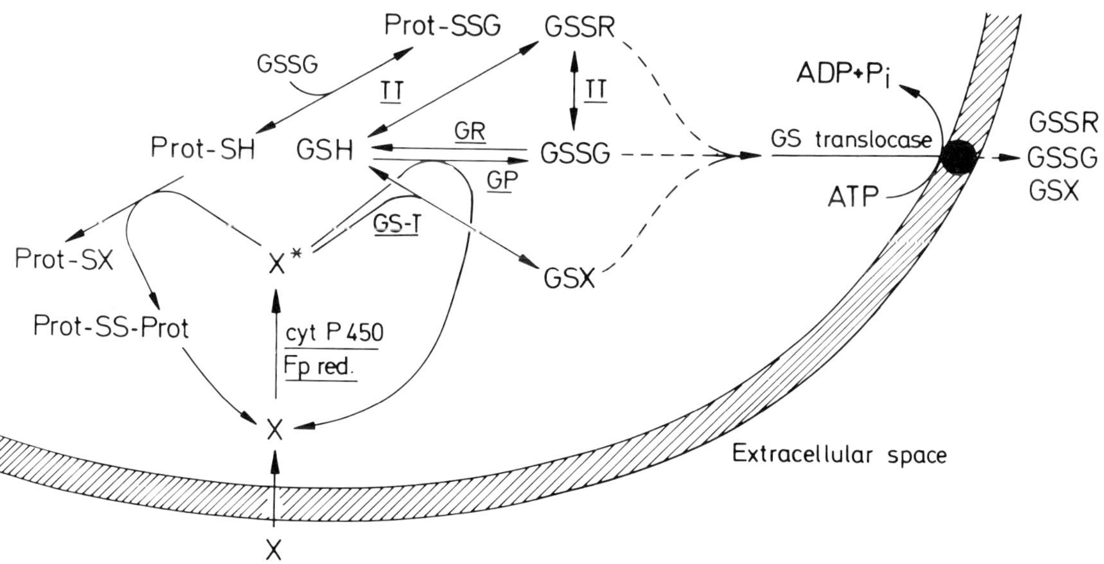

Figure 7-2. Glutathione metabolism during oxidative stress. GSSR = glutathione mixed disulfides; GSX = glutathione S-conjugates; Prot-SSG = protein glutathione disulfide; Prot-SX = arylated/alkylated protein; GS-T = glutathione S-transferase; TT = thiol transferases.

process; however, there is now evidence that some S-conjugates can also contribute to redox cycling.[25] Thus, the conjugate itself may not always be less toxic than the metabolic intermediate. However, the conjugate is more readily excreted from the cell, and so its ability to redox cycle is diminished by its extrusion.

Direct oxidation of GSH by free radical intermediates to form GSSG has been shown to occur:

$$GSH + X^{\cdot} \rightarrow GS^{\cdot} + X$$
$$2\ GS^{\cdot} \rightarrow GSSG$$

This postulates the existence of the thiyl radical, $GS^{\cdot}$. Although the formation of this thiyl radical by the interaction of GSH with transition metals has long been known, the existence of this radical has only recently been directly demonstrated in biological systems.[26] $GS^{\cdot}$ formation during horseradish peroxidase- or prostaglandin synthase-mediated acetaminophen metabolism has been detected by Electron Paramagnetic Resonance (EPR) spectroscopy.[27] Thus, some free radical intermediates can be directly reduced back to the parent molecule by GSH by a mechanism analogous to their interaction with molecular oxygen. Although it has been shown to occur *in vitro*, it is presently unknown whether $GS^{\cdot}$ interaction with O_2 to produce oxygen-containing products of GSH is a biologically relevant phenomenon.[28]

Aside from dimerization of the thiyl radical, GSSG can also be generated during the glutathione peroxidase-mediated reduction of inorganic and organic peroxides (for review, see ref. 29). The overall reaction is summarized in Figure 7-1.

In addition to the metabolism of both H_2O_2 and ROOH by glutathione peroxidase, a selenium-independent glutathione peroxidase activity that can utilize only ROOH as a substrate has been identified by Lawrence and Burk and was subsequently shown to be a property of the glutathione S-transferases.[30] Glutathione peroxidase activity can be followed by the peroxide-in-duced GSSG efflux from the cell or the NADPH oxidation that is due to the re-reduction of GSSG back to GSH (see Figure 7-2). It was previously thought that low NADPH levels directly triggered the efflux of GSSG from the cell; however, recent evidence from our laboratory has demonstrated that it is the intracellular GSSG accumulation rather than NADPH oxidation that stimulates the efflux of the disulfide.[31]

The ATP-dependent extrusion of GSSG was first demonstrated in the eye lens and later in perfused liver and isolated hepatocytes.[32-34] Recently, we identified and characterized[34,35] an ATPase activity associated with the hepatocyte plasma membrane that is stimulated by micromolar concentrations of GSSG as well as by various glutathione S-conjugates.[34-36] In a more recent study, we have also shown that transport of GSSG and glutathione conjugates in liver plasma membranes is stimulated by ATP, suggesting that the stimulated ATPase activity may be responsible for the active efflux of GSSG observed in hepatocytes under conditions of oxidative stress.[36]

The rapid removal of GSSG and glutathione conjugates would be beneficial to the cell as high intracellular levels of GSSG may contribute to the oxidation of protein thiols (see Figure 7-2). It also provides a mechanism for the removal of the glutathione conjugates of electrophilic metabolites which, as noted previously, may also contribute to redox cycling. However, GSSG and glutathione conjugate efflux will also result in a depletion of substrate and, after prolonged oxidation, the resynthesis of glutathione from its constituent amino acids would be necessary. This may explain in part the protective effect of administration of glutathione precursors to cells exposed to "oxidative stress."

HEPATOTOXICITY MEDIATED BY REACTIVE INTERMEDIATE PRODUCTION

Bromobenzene and Acetaminophen

The covalent binding of certain carcinogens to tissue protein was first observed in the 1940s and 1950s, and the similar binding pattern of a variety of hepatotoxins to liver protein led to the development of a theory relating the cytotoxicity of an agent to the extent of its protein binding.[37,38] Such a mechanism was initially postulated as the basis of cytotoxicity caused by the halogenated hydrocarbon, bromobenzene, and the analgesic drug, acetaminophen.

Bromobenzene is metabolized by microsomal cytochrome P_{450} to a variety of metabolites of which the highly reactive 3,4-epoxide is thought to be the most toxic.[39] Contribution of other metabolic products, the bromophenols, to bromobenzene-induced liver necrosis is probably less important.[39,40] ^{14}C-labeled metabolites of bromobenzene have been shown to bind to protein, and pretreatments that altered the amount of covalently bound metabolites correlated with the severity of tissue damage.[41-43] Results from our laboratory have shown that the increase in covalently bound metabolites was preceded by a loss of soluble GSH due to conjugation of glutathione with the reactive epoxide.[44] Bromobenzene metabolite binding also decreased the activities of a number of enzymes, including glucose-6-phosphate dehydrogenase and glutathione reductase, presumably by arylating free SH-groups required for enzyme activity.[45]

Redox cycling has not been shown to be important in bromobenzene toxicity. Studies in our laboratory have established that bromobenzene does not stimulate O_2 uptake or affect mitochondrial function during the early phase of development of toxicity in hepatocytes.[45] In addition, no lipid peroxidation, an indirect measure of ·OH production, was induced by incubation of hepatocytes with bromobenzene.[46] In contrast, treatment of mice with bromobenzene has recently been shown to produce both lipid peroxidation and liver necrosis; however, lipid peroxidation was only in part correlated with the onset of liver necrosis. A dual mode for bromobenzene toxicity was therefore proposed in this study: one involving the peroxidative decomposition of membrane lipids and the other involving a disruption of thiol and Ca^{2+} homeostasis.[47]

Similarly, the metabolism of acetaminophen by cytochrome P_{450} produces an intermediate that is thought to be responsible for the observed hepatotoxicity.[48] It appears now that the toxicity of acetaminophen is mediated by N-acetyl-p-benzoquinone imine (NAPQI).[49,50] Indeed, administration of the quinone imine itself has been shown to produce toxicity sufficient to account for the hepatic necrosis caused by acetaminophen overdose.[51-53] Early investigations with acetaminophen demonstrated that, as with bromobenzene, the extent of covalently bound drug was affected by the degree of cytochrome P_{450} metabolism and that this was closely associated with the level of hepatotoxicity.[54] GSH and protein thiol depletion was also observed after incubation of hepatocytes with either acetaminophen or NAPQI.[53] As NAPQI can directly oxidize GSH and NADPH to produce GSSG, the observed decrease in GSH is produced by both oxidation and conjugation.[55,52] Moreover, thiol depletion always precedes cytotoxicity, demonstrating a close association between thiol redox status and hepatotoxicity. To investigate the relative contribution of arylation and oxidation to NAPQI-induced cell killing, the effects of two analogues of NAPQI, the 3,5-dimethyl-NAPQI, which primarily oxidizes thiols, and the 2,6-dimethyl-NAPQI, which primarily binds to thiols, have recently been evaluated in isolated hepatocytes.[55] Results from this study have shown that both analogues can cause a disruption of intracellular thiol homeostasis and cell death. However, the arylating analogue, 2,6-dimethyl-NAPQI, was significantly more efficient than the oxidative analogue in causing thiol depletion and cell killing, suggesting that arylation of protein thiols is the most important mechanism involved in NAPQI cytotoxicity.[56]

The importance of oxygen radical formation in acetaminophen toxicity is presently a topic of debate. Lipid peroxidation has been observed by some authors after acetaminophen administration to experimental animals; however, this may be a later event that is not associated with the initiation of cell death. For example, studies with the putative metabolic intermediate, NAPQI, have not revealed any appreciable redox cycling with O_2 (M. Moore and H. Thor, unpublished observations), and the cytotoxicity of NAPQI has been dissociated from lipid peroxidation.[50,51,57] Thus, the toxicity of acetaminophen is less likely to be due to the formation of active oxygen species than to the arylation and/or oxidation of cellular protein. A summary of the effects of acetaminophen and bromobenzene on isolated hepatocytes is shown in Table 7-1.

Menadione and Diquat

Two other well-studied compounds that are metabolized by the cytochrome P_{450} system to reactive intermediates are the quinone, menadione, and the bipyridilium herbicide, diquat (1,1'-ethylene,-2,2'-bipyridilium dichloride). Menadione can be metabolized by one-electron reduction by NADPH-cytochrome P_{450} reductase ER or NADH-ubiquinone oxidoreductase (mitochondria) to produce the semiquinone radical.[5,58] This radical has been shown to redox cycle extensively with molecular oxygen, as demonstrated by the marked stimulation of O_2 consumption. In addition to redox cycling, a certain proportion of the semiquinone radical can also be bound to protein thiols, although this constitutes a comparatively small percentage of the loss of cellular sulfhydryl groups ($\simeq 20\%$), suggesting that the observed loss of protein thiols is primarily the result of oxidation rather than covalent binding.[59] Nevertheless, in contrast to bromobenzene and acetaminophen, menadione can initiate oxidative stress both by redox cycling and by covalent binding to cellular proteins.

The herbicide diquat, a compound closely related to paraquat, can also undergo one-electron reduction by flavoenzymes to form stable cation radicals that can interact with O_2 to produce O_2^-, a feature that is thought to be essential for its herbicidal activity.[60] This agent can be considered a "pure" redox cycling compound because the relatively stable diquat radical ($DQ^{\cdot+}$) has not been shown to bind to macromolecules in mammalian cells.[61] The mechanism of toxicity of diquat is thought to be analogous to its herbicidal action and, therefore, dependent on the production of the superoxide radical.[61,62]

Evidence for the formation of toxic oxygen species by both menadione and diquat comes from experiments in which $\cdot OH$ formation by these agents was demonstrated by dimethylsulfoxide demethylation as well as by inhibition of the resulting lipid peroxidation by the iron chelator, desferrioxamine.[64]

TABLE 7-1
The Effects of Various Hepatotoxins on Isolated Hepatocytes

Agent	Covalent Binding	O_2 Consumption	GSH Depletion	Protein SH Depletion	Protection by DTT[a]
Bromobenzene	+++	−	++	++	−
Acetaminophen	++	−	++	++	++
Menadione	+	++[b]	+++	++	++
Diquat	−	++	+	(+)	+

[a]DTT protection is defined as a reduction in cell death induced by drug administration. DTT concentration = 5 mmol.

[b]Menadione can redox cycle in both the cytosolic and mitochondrial compartments. Diquat can redox cycle only in the cytosol.

The greater cytotoxic potency of menadione, when compared to diquat, may be attributable in part to the ability of menadione to redox cycle within the mitochondria, thereby producing a disruption of mitochondrial function, especially the release of CA^{2+} (see "Thiol Oxidation/Arylation and Disruption of Ca^{2+} Homeostasis"). Diquat cannot penetrate the mitochondrial membrane, and this restriction of its redox cycling to the cytosolic compartment may explain its lesser toxicity when compared to menadione.[64,65]

A comparative scheme of the effects of these four agents on isolated hepatocytes is presented in Table 7-1. All four agents produce cytotoxicity, although the mechanisms are not precisely the same. For example, bromobenzene and acetaminophen and, to a lesser extent, menadione all bind to cellular proteins. However, despite the attractiveness of the covalent binding theory, objections have been raised against the existence of a direct association of drug binding and hepatic injury. Monks et al. have speculated that a proportion of the covalent binding of bromobenzene intermediates may be due to binding of less toxic metabolites (such as p-bromophenol).[39] Nevertheless, covalent binding to cellular protein remains the most convincing explanation for bromobenzene-induced cell damage.

There is still debate about the applicability of this theory to acetaminophen-induced hepatotoxicity. Results from several laboratories have indicated a dissociation between the extent of covalent binding and cell death.[66-68] If covalent binding is not important in acetaminophen cytotoxicity, then, in view of its poor redox cycling ability, another plausible mechanism could be the oxidation of cellular nucleophiles.

In fact, the ability of this agent to oxidize both soluble and protein-bound SH-groups has led us and others to suggest that protein thiol oxidation may trigger irreversible hepatocyte injury.[52,53] However, this does not exclude the possible involvement of protein arylation in the mechanism of action of acetaminophen. More recent studies in our and other laboratories have, in fact, shown that oxidation contributes to a lesser extent than arylation to the toxicity of either NAPQI or paracetamol.[56,69] That the overall degree of covalent binding does not necessarily correlate with the extent of drug-induced liver necrosis can be explained by the fact that only a small number of the complexed sites may be critical for the initiation of cytotoxicity. Quinones also appear to exert their toxic effects through arylation and/or oxidation of thiols in a fashion similar to that observed with NAPQI. In a recent study comparing different substituted benzoquinone analogues, we have shown that arylating analogues are more cytotoxic than purely oxidizing analogues.[70]

In contrast to bromobenzene and acetaminophen, menadione and diquat are also able to stimulate O_2 consumption by redox cycling. For menadione, the enhanced H_2O_2 production results in GSH depletion due to glutathione peroxidase activity; GSH conjugation also contributes to the loss of reduced glutathione. On the other hand, diquat-induced GSH consumption appears to be entirely due to GSSG formation coupled to glutathione peroxidase-catalyzed reduction of H_2O_2 and lipid hydroperoxides.[64]

As mentioned previously, redox cycling may result in the production of ·OH which can initiate the chain of events leading to lipid peroxidation and membrane damage. The classical mechanism proposed for lipid peroxidation is as follows: A sufficiently reactive radical (not O_2^-) abstracts a hydrogen atom from the lipid molecule, and the resulting carbon-centered radical stabilizes by forming a diene; interaction of the diene with molecular oxygen forms the hydroperoxy-radical, and lipid hydroperoxides plus more radical species are generated in the presence of transition metals. Thus, once initiated, the cycle is self-sustaining. As stated earlier, glutathione depletion can be shown to result from the metabolism of the lipid hydroperoxides by glutathione peroxidase.[64]

The hydroxyl radical is commonly thought to be responsible for the initiation of lipid peroxidation during oxidative stress; however, demonstration of microsomal lipid peroxidation, which is insensitive to hydroxyl radical scavengers, by our laboratory in collaboration with Ernster et al., and by others, has induced speculation regarding other possible initiators of lipid peroxidation.[69-71] We, as well as Aust and Svingen, have suggested that the perferryl radical may be involved, although Halliwell has suggested that the low reactivity of this complex makes the ferryl radical a more likely candidate.[17,71-73]

Regardless of which radical initiates the peroxidation of membrane lipids, their formation could be expected to produce cell death by a variety of mechanisms. First, destabilization of the membrane may cause an increase in cell permeability *per se* or, alternatively, may inhibit critical membrane-associated enzymes by disrupting the annular lipid structure. The breakdown products of lipid hydroperoxides are also known to be toxic: The most familiar compound is malondialdehyde, the measurement of which forms the basis of the most simple (albeit indirect) method of monitoring the extent of lipid peroxidation.[74] Malondialdehyde may be metabolized by aldehyde dehydrogenases or may cause the formation of Schiff bases that can cross-link (and thereby inactivate) enzymes.[75]

The identification of other toxic products of lipid peroxidation by Benedetti and co-workers has provided an alternative mechanism whereby lipid hydroperoxides may cause cytotoxicity. These compounds, the 4-hydroxyalkenals, have been isolated from peroxidized microsomal lipids, and both the endogenous and synthetic alkenals have been shown to induce a variety of cytotoxic events, including erythrocyte lysis and inhibition of glucose-6-phosphatase activity.[76] In addition, they can interact with low molecular weight as well as protein-bound SH-groups, and it is this mechanism that is postulated to account for the observed inhibition of microsomal Ca^{2+} uptake by the 4-hydroxyalkenals.[76]

However, whether lipid peroxidation *per se* can initiate oxidative cell death is presently in doubt. We have observed that high levels of lipid peroxidation in control hepatocyte incubations produce no concomitant increase in cell death. The protective effect of the iron chelator, desferrioxamine, on diquat-induced cytotoxicity seems to indicate that ·OH is responsible for its toxicity. However, the salutary effects of desferrioxamine are not observed when hepatocytes are incubated with toxic doses of menadione, and we have previously observed that several radical scavengers (e.g., dimethylsulfoxide, thiourea, alpha-keto-gamma-methiolbutyric acid) offer scant protection against menadione-induced cytotoxicity.[79] Thus, GSH and protein thiol depletion, as well as mitochondrial redox cycling, may be more important for the hepatotoxicity of the quinone. In contrast, desferrioxamine invariably prevents diquat cytotoxicity, and radical scavengers are able to reduce the extent of cell death to varying degrees, suggesting that reactive oxygen species are critically involved in initiating the toxicity of diquat. Whether it is the lipid peroxidation *per se* that is responsible for the cell death due to diquat is unclear. Shu and co-workers examined paraquat toxicity and found that although lipid peroxidation of mouse lung microsomes could be prevented by DPPD (N,N'-diphenyl-*p*-phenylenediamine) or high carbohydrate diet, the lethal effects were not prevented.[80] In addition, conjugated diene formation was not elevated *in vivo* even after administration of a dose of paraquat twice the LD_{50}, from which they concluded that the *in vitro* effects of the bipyridyl herbicides may be unrelated to the toxic effects *in vivo*.[80] Thus, the toxicity of diquat appears to be related to ·OH formation, but it may involve other mechanisms in addition to lipid peroxidation.[81,82]

Despite the variation in covalent binding and redox cycling, a common feature of the agents just discussed is the depletion of both soluble and protein-bound SH-groups. Furthermore, the cytoprotective effect of agents that elevate the intracellular glutathione pool (such as methionine, N-acetylcysteine) or act as thiol reductants (e.g., dithiothreitol, dithioerythritol, beta-mercaptoethanol) as well as the enhancement of cytotoxicity by pretreatments that deplete the glutathione pool (e.g., diethylmaleate or glutathione reductase inhibition) has been observed for all of the above agents; bromobenzene, acetaminophen, menadione, and diquat (see Table 7-1).[45,83-87] We have demonstrated that the loss of glutathione preceded the loss of protein-bound SH-groups after incubation of hepatocytes with either menadione or NAPQI, suggesting that GSH may play a role in maintaining the reduced state of protein thiols.[53,59] Furthermore, the loss of protein thiol groups due to oxidation and/or arylation was closely associated with the onset of toxicity for both of these agents. Considering the important role played by SH-groups in a variety of enzymes, both structurally and as part of the catalytic site (for review see ref. 88), any interference with protein SH-groups would be expected to have dramatic consequences on cellular function. Evidence from our laboratory suggests that one of the critical alterations produced by thiol depletion is the disruption of hepatocyte Ca^{2+} homeostasis which results in an uncontrolled sustained increase in cytosolic Ca^{2+} concentration.

THIOL OXIDATION/ARYLATION AND DISRUPTION OF Ca^{2+} HOMEOSTASIS

The level of cytosolic Ca^{2+} is maintained at a very low concentration ($\approx 0.2\ \mu$mol) against an extracellular concentration of ≈ 1.3 mmol. This is achieved by several different mechanisms: Uptake of Ca^{2+} into mitochondria by an energy-dependent process, Ca^{2+} sequestration by the ER via a Ca^{2+}-ATPase activity, and extrusion of Ca^{2+} from the cell by a distinct plasma membrane Ca^{2+}-translocase. Binding to intracellular proteins (such as calmodulin) also buffers the cytosolic Ca^{2+} concentration. We have shown that thiol depletion can affect all of these processes and that this can lead to a rise in the level of free Ca^{2+} within the cytosol (see following). Evidence of an increase in cytosolic Ca^{2+} has been obtained by the use of several techniques: (1) sequential release of intracellular Ca^{2+} pools by the mitochondrial uncoupler carbonyl-cyanide-*m*-chlorophenylhydrazone (CCCP) and the Ca^{2+} ionophore (A23187) in the presence of the Ca^{2+}-sensitive dye, arsenazo III; (2) measurement of phosphorylase activation by Ca^{2+}-dependent phosphorylase *b* kinase, a process sensitive to micromolar concentrations of Ca^{2+} and; (3) fluorescence changes of other Ca^{2+}-sensitive indicators, such as Quin 2-AM and Fura 2-AM.[89-91]

We have observed that thiol depletion always precedes the disruption of Ca^{2+} homeostasis and this, in turn, always occurs prior to cell death. The rise in the free Ca^{2+} concentration is therefore closely associated with the onset of cell death, possibly due to an irreversible activation of degradative processes. (see "Possible Mechanisms of Oxidant-Induced Cell Killing"). A more detailed account of the effects of thiol depletion on the various Ca^{2+} pools follows.

Microsomes

ATP-dependent Ca^{2+} sequestration in liver microsomes was first described by Moore et al. and has been studied extensively in the following years because of its postulated role in regulating the physiological Ca^{2+} concentration.[92-94] Microsomes have a fairly low capacity to accumulate Ca^{2+} as compared to mitochondria, although the release of the ER pool has been shown to mediate the actions stimulated by alpha-adrenergic hormones.[95,96] Studies in our laboratory have demonstrated the sensitivity of microsomal Ca^{2+} uptake to thiol depletion by agents such as *t*-butylhydroperoxide, NAPQI p-chloromercuribenzoate (PCMB), and diamide.[53,97,98] Thus, Ca^{2+}-sequestration by microsomes is inhibited by alkylating (PCMB) as well as oxidizing (diamide) agents. In addition, the ability of GSH or DTT to prevent the loss of Ca^{2+} uptake induced by t-butylhydroperoxide further strengthens the association between thiol redox status and microsomal Ca^{2+} sequestration.[97]

Mitochondria

Mitochondria form a quantitatively greater pool of Ca^{2+} within the hepatocyte and have been shown to be more important in regulating cytosolic Ca^{2+} concentration compared with the ER under conditions in which plasma membrane Ca^{2+} extrusion is inhibited.[99] Ca^{2+} uptake into this organelle is dependent upon the electrical component of the proton-motive force, and release occurs via a distinct $Ca^{2+}/2H^+$ anti-porter.[100,101]

The original observation by Lehninger and associates that mitochondrial Ca^{2+} release is regulated by the oxidation state of the pyridine nucleotides, has been confirmed by other laboratories.[102–104] We have observed such an association between pyridine nucleotide oxidation and mitochondrial Ca^{2+} efflux during t-butylhydroperoxide metabolism in isolated hepatocytes.[89] Furthermore, when NADPH oxidation was prevented by selective inhibition of glutathione reductase, no loss of mitochondrial Ca^{2+} was induced by t-butylhydroperoxide.[89] Studies with NAPQI have also demonstrated the ability of this metabolite to directly cause Ca^{2+} release from isolated rat liver mitochondria, an effect that was accompanied by pyridine nucleotide oxidation and that preceded nonspecific mitochondrial membrane damage.[53] The subsequent irreversible damage to the organelle appears to be related to the continued cycling of Ca^{2+} after the efflux has occurred, as membrane damage is prevented if reuptake of Ca^{2+} is blocked by ruthenium red.[105]

Plasma Membrane

A high-affinity Mg^{2+}-dependent Ca^{2+}-ATPase activity of rat liver plasma membrane fraction was identified by Lotersztajn et al. in 1981.[106] It has optimal substrate and pH conditions that are distinct from the Ca^{2+}-ATPase of the ER membrane. However, like the microsomal Ca^{2+}-uptake system, the ATP-dependent membrane Ca^{2+} translocase is sensitive to agents that oxidize or alkylate critical thiol groups. Studies in our laboratory using diamide and t-butylhydroperoxide have demonstrated that these compounds are able to inhibit energy-dependent Ca^{2+} sequestration by rat liver plasma membrane vesicles and that DTT and glutathione are able to prevent the inhibitory effects of these agents.[107]

Subsequent experiments using a highly purified plasma membrane fraction from rat liver showed that incubation of hepatocytes with menadione (200 μmol) decreased the high-affinity Ca^{2+}-ATPase activity as compared to the control. Restoration of activity could be effected by addition of either GSH or DTT.[86] Interestingly, the pattern of enzyme inhibition was mimicked by treatment of the cells with the thiol-complexing agent, N-ethylamaleimide, suggesting that the observed inhibition of enzyme activity was the result of a loss of critical thiol group(s). Because of its high affinity for Ca^{2+}, the plasma membrane Ca^{2+}-translocase plays a major role in the maintenance of the concentration gradient between extra- and intracellular environment. Thus, inhibition of the Ca^{2+} efflux pathway can ultimately result in the accumulation of Ca^{2+} in the intracellular compartments and elevation of its cytosolic levels. Inhibition of the plasma membrane Ca^{2+}-ATPase and Ca^{2+} efflux caused by a reactive disulfide, cystamine, has recently been shown to cause intracellular accumulation of Ca^{2+} and a sustained increase of cytosolic Ca^{2+} levels. These events were preceding the appearance of morphological and biochemical index of cell death. Inhibition of the plasma membrane Ca^{2+}-ATPase has now been observed with several toxic agents, including NAPQI and its dimethylated analogues, acetaminophen, CCl_4 and the parkinsonian-inducing compound, 1 - methyl -4- phenyl -1,2,3,6 - tetrahydropyridine (MPTP).[53,56,108,109]

Cytosolic Ca²⁺

As summarized in Figure 7-3, thiol oxidation can produce a rise in intracellular Ca^{2+} concentration in a number of ways. Depletion of GSH and generation of GSSG results in the oxidation of pyridine nucleotides due to stimulation of glutathione reductase activity. (In addition, some electrophiles, for example, NAPQI, are capable of nonenzymatically oxidizing pyridine nucleotides.) A decrease in the mitochondrial ratio of NADPH/NADP⁺ favors the release of Ca^{2+} into the cytosol. Oxidation or arylation of critical thiol(s) of the microsomal Ca^{2+}-ATPase prevents reuptake into this pool. The subsequent increase in intracellular Ca^{2+} can be transiently compensated for by the extrusion of Ca^{2+} by the plasma membrane Ca^{2+} pump; however, as this enzyme is also inactivated by the modification of protein thiols, the elevation of cytosolic Ca^{2+} becomes sustained and cell death ensues.

A sustained increase in cytosolic Ca^{2+} concentration has been suggested as a final mediator in the development of oxidative cell injury.[110–112] Under physiological conditions the cytosolic Ca^{2+} concentration in hepatocytes is modulated by various hormones and other Ca^{2+}-mobilizing agents.

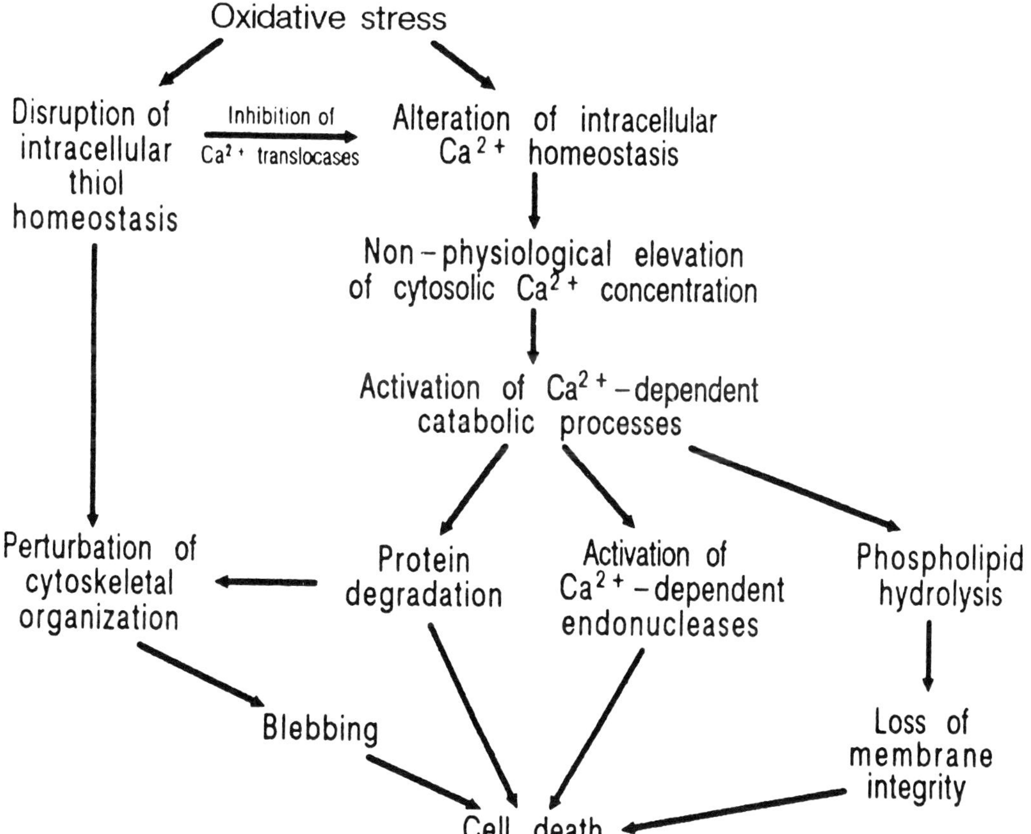

Figure 7-3. Proposed pathway of cell death mediated by thiol depletion and disruption of Ca^{2+} homeostasis.

To investigate the differences between physiological and toxicological fluctuations in cytosolic Ca^{2+} level, we recently compared the effects of an alpha-adrenergic agonist, phenylephrine, with those of both nontoxic and toxic concentrations of menadione in Quin-2-loaded hepatocytes.[113] Our study has shown that both the extent and the duration of the increase in cytosolic Ca^{2+} observed in menadione-treated hepatocytes were greater than those observed in cells stiumulated with phenylephrine. Furthermore, the thiol-reducing agent, dithiothreitol, prevented both the increase in cytosolic Ca^{2+} and toxicity in menadione-treated cells. Thus, it is apparent that oxidative stress causes an alteration of protein thiols which results in a disruption of Ca^{2+} homeostasis, and that this is critically involved in the events which lead to cell death.

POSSIBLE MECHANISMS OF OXIDANT-INDUCED CELL KILLING

Cytoskeletal Changes

The appearance of surface blebs is a prominent feature of isolated hepatocytes that have been subjected to oxidative stress. Extensive blebbing has been observed in our laboratory with bromobenzene, NAPQI, menadione and, to a lesser extent, with diquat. Moreover, other thiol oxidants, such as diamide and t-butylhydroperoxide, also produce comparable changes in surface morphology. The similarity between the xenobiotic-induced shape changes and those resulting from treatment with two classes of compounds that are known to produce alterations in cytoskeletal structure, the cytochalasins and phalloidins, prompted us to speculate that a disruption of the cytoskeleton formed the basis of blebbing.[114,115] Furthermore, the association of surface blebbing with the increase in cytosolic Ca^{2+} induced by oxidative stress is also a consistent finding, and we have observed that the

rise in intracellular Ca^{2+} levels produced by menadione, NAPQI, or the Ca^{2+}-ionophore, A23187, is closely associated with the alteration in hepatocyte morphology.

The hepatocyte cytoskeleton consists of bundles of actin filaments, associated with myosin, which can alter their degree of polymerization [filamentous (F) actin $\longleftrightarrow$ globular (monomeric) (G) actin] depending on a variety of conditions, including ionic strength and the presence of actin-binding proteins. Actin-binding proteins produce different effects on cytoskeletal structure, mostly by altering the ratio of F:G actin or by reducing the size of the F actin fragments. The list of proteins that have been shown to interact with actin is rapidly increasing (see refs. 116, 117), and some of those that have been so characterized appear to be activated by Ca^{2+}.

Ca^{2+} alone may directly activate these proteins, but there is also evidence that Ca^{2+}-calmodulin interactions can affect the polymerization state of actin.[118,119] Alternatively Ca^{2+}-activated catabolic enzymes, such as cytosolic proteases, can cleave proteins associated with the cytoskeleton and thus may cause alteration of the microfilament network and blebbing.[120-122]

The effects of diamide and N-ethylmaleimide (a thiol oxidant and SH-complexing agent, respectively) are similar to those produced by compounds that are known to increase the intracellular Ca^{2+} concentration (e.g., A23187, NAPQI, menadione). It is possible that the effects of these agents may be mediated by Ca^{2+} (resulting from the drug-induced disruption of Ca^{2+} homeostasis), but the sensitivity of actin as well as certain actin-binding proteins to thiol modifications suggests that an interaction with protein SH-groups on either of these proteins may also affect the cytoskeletal network.[123,124] Recent studies have shown that the thiol oxidation caused by menadione in hepatocytes can perturb microfilament organization through S-S cross-linking of actin molecules.[125] Both cross-linking of actin and plasma membrane blebbing can be prevented under conditions in which protein thiols are chemically maintained in a reduced state, suggesting that protein thiol modification can play a major role in the onset of blebbing in this system. Figure 7-4 summarizes the potential mechanisms by which modification of cytoskeletal organization can cause blebbing.

Phospholipase, Protease and Endonuclease Activation

The discovery of a relationship between a sustained increase in cytosolic Ca^{2+} concentration and the toxicity of menadione and several other agents in hepatocytes has led to a search for mechanism(s) by which the increased cytosolic Ca^{2+} could trigger cytotoxicty. It has been proposed that a sustained increase in Ca^{2+} concentration may cause abnormal stiumlation of physiological functions, including various Ca^{2+}-dependent degradative processes (Figure 7-5).

Ca^{2+}-activated phospholipases are widely distributed in mammalian cells, and it has been suggested that an enhanced rate of phospholipid hydrolysis may result in irreversible cell damage. This assumption is supported by the observation that inhibitiors of phospholipase activity can prevent ischemic cell death in liver and heart and by the finding that phospholipid breakdown is enhanced during tissue injury.[126,127]

Intracellular protein degradation is also known to be stimulated by a rise in cytosolic Ca^{2+} levels; a group of cytosolic, thiol-sensitive proteases, whose activity depends on micromolar concentrations of Ca^{2+}, seems to be involved in this process.[128] These proteases function in the cleavage of membrane receptors, cytoskeletal proteins, and protein kinase C.

Recent studies in our laboratory have provided evidence for the involvement of both phospholipases and proteases in different experimental models. Activation of phospholipases caused by a toxic elevation of cytosolic Ca^{2+} seems to play an important

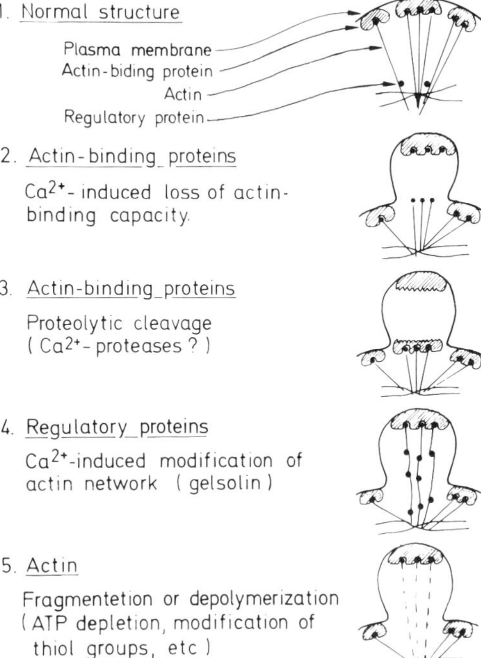

Figure 7-4. Mechanisms by which xenobiotics can affect the cytoskeletal network of hepatocytes to cause blebbing of the plasma membrane. Some agents may have multiple effects; e.g., menadione may disrupt the cytoskeleton by thiol magnification as well as by elevation of cytosolic Ca^{2+} levels.

role in the toxicity of the reactive metabolite of acetaminophen, NAPQI.[56] Although Ca^{2+}-activated proteases are also involved in NAPQI toxicity, their contribution seems to be of less importance.[56] In contrast, activation of cytosolic Ca^{2+} proteases appears to play a major role in the toxicity caused by the reactive disulfide cystamine and to be involved, together with phospholipases, in the toxicity of the herbicide diquat.[121,129]

A third catabolic process found to be sensitive to a nonphysiological elevation of cytosolic Ca^{2+} concentration during oxidative stress is that of an endogenous endonuclease. Since Ca^{2+}-dependent endonuclease activity has previously been associated with "programmed" or physiological cell death, the identification of its contribution to oxidative cell killing provides convincing evidence for the recruitment of physiological processes under certain toxic conditions.

Involvement of the endonuclease in programmed cell death (apoptosis) has been most extensively studied in cells from the immature thymus, a tissue in which up to 95% of cells are eliminated by this process. Glucocorticoid hormones and cytotoxic T-lymphocytes cause widespread chromatin condensation and endonuclease activation characteristic of apoptosis.[130–132]

Previous investigations into the mechanism of endonuclease activation have provided strong evidence for the involvement of Ca^{2+} in the process. Incubation of hepatocyte or thymocyte[134] nuclei in a Ca^{2+}-and Mg^{2+}-containing buffer activates the enzyme, while incubation of nuclei in a buffer lacking one (or both) cations does not.[133,134] Extensive plasma membrane blebbing accompanies induction to apoptosis in thymocytes, an effect which can be related to a sustained increase in cytosolic Ca^{2+} concentration as previously noted. In fact, the Ca^{2+} ionophore A23187 can induce both the morphological changes and -the endonuclease activation characteristic of apoptosis in thymocytes.[134]

Since menadione metabolism in hepatocytes causes a disruption in intracellular Ca^{2+} homeostasis and results in extensive plasma membrane blebbing, we recently studied its effects on endonucleaase activity in these cells.[135] In the presence of moderately toxic concentrations of menadione (50–100 μmol), a time-dependent stimulation of endonuclease activity was detectable after approximately two hours of incubation with maximal DNA cleavage after four hours. Higher concentrations of menadione failed to activate the endonuclease. The observation that extensive protein thiol depletion occurs in hepatocytes exposed to high concentrations of menadione suggests that cell killing is, in this case, caused by other mechanisms (e.g., disruption of cytoskeletal structure by thiol modification). Endonuclease activation preceded a detectable loss of cell viability, and was therefore not a conse-

quence of cell death. Interestingly, morphological alterations associated with menadione treatment of hepatocytes follow those observed in apoptotic thymocytes quite closely. The formation of surface blebs has been mentioned but chromatin condensation is also seen, and a close similarity exists between DNA cleavage products isolated fromk apoptotic thymocytes and from menadione-treated hepatocytes. Thus, it appears from this study that conditions of oxidative stress can cause not only endonuclease activation but also apoptosis.

To more clearly define the role of Ca^{2+} in apoptosis, we investigated the effects of the synthetic glucocorticoid methylprednisolone on cytosolic Ca^{2+} concentration in isolated thymocytes.[136] In these cells, methylprednisolone caused a sustained increaase in cytosolic Ca^{2+} concentration that was dependent both on glucocorticoid receptor binding and on protein synthesis. Buffering of the methylprednisolone-induced elevation in cytosolic Ca^{2+}, using the Ca^{2+}-dye Quin-2, completely abolished the stimulation of the endonuclease. These findings not only provide a direct demonstration for the Ca^{2+}-dependent activation of the endonuclase but also provide an interesting basis for comparison of toxic cell killing with physiological cell killing.

The role that Ca^{2+} plays in DNA damage caused by oxidative stress has also been proposed in recent work, where active oxygen-induced DNA strand breakage could be prevented by buffering intracellular Ca^{2+} with Quin-2.[137] Although long-term consequences of extensive DNA cleavage appear evident, it is still unclear how DNA damage may acutely cause other disruptive events in the cell. It has been proposed that overwhelming damage of DNA may exhaust cellular repairing capacities, particularly those involving the activation of the NAD^+, protein ADP-ribosyltransferase.[138] This enzyme, otherwise referred to as poly(ADP-ribose) polymerase, can cleave coenzyme NAD^+ into nicotinamide and ADP-ribose and covalently bind the ADP-ribose moiety to various proteins, including histones and enzymes involved in DNA repair and replication. The enzyme activity is activated by DNA strand breaks that may result either by treatment of cells with toxic agents or by endogenous processes involving gene rearrangement.[139,140] Regardless of the mechanism of activation, an increased turnover of poly(ADP-ribose) may, under certain conditions, cause the depletion of intracellular pool of NAD^+. This would ultimately cause a depletion of ATP and disruption of intracellular Ca^{2+} homeostasis, resulting in cell death. Such a mechanism has been proposed for toxic cell killing caused by H_2O_2 in macrophage-like $P388D_1$ cells.[139] It is unclear whether this process is equally important in other cell types. Ongoing studies in our laboratory involve the investigation of the role of DNA damage in acute and subacute toxic injury.

CONCLUSIONS

As can be seen from this discussion, the term *oxidative cell injury* is a general one, and the oxidation and/or alkylation of cellular macromolecules can proceed via more than one mechanism. Furthermore, the metabolic intermediate need not be a free radical to produce cytotoxicity (e.g., bromobenzene), as long as it is sufficiently electrophilic.

It is apparent that the cell possesses efficient defense mechanisms to deal with production of $O_2^{\cdot-}$ and H_2O_2 and to detoxify electrophilic metabolites, and it is only when these defenses are overwhelmed that the toxic effects are evident. Of these defense systems, glutathione appears to be critical in preventing the cell damage caused by oxidative xenobiotic metabolism. For example, even with a "pure" redox-cycling agent such ad diquat, toxicity is not observed until GSH is depleted to less than ≈ 5 nmol/10^6 cells, regardless of the degree of lipid peroxidation.[121]

We have established that thiol depletion by oxidation and/or arylation can disrupt hepatocyte Ca^{2+} homeostasis and that the rise in the intracellular Ca^{2+} concentration is closely associated with the development of cytotoxicity induced by a variety of hepatotoxins. Thus, it appears that despite the differences in the initiation of oxidative stress by different agents Ca^{2+} may represent a common endpoint in their mechanism of cell damage. Activation of degradative enzymes, such as phospholipases, proteases and endonucleases, appears to mediate the toxic action of the calcium ion. However, the association of cell death with changes in Ca^{2+} homeostasis does not necessarily preclude the involvement of protein thiols in the cytotoxicity induced by oxidative stress, nor does it rule out the contribution of additional mechanisms.

ACKNOWLEDGMENTS

We would like to acknowledge the financial support of the Swedish Medical Research Council and the Canadian Liver Foundation.

REFERENCES

1. Smith MT, Orrenius S. Studies on drug metabolism and drug toxicity in isolated mammalian cells. In: Mitchell JR, Horning MG, eds. *Drug Metabolism and Drug Toxicity.* New York: Raven Press; 1984.
2. Orrenius S, Ernster L. Microsomal cytochrome P-450-linked mono-oxygenase systems in mammalian tissues. In: Hayaishi O, ed. *Molecular Mechanisms of Oxygen Activation.* New York: Academic Press; 1974:215.
3. Coon MJ, Persson AV. Microsomal cytochrome P-450: a central catalyst in detoxication reactions. In: Jakoby WB, ed. *Enzymatic Basis of Detoxication,* Vol. I. New York: Academic Press; 1980:6.
4. Thor H, Smith MT, Hartzell P, Bellomo G, Jewell SA, Orrenius S. The metabolism of menadione (2-methyl-1,4-naphthoquinone) by isolated hepatocytes: a study of the implications of oxidative stress in intact cells. *J Biol Chem.* 1982;257:12419.
5. Iyanagi T, Yamasaki I. One-electron-transfer reactions in biochemical systems, V: difference in

the mechanism of the NADH dehydrogenase and the NAD(P)H dehydrogenase (DT-diaphorase). *Biochem Biophys Acta.* 1970;216:282.

6. Powis G, Svingen B, Appel P. Quinone-stimulated superoxide formation by subcellular fractions, isolated hepatocytes and other cells. *Molec Pharmacol.* 1981;20:387.

7. Oshino N, Imai Y, Sato R. Fatty acid desaturation by liver microsomes: possible involvement of cytochrome b_5. *7th International Congress of Biochemistry.* Tokyo, 1967 (abstract).

8. Depierre JW, Dallner G. Structural aspects of the membrane of the endoplasmic reticulum. *Biochem Biophys Acta.* 1975;415:411.

9. Eling T, Boyd J, Reed G, Mason R, Sivarajah K. Xenobiotic metabolism by prostaglandin endoperoxide synthetase. *Drug Metab Rev.* 1983; 14:219.

10. Moldéus P, O'Brien PJ, Thor H, Berggren M, Orreniuis S. Oxidation of glutathione by free radical intermediates formed during peroxidase-catalyzed N-demethylation reactions. *FEBS Lett.* 1983;162:411.

11. Ross D, Larsson R, Norbeck K, Ryhage R, Moldéus P. Characterization and mechanism of formation of reactive products formed during peroxidase-catalyzed oxidation of p-phenetidine. *Molec Pharmacol.* 1985;27:277.

12. Christ EJ, van Dorp DA. Comparative aspects of prostaglandin biosynthesis in animal tissues. *Biochem Biophys Acta.* 1972;270:537.

13. Jones DP, Mason HS. Gradients of O_2 concentration in hepatocytes. *J Biol Chem.* 1978;253:4874.

14. Fridovich I. Superoxide radical: an endogenous toxicant. *Annu Rev Pharmacol Toxicol.* 1983; 23:239.

15. Chance B, Sies H, Boveris A. Hydrogen peroxide metabolism in mammalian organs. *Physiol Rev.* 1979;59:527.

16. Jones DP, Eklöw L, Thor H, Orrenius S. Metabolism of hydrogen peroxide in isolated hepatocytes: relative contributions of catalase and glutathione peroxidase in decomposition of endogenously-generated H_2O_2. *Arch Biochem Biophys.* 1981;210:505.

17. Halliwell B, Gutteridge JMC. Oxygen toxicity, oxygen radicals, transition metals and disease. *Biochem J.* 1984;219:1.

18. DiGiuseppi J, Fridovich I. Oxygen toxicity in *Streptococcus sanguis:* the relative importance of superoxide and hydroxyl radicals. *J Biol Chem.* 1982;257:4046.

19. Azzi A, Loschen G, Flohé L. Structural and functional aspects of H_2O_2 formation in the mitochondrial membrane. In: Flohé L, Benohr HC, Sies H, Waller HD, Wendel A, eds. *Glutathione.* Stuttgart: Thieme; 1974:237.

20. Zimmerman R, Flohé L, Weser M, Hartmann HJ. Inhibition of lipid peroxidation in isolated inner membranes of rat liver mitochondria by superoxide dismutase. *FEBS Lett.* 1973;29:117.

21. Cadenas E, Sies H, Graf H, Ullrich V. Oxene donors yield low-level chemiluminescence with microsomes and isolated cytochrome P-450. *Eur J Biochem.* 1973;130:117.

22. Lynch RE, Fridovich I. Autoinactivation of xanthine oxidase: the role of superoxide radical and hydrogen peroxide. *Biochem Biophys Acta.* 1979;571:195.

23. Meister A, Tate S. Glutathione and related γ-glutamyl compounds: biosynthesis and utilization. *Annu Rev Biochem.* 1976;45:559.

24. Jakoby WB. The glutathione S-transferases: a group of multifunctional detoxification proteins. *Adv Enzymol.* 1978;46:383.

25. Wefers H, Sies H. Hepatic low-level chemiluminescence during redox cycling of menadione and menadione–glutathione conjugate: relation to NAD(P)H:quinone reductase (DT diaphorase) activity. *Arch Biochem Biophys.* 1983;224:568.

26. Peisach J, Blumberg WE. A mechanism for the action of penicillamine in the treatment of Wilson's disease. *Molec Pharmacol.* 1969;5:200.

27. Ross D, Albano E, Nilsson U, Moldéus P. Thiyl radicals-formation during peroxidase-catalyzed metabolism of acetaminophen in the presence of thiols. *Biochem Biophys Res Commun.* 1984; 125:109.

28. Moldéus P, Jernström B. Interaction of glutathione with reactive intermediates. In: Larsson A, Orrenius S. Holmgren A, Mannervik B, eds. *Functions of Glutathione: Biochemical, Physiological, Toxicological and Clinical Aspects.* New York: Raven Press; 1983;99.

29. Burk RF, Lawrence RA. Non-selenium-dependent glutathione peroxidase. In: Sies H, Wendel A, eds. *Functions of Glutathione in Liver and Kidney.* Berlin: Springer-Verlag; 1978;114.

30. Lawrence RA, Burk RG. Glutathione peroxidase activity in selenium-deficient rat liver. *Biochem Biophys Res Commun.* 1976;71:152.

31. Eklöw L, Thor H, Orrenius S. Formation and efflux of glutathione disulfide studied in isolated rat hepatocytes. *FEBS Lett.* 1981;127:125.

32. Srivastava SK, Beutler E. Permeability of normal and cataractous rabbit lenses to glutathione. *Proc Soc Exp Biol Med.* 1968;127:512.

33. Sies H, Gerstenecker C, Menzel H, Flohé L. Oxidation in the NADP system and release of GSSG from hemoglobin-free perfused rat liver during peroxidatic oxidation of glutathione by hydroperoxides. *FEBS Lett.* 1972;27:171.

34. Nicotera P, Moore M, Bellomo G, Mirabelli F, Orrenius S. Demonstration and partial characterization of glutathione disulfide-stimulated ATPase activity in the plasma membrane fraction from rat hepatocytes. *J Biol Chem.* 1985;260: 1999.

35. Nicotera P, Baldi C, Svensson SÅ, Larsson R, Bellomo G, Orrenius S. Glutathione S-conjugates stimulate ATP hydrolysis in the plasma membrane fraction of rat hepatocytes. *FEBS Lett.* 1985;187:121.

36. Kobayashi K, Sogame Y, Hayashi K, Nicotera N, Orrenius S. ATP stimulates the uptake of dinitrophenylglutathione from rat liver plasma membrane vesicles. *FEBS Lett.* 1988;240:55.

37. Miller EC, Miller JA. Presence and significance of bound aminoazo dyes in livers of rats fed p-dimethylamino-azobenzene. *Cancer Res.* 1947;7:468.

38. Bhargava PM, Heidelberger C. Studies in the

structure of the skin protein-bound compounds following topical application of 1,2,5,6-dibenzanthracene-9,10-^{14}C, II: nature of the 2-phenylphenanthracene-3,2'-dicarboxylic acid-protein bond. *J Am Chem Soc.* 1956;78:3671.

39. Monks TJ, Hinson JA, Gillette JR. Bromobenzene and *p*-bromophenol toxicity *in vivo. Life Sci.* 1982;30:841.

40. Thor H, Svensson S-Å, Hartzell P, Orrenius S. Biotransformation of bromobenzene to reactive metabolites by isolated hepatocytes. In: Snyder R, Parke DV, Kocsis JJ, Jollow DJ, Gibson CG, Witmer CM, eds. *Biological Reactive Intermediates,* Vol. II. New York: Plenum Press; 1982:287.

41. Brodie BB, Reid W, Cho AK, Sipes G, Krishna G, Gillette JR. Possible mechanisms of liver necrosis caused by aromatic organic compounds. *Proc Natl Acad Sci USA.* 1971;68:160.

42. Dent JG, Sun JD. Development of a novel method for measuring covalent binding and its application to investigation of bromobenzene hepatotoxicity. In: Snyder R, Parke DV, Kocsis JJ, Jollow DJ, Gibson CG, Witmer CM, eds. *Biological Reactive Intermediates,* Vol. II. New York: Plenum Press; 1982:275.

43. Reid WD, Christie B, Krishna G, Mitchell JR, Moskowitz J, Brodie BB. Bromobenzene metabolism and hepatic necrosis. *Pharmacology.* 1971;6:41.

44. Thor H, Moldéus P, Kristoferson A, Högberg J, Reed DJ, Orrenius S. Metabolic activation and hepatotoxicity: metabolism of bromobenzene in isolated hepatocytes. *Arch Biochem Biophys.* 1978;188:114.

45. Thor H, Orrenius S. The mechanism of bromobenzene-induced cytotoxicity: studies with isolated hepatocytes. *Arch Toxicol.* 1980;44:31.

46. Orrenius S, Thor H, Eklöw L, Moldéus P, Jones DP. Drug-stimulated H_2O_2 formation in hepatocytes: possible toxicological implications. In: Snyder R, Parke DV, Kocsis JJ, Jollow DJ, Gibson CG, Witmer CM, eds. *Biological Reactive Intermediates,* Vol. II. New York: Plenum Press; 1982:395.

47. Casini AF, Maellaro E, Pompella A, Ferrali M, Comporti M. Lipid peroxidation, protein thiols and calcium homeostasis in bromobenzene-induced liver damage. *Biochem Pharmacol.* 1987;36:3689.

48. Mitchell JR, Jollow DJ, Potter WZ, Davis DC, Gillette JR, Brodie BB. Acetaminophen-induced hepatic necrosis, I: role of drug metabolism. *J Pharmacol Exp Ther.* 1973;187:185.

49. Dahlin DC, Miwa GJ, Lu AYH, Nelson SD. N-acetyl-*p*-benzoquinone imine: a cytochrome P_{450} product of acetaminophen. *Proc Natl Acad Sci USA.* 1984;81:1327.

50. Holme JA, Dahlin DC, Nelson SD, Dybing E. Cytotoxic effects of N-acetyl-*p*-benzoquinone imine, a common arylating intermediate of paracetamol and N-hydroxyparacetamol. *Biochem Pharmacol.* 1984;33:2367.

51. Prowis G, Svingen BA, Dahlin CC, Nelson SD. Enzymatic and non-enzymatic reduction of N-acetyl-*p*-benzoquinone imine and some properties of the N-acetyl-*p*-benzosemiquinone imine radical. *Biochem Pharmacol.* 1984;33:2367.

52. Albano E, Rundgren M, Harvison PJ, Nelson SD, Moldéus P. Mechanisms of N-acetyl-*p*-benzoquinone imine (NAPQI) cytotoxicity. *Molec Pharmacol.* 1985;28:306.

53. Moore MM, Thor H, Moore GA, Nelson SD, Moldéus P, Orrenius S. The toxicity of acetaminophen and N-acetyl-*p*-benzoquinone imine (NAPQI) in isolated hepatocytes is associated with thiol depletion and increased cytosolic Ca^{2+}. *J Biol Chem.* 1985;230:13035.

54. Jollow DJ, Mitchell JR, Potter WZ, Davis DC, Gillette JR, Brodie BB. Acetaminophen-induced hepatic necrosis, II: role of covalent binding *in vivo. J Pharmacol Exp Ther.* 1973;187:211.

55. Rosen GM, Rauckman EJ, Ellington SP, Dahlin DC, Christie JL, Nelson SD. Reduction and glutathione conjugation reactions of N-acetyl-*p*-benzoquinone imine and two dimethylated analogues. *Molec Pharmacol.* 1984;25:151.

56. Nicotera P, Rundgren M, Porubeck D, Cotgreave I, Moldéus P, Orrenius S, and Nelson S. Implications of arylation and oxidation in the cytotoxicity of quinoneimines in isolated hepatocytes: effects on thiol and Ca^{2+} homeostasis. *Arch Biochem Biophys.* (Submitted).

57. Younes M, Siegers CP. The role of iron in the paracetamol- and CCl_4-induced lipid peroxidation in heapatotoxicity. *Chem Biol Interact.* 1985;55:327.

58. Powis G, Appel PL. Relationship of the single-electron reduction potential of quinones to their reduction by flavoproteins. *Biochem Pharmacol.* 1980;29:2567.

59. DiMonte D, Ross D, Bellomo G, Eklöw L, Orrenius S. Alterations in intracellular thiol homeostasis during the metabolism of menadione by isolated rat hepatocytes. *Arch Biochem Biophys.* 1984;235:334.

60. Dodge AD. The mode of action of the bipyridilium herbicides, paraquat and diquat. *Endeavour* 1971;30:130.

61. Ilett KF, Stripp B, Menard RH, Reid WD, Gillette JR. Studies on the mechanism of the lung toxicity of paraquat: comparison of tissue distribution and some biochemical parameters in rats and rabbits. *Toxicol Appl Pharmacol.* 1974; 28:216.

62. Farrington JA, Ebert M, Land EJ, Fletcher K. Bipyridilium quaternary salts and related compounds, V: pulse radiolysis studies of the reaction of paraquat radical with oxygen; implications for the mode of action of bipyridyl herbicides. *Biochem Biophys Acta.* 1972;314:372.

63. Hassan HM, Fridovich I. Superoxide radical and oxygen enhancement of the toxicity of paraquat in *Escherichia coli. J Biol Chem.* 1978;253:8143.

64. Orrenius S, Rossi L, Eklöw-Lästbom L, Thor H. Oxidative stress in intact cells: a comparison of the effects of menadione and diquat in isolated hepatocytes. In: Poli G, Cheesman KH, eds. *Free Radicals in Liver Injury.* Oxford: IRL Press Ltd; 1986:99.

65. Gage JC. The action of paraquat and diquat on

the respiration of liver cell fractions. *Biochem J.* 1968;109:757.

66. Gerber JG, McDonald JS, Harbison RD, Villeneuve J-P, Wood AJJ, Nies AS. Effect of N-acetylcysteine on hepatic covalent binding of paracetamol (acetaminophen) *Lancet* 1977;I:657.

67. Labadarios D, Davis M, Portmann B, Williams P. Paracetamol-induced hepatic necrosis in the mouse: relationship between covalent binding, hepatic glutathione and the protective effect of α-mercaptopropionylglycine. *Biochem Pharmacol.* 1977;26:31.

68. Devalia JL, Ogilvie RC, McLean AEM. Dissociation of cell death from covalent binding of paracetamol by flavones in a hepatocyte system. *Biochem Pharmacol.* 1982;31:3745.

69. Van De Straat R, De Vries J, Debets AJJ, Vermeulen PE. The mechanism of prevention of paracetamol-induced hepatotoxicity by 3,5-dialkyl substitution. *Biochem Pharmacol.* 1987;36:2065.

70. Rossi L, Moore GA, Orrenius S, O'Brien PJ. Quinone toxicity in hepatocytes without oxidative stress. *Arch Biochem Biophys.* 1986;251:25.

71. Ernster L, Lind C, Nordenbrand K, Thor H, Orrenius S. NADPH-cytochrome P_{450} reductase as a superoxide radical generator. In: Nozacki M, Yamamoto S, Ishimura Y, Coon MJ, Ernster L, Estabrook RW, eds. *Oxygenases and Oxygen Metabolism,* New York: Academic Press; 1982:357.

72. Aust SD, Svingen BA. The role of iron in enzymatic lipid peroxidation. In: Pryor WA, ed. *Free Radicals in Biology,* Vol. 5. New York: Academic Press; 1982:1.

73. Gutteridge JMC, Rowley DA, Halliwell B. Superoxide-dependent formation of hydroxyl radicals and lipid peroxidation in the presence of iron salts. *Biochem J.* 1982;206:605.

74. Bernheim F, Bernheim MLC, Wilbur KM. The reaction between thiobarbituric acid and the oxidative products of certain lipids. *J Biol Chem.* 1948;174:257.

75. Tappel AL. Lipid peroxidation damage to cell components. *Fed Proc.* 1973;32:1870.

76. Benedetti A, Comporti M, Esterbauer H. Identification of 4-hydroxynonenal as a cytotoxic product originating from the peroxidation of liver microsomal lipids. *Biochem Biophys Acta.* 1980;620:281.

77. Cadenas E, Müller A, Brigelius R, Esterbauer H, Sies H. Effects of 4-hydroxynonenal on isolated hepatocytes: studies on chemiluminescence response, alkane production and glutathione status. *Biochem J.* 1983;214:479.

78. Benedetti A, Fulceri R, Comporti M. Inhibition of Ca^{2+} sequestration activity of liver microsomes by 4-hydroxy-alkenals originating from the peroxidation of liver microsomal lipids. *Biochem Biophys Acta.* 1984;793:489.

79. Orrenius S, Thor H, DiMonte D. Metabolic activation and inactivation: a critical balance in toxicity. In: Caldwell J, Paulson GD, eds. *Foreign Compound Metabolism.* London: Taylor and Francis; 1984:245.

80. Shu H, Talcott RE, Rice SA, Wei ET. Lipid peroxidation and paraquat toxicity. *Biochem Pharmacol.* 1979;28:327.

81. Eklöw-låstbom L, Rossi L, Thor H, Orrenius S. Effects of oxidative stress caused by hyperoxia and diquat: a study in isolated hepatocytes. *Free Rad Res Commun.* 1986;2:57.

82. Sandy MS, Moldéus P, Ross D, Smith MT. Cytotoxicity of the redox cycling compound diquat in isolated hepatocytes: involvement of hydrogen peroxide and transition metals. *Arch Biochem Biophys.* 1987;259:29.

83. Thor H, Moldéus P, Hermanson R, Högberg J, Reed DJ, Orrenius S. Metabolic activation and hepatotoxicity: toxicity of bromobenzene in hepatocytes isolated from phenobarbital- and diethylmaleate-treated rats. *Arch Biochem Biophys.* 1978;188:122.

84. Hinson JA. Biochemical toxicology of acetaminophen. In: Hodgson E, Bend JR, Philpot RM, eds. *Reviews in Biochemical Toxicology,* Vol. 2. New York: Elsevier; 1980:103.

85. Miners JO, Drew R, Birkett DJ. Mechanism of action of paracetamolprotective agents in mice *in vivo. Biochem Pharmacol.* 1984;33:2995.

86. Nicotera P, Moore M, Mirabelli F, Bellomo G, Orrenius S. Inhibition of hepatocyte plasma membrane Ca^{2+}-ATPase activity and its restoration by thiols. *FEBS Lett.* 1985;181:149.

87. DiMonte D, Bellomo G, Thor H, Nicotera P, Orrenius S. Menadione-induced cytotoxicity is associated with protein thiol oxidation and alteration in intracellular Ca^{2+} homeostasis. *Arch Biochem Biophys.* 1984;235:343.

88. Boyer PD. Sulfhydryl and disulfide groups of enzymes. In: Boyer PD, Lardy H, Myrbäck K, eds. *The Enzymes,* Vol. 1. 2nd ed. New York: Academic Press; 1959:511.

89. Bellomo G, Jewell SA, Thor H, Orrenius S. Regulation of intracellular calcium compartmentation: studies with isolated hepatocytes and t-butylhydroperoxide. *Proc Natl Acad Sci USA.* 1982; 79:6842.

90. Malencik DA, Fisher EH. Structure function and regulation of phosphorylase kinase. In: Cheung WY, ed. *Calcium and Cell Function.* New York: Academic Press; 1982:161.

91. Grynkiewicz G, Poenie M, Tsien RY. A new generation of Ca^{2+} indicators with greatly improved fluorescence properties. *J Biol Chem.* 1985; 260:3440.

92. Moore L, Chen T, Knapp HR Jr, Landon EJ. Energy-dependent Ca^{2+} sequestration activity in rat liver microsomes. *J Biol Chem.* 1975; 250:4562.

93. Becker GL, Fiskum G, Lehninger AL. Regulation of free Ca^{2+} by liver mitochondria and endoplasmic reticulum. *J Biol Chem.* 1980;255:9009.

94. Dawson AP, Fulton DV. Some properties of the Ca-stimulated ATPase of rat liver microsomal fraction. *Biochem J.* 1983;210:405.

95. Murphy E, Coll K, Rich JL, Williamson JR. Hormonal effects on Ca^{2+} homeostasis in isolated hepatocytes. *J Biol Chem.* 1980;255:6600.

96. Berridge MJ, Irvine RZ. Inositol triphosphate, a novel second messenger in cellular signal transduction. *Nature* (Lond). 1984;312:315.

97. Jones DP, Thor H, Smith MT, Jewell SA, Orrenius S. Inhibition of ATP-dependent micro-

somal Ca^{2+} sequestration during oxidative stress and its prevention by glutathione. *J Biol Chem.* 1983;258:6390.

98. Thor H, Hartzell P, Svensson S-Å, et al. On the role of thiol groups in the inhibition of liver microsomal Ca^{2+} sequestration by toxic agents. *Biochem Pharmacol.* 1987;36:1313.

99. Bellomo G, Nicotera P, Orrenius S. Alterations in intracellular Ca^{2+} compartmentation following inhibition of calcium efflux from isolated hepatocytes. *Eur J Biochem.* 1984;144:19.

100. Carafoli E, Crompton M. The regulation of intracellular Ca^{2+}. *Curr Top Membr Transp.* 1978; 10:151.

101. Fiskum G, Lehninger AL. Regulated release of Ca^{2+} from respiring mitochondria by Ca^{2+}/2H$^+$ antiporter. *J Biol Chem.* 1979;254:6236.

102. Lehninger AL, Vercesi A, Bababummi EA. Regulation of Ca^{2+} release from mitochondria by the oxidation-reduction state of pyridine nucleotides. *Proc Natl Acad Sci USA.* 1978;75:1690.

103. Lötscher HR, Winterhalter K, Carafoli E, Richter C. Hydroperoxide-induced loss of pyridine nucleotides and release of calcium from rat liver mitochondria. *J Biol Chem.* 1980;255:9325.

104. Prpić V, Bygrave FL. On the inter-relationship between glucagon action, the oxidation-reduction state of pyridine nucleotides, and calcium retention by rat liver mitochondria. *J Biol Chem.* 1980;255:6193.

105. Bellomo G, Martino A, Richelmi P, Moore GA, Jewell SA, Orrenius S. Pyridine nucleotide oxidation, Ca^{2+} cycling and membrane damage during t-butyl hydroperoxide metabolism by rat liver mitochondria. *Eur J Biochem.* 1984;140:1.

106. Lotersztajn S, Hanoune J, Pecker F. A high-affinity Ca^{2+}-stimulated magnesium-dependent ATPase in rat liver plasma membranes: dependence on an endogenous protein activator distinct from calmodulin. *J Biol Chem.* 1981;256: 11209.

107. Bellomo G, Mirabelli F, Richelmi P, Orrenius S. Critical role of sulfhydryl group(s) in ATP-dependent Ca^{2+}sequestration by the plasma membrane fraction from rat liver. *FEBS Lett.* 1983; 163:136.

108. Tsokos-Kuhn JO, Todd EL, McMillin-Wood JB, Mitchell JR. ATP-dependent calcium uptake by rat liver plasma membrane vesicles: effect of alkylating hepatotoxins *in vivo. Mol Pharmacol.* 1985;28:56.

109. Kass GEN, Wright J, Nicotera P, Orrenius S. The mechanism of 1-methyl-4-phenyl-1,2,3,6-tetrahydropyridine toxicity: role of intracellular calcium. *Arch Biochem Biophys.* 1988;260:189.

110. Schanne FAX, Kane A, Young EE, Farber JL. Calcium-dependence of toxic cell death: a final common pathway. *Science.* 1979;206:700.

111. Smith MT, Thor H, Orrenius S. Toxic injury to isolated hepatocytes is not dependent on extracellular calcium. *Science* 1981;213:1257.

112. Jewell SA, Bellomo G, Thor H, Orrenius S. Bleb formation in hepatocytes during drug metabolism is caused by disturbances in thiol and calcium ion homeostasis. *Science.* 1982;214:1257.

113. Nicotera P, McConkey D, Svensson S-Å, Bellomo G, Orrenius S. Correlation between cytosolic Ca^{2+} concentration and cytotoxicity in hepatocytes exposed to oxidative stress. *Toxicology.* 1988;52:55.

114. Lengsfeld AM, Low J, Willand T, Dancker P, Hasselbach W. Interaction of phalloidin with actin. *Proc Natl Acad Sci USA.* 1974;71:2803.

115. Schliwa M. Action of cytochalasin D on cytoskeletal networks. *J Cell Biol.* 1982;92:79.

116. Schliwa M. Proteins associated with cytoplasmic actin. *Cell.* 1981;25:587.

117. Korn ED. Actin polymerization and its regulation by proteins from non-muscle cells. *Physiol Rev.* 1982;62:672.

118. Kakiuchi S, Sobue K. Control of the cytoskeleton by calmodulin and calmodulin-binding proteins. *Trends Biochem Sci.* 1983;8:59.

119. Piazza GA, Wallace RW. Calmodulin accelerates the rate of polymerization of human platelet actin and alters the characteristics of actin filaments. *Proc Natl Acad Sci USA.* 1985;82:1683.

120. Collier HL, Wang K. Purification and properties of human platelet, P-235. *J Biol Chem.* 1982; 257:6937.

121. Nicotera P, Hartzell P, Svensson S-Å, Bellomo G, Orrenius S. Cystamine induces toxicity in hepatocytes through the elevation of cytosolic Ca^{2+} and the stimulation of a non-lysosomal proteolytic system. *J Biol Chem.* 1986;261:14628.

122. Nicotera P, Hartzell P, Davis G, Orrenius S. The formation of plasma membrane blebs in hepatocytes exposed to agents that increase cytosolic Ca^{2+} is mediated by the activation of a non-lysosomal proteolytic system. *FEBS Lett.* 1986; 209:139.

123. Tait JF, Frieden C. Chemical modification of actin: acceleration of polymerization and reduction of network formation with N-ethylmaleimide, (Iodoacetamido) tetramethylrhodamine or 7-chloro-4-nitro-2,1,3,benzooxadiazole. *Biochemistry.* 1982;24:6046.

124. Las AA, Korn ED. Reinvestigation of the inhibition of actin polymerization by profilin. *J Biol Chem.* 1985;260:10132.

125. Mirabelli F, Salis A, Marinoni V, Finardi G, Bellomo G, Thor H, Orrenius S. On the mechanism of surface bleb formation in hepatocytes exposed to menadione induced oxidative stress. Submitted for publication.

126. Chien KR, Abrams J, Ptau RG, Farber JL. Ischemic myocardial cell injury: prevention by chlorpromazine of an accelerated phospholipid degradation and associated membrane dysfunction. *Am J Pathol.* 1979;94:505.

127. Shier WT, DuBourdieu DT. Stimulation of phospholipid hydrolysis and cell death by mercuric chloride: evidence for mercuric ion acting as a calcium mimetic agent. *Biochem Biophys Res Commun.* 1983;110:758.

128. Murachi T, Tanaka K, Hatanaka M, Murakami T. Intracellular Ca^{2+}-dependent protease (calpain) and its high molecular weight endogenous inhibitor (calpastatin). *Adv Enz Regul.* 1981; 19:407.

129. Tsokos-Kuhn JO, Smith CV, Mitchell JR. Diquat cytotoxicity in isolated Fisher-344 rat hepatocytes. In: Fiskum G, ed. *Cell Calcium Metabo-*

lism '87, 7th Washington Spring Symposium. 1987;143.

130. Wyllie AH. Glucocorticoid-induced thymocyte apoptosis is associated with endogenous endonuclease activation. *Nature.* 1980;284:555.

131. Duke RC, Chervenak R, Cohyen JJ. Endogenous endonuclease-induced DNA fragmentation: an early event in cell-mediated cytolysis. *Proc Natl Acad Sci USA.* 1983;80:6361.

132. Ucker DS. Cytotoxic T lymphocytes and glucocorticoids activate an endogenous suicide process in target cells. *Nature.* 1987;327:32.

133. Vanderbilt JN, Bloom KS, Andersson JN. Endogenous nuclease: properties and effects of transcribed genes in chromatin. *J Biol Chem.* 1982; 257:13009.

134. Wyllie AH, Morris RG, Smith AL, Dunlop D. Chromatin cleavage in apoptosis: association with condensed chromatin morphology and dependence on macromolecular synthesis. *J Pathol.* 1984;14:67.

135. McConkey DJ, Hartzell P, Nicotera P, Wyllie AH, Orrenius S. Stimulation of endogenous endonuclease activity in hepatocytes exposed to oxidative stress. *Toxicol Lett.* 1988;42:123.

136. McConkey DJ, Nicotera P, Hartzell P, Bellomo G, Wyllie AH, Orrenius S. Glucocorticoids activate a suicide process in thymocytes through an elevation of cytosolic Ca^{2+} concentration. *Arch Biochem Biophys.* 1989;269:365.

137. Muehlematter D, Larsson R, Cerutti P. Active oxygen induced DNA strand breakage and poly ADP-ribosylation in promotable and non-promotable J B6 mouse epidermal cells. *Carcinogenessis.* 1988;9:239.

138. Gaal JC, Smith KR, Pearson CK. Cellular euthanasia mediated by a nuclear enzyme: a central role for nuclear ADP-ribosylation in cellular metabolism. *Trends Biochem Sci.* 1987; 12:129.

139. Schraufstatter IV, Hyslop PA, Hinshaw DB, Spragg RG, Sklar LA, Cochrane CG. Hydrogen peroxide-induced injury of cells and its prevention by inhibitors of poly(ADP-ribose) polymerase. *Proc Natl Acad Sci USA.* 1986;83:4908.

140. Farzaneh F, Zalin R, Brill D, Shall S. DNA strand breaks and ADP-ribosyl transferase activation during cell differentiation. *Nature.* 1982;300: 362.

8

Phospholipids, Phospholipases, and Cell Injury

Yasaman Shirazi
Wolfgang J. Mergner

THE ROLE OF PHOSPHOLIPASES AND PHOSPHOLIPIDS IN MEMBRANE TURNOVER AND CELL INTEGRITY

Several recent studies have emphasized the role of an altered phospholipid metabolism in cell injury.[1-16] These studies have proposed new concepts regarding the role of phospholipases in membrane damage following ischemia, toxic or physical cell injury. They propose that a disturbed phospholipid metabolism is likely to affect membrane integrity. Such disturbances can be directed either at the plasma membrane or at intracellular membranes. Alterations in the plasma membrane cause equilibrium of the gradients inside and outside the cells, while those in the intracellular membranes modify specialized cell functions such as energy metabolism, protein synthesis, and lysosomal digestion. The alteration of phospholipids by either phospholipases alone or by phospholipases plus free radicals and other factors such as electrolytes is thought to be an extension of the normal catalytic function of phospholipases. This chapter, therefore, will summarize the normal function of phospholipids and phospholipases in cells and then extrapolate from the fragments of information that are available regarding their role in cell injury.

Membrane Lipids

Phospholipids present in the lipid bilayer are heterogeneous mixtures of molecules, most of which are amphiphilic and contain both hydrophilic (polar) and hydrophobic (apolar) moieties (see Table 8-1). A significant additional modification of the function of phospholipids in membranes is derived from their admixture with other lipid moieties such as cholesterol, from the incorporation of different fatty acids, and from the association with ions. Phospholipids present essential components of biomembranes and characterize, to a large extent, their biological behavior (for review, see Chapter 1).

Topography of Phospholipids in the Membranes

Van Deenen investigated the topography of phospholipids in the plasma membrane.[17] He stressed the asymmetry of phospholipid arrangement in the plasma membrane: Phosphatidylserine and phosphatidylethanolamine are located mostly in the inner leaflet of the plasma membrane, whereas phosphatidylcholine is arranged in the outer leaflet. Similarly, the preferred hydrolytic sites of phospholipases may be at either the inner or outer surface of cellular membranes. Phospholipase C isolated from *Bacillus cereus* for example, acts only at the inner surface of erythrocyte membranes.[18]

TABLE 8-1
The Main Groups of Phospholipids

Head Group	Name of Phospholipid	Abbreviation	Charge
Choline	Phosphatidylcholine	PC	Zwitterionic
Serine	Phosphatidylserine	PS	Anionic
Ethanolamine	Phosphatidylethanolamine	PE	Anionic
Glycerol	Diphosphatidylglycerol (Cardiolipin)	DPG	Anionic
Inositol	Phosphatidylinositol	PI	Anionic

Modified from Katz and Messino, 1981.

Phospholipids and Enzymes

Phospholipids are essential components of many enzymes. This applies to both the plasma membrane and to organelle membranes. For example, Bruni and Bigon described the interaction of diphosphatidylglycerol (cardiolipin) with the mitochondrial ATPase.[19] Cardiolipin comprises approximately 20% of the mitochondrial phospholipids. It produces a distinct stimulation of the oligomycin-sensitive ATPase. When cardiolipin was incubated in surplus with submitochondrial particles deprived of magnesium, the compound caused solubilization of mitochondrial ATPase and a visual decline of inner membrane spheres in negative staining preparations of mitochondria.

Mobility of Proteins and Lipids in Membranes

Membranes are lipid bilayers and retain great fluidity at the physiologic temperatures of 37°C. This means that lipid molecules in the membrane have great lateral mobility and diffuse laterally fast. A member of one lipid bilayer exchanges places with its neighbor 107 times per second. In addition, the fatty acids have flexion and rotational mobility. However, there are restrictions on mobility. Such restrictions are induced by packing phospholipids with cholesterol. Flip-flop mobility between the two lipid layers is rare. Membrane phospholipid methylation in response to signals has been linked to increased flip-flop movements.

The membrane lipids act as solvent for the suspended proteins, which equal the lipids in mass but not in size. Cell surface proteins interact with each other and with structures inside and outside the cell to carry out physiological functions. These processes also require that proteins move laterally in the plane of the plasma membrane.[20] Schlessinger described the clustering of polypeptide hormone receptors necessary to trigger cellular reactions and the subsequent aggregation to initiate endocytosis.[21]

Rotational and Lateral Diffusion

Rotational diffusion differs for lipid molecules and proteins. Lipid molecules move rapidly, about 0.1 nanosecond, while protein molecules move more slowly in the range of microseconds. The measuring system, therefore, differs for both. Steady-state measurement of transient dichroism has been used to measure protein rotation in the erythrocyte membranes.[22,23] Lateral diffusion can be measured by fluorescence photo-bleaching recovery methods. While rotational diffusion is greatly influenced by size, lateral diffusion is much less so. Thus, rotational diffusion can be used as a measure of aggregation of membrane proteins.

Beta-Receptor Adenylyl Cyclase Interaction

In the case of beta-adrenergic receptors, the primary interaction of the hormone-occupied receptor occurs with the guanine nucleotide binding proteins (G proteins) in the membrane. This initiates a sequence of reactions that stimulate adenylyl cyclase. The first steps in this sequence are the binding of GTP, the hydrolysis of GTP, and the release of GDP. The G protein interacts with the catalytic subunit of adenylyl cyclase to form the G-protein adenyl cyclase complex. Thus, hormone activation of adenylyl cyclase is governed by protein–protein interaction in the membrane between spatially separated proteins. This could mean that the mobility of the interacting polypeptides is the rate-limiting step in the reaction. Diffusion of cell surface proteins also plays a role in immune responses. Membrane proteins cluster into patches when they are cross-linked with antibodies.[24] Cross-linked membrane proteins are actively moved to one pole of the cell, a process called *capping*. From this pole, receptor-mediated endocytosis may be initiated.

Phospholipases

The discussion of phospholipases within the theme of *cell injury* may unduly emphasize their catabolic and lytic function, which can be observed in the process of autolysis. The activity of phospholipases in normal cellular metabolism, however, appears as a tightly controlled mechanism that by checks and balances is integrated into complex cellular metabolic functions. The reactions of phospholipases may be subdivided as shown in Table 8-2.

Classes of Phospholipases

Phosphoglycerides are hydrolyzed by a number of phospholipases. Phospholipase A_1 specifically removes fatty acid from the 1 position and phospholipase A_2 removes

TABLE 8-2
Proposed Function of Phospholipases

Phospholipid turnover
Phospholipid synthesis
Autolysis and degradation
Generation of prostaglandins
Ca^{2+} messenger
Protein kinase activation

fatty acid from the 2 position. Removal of one fatty acid molecule from the phosphoglyceride yields a lysophosphoglyceride. As intermediary products of phosphoglyceride metabolism, lysophosphoglycerides exist in the tissue only in very small amounts and are rapidly removed by lysophospholipases (or by acyl-CoA-lyso-phospholipid-acyltransferases). Lysophosphoglycerides are injurious to membranes at higher concentrations. Phospholipase B, the existence of which is debated, removes two fatty acids from phosphoglyceride; while phospholipase C attacks the bond between phosphoric acid and glycerol. Phospholipase D removes the polar head groups to yield diacylglycerol and phosphatidic head group. Acute cell injury has been mainly studied as an event that activates phospholipase A_1, phospholipase A_2, or phospholipase C.

Phospholipase A_1 and A_2. Phospholipase A_1 activities have been found in almost every cell in which they have been sought.[25] In the rat liver, it was shown that phospholipase A_1 and A_2 exist in the plasma membrane and Golgi membranes; both were stimulated by calcium and activated at a pH of 8.0 to 9.5. The purified phospholipases A_2 obtained from different sources display considerable variation of enzyme activity and molecular weight, ranging from 12 to 14 kd. All preparations of phospholipase A_2 are calcium dependent and produce equimolar amounts of 1-acyl lysophospholipid and free fatty acid. A very detailed and classic study by Nachbaur et al. determined that the soluble phospholipase A in rat liver occurred exclusively in the lysosomes, while the membrane-bound phospholipase A_2 existed in mitochondria, plasma membrane, and microsomes.[25] The soluble lysosomal phospholipase had a optimum pH of 4.0 to 5.0, was not calcium dependent, and was inhibited by KCl and $CaCl_2$, while the membrane-bound phospholipase is activated by the same chemicals. It was determined that the soluble phospholipases have both A_1 and A_2 activities. The mitochondria have a phospholipase A_2 that is stimulated by calcium at a pH of 8.0 to 9.0. The alkaline phospholipase of microsomes and plasma membrane has predominantly A_1 activity in contrast to that in the mitochondria. Nachbaur et al. also characterized membrane-bound phospholipases as displaying acyl chain specificity in addition to specificity for exogenous substrate.[25] For example, it was shown that lysis of exogenously added phosphatidylethanolamine was higher in the plasma membrane and outer mitochondrial membrane than in the inner membrane and rough endoplasmic reticulum.

Phospholipase C. Phosphatidylinositol (PI)–specific phospholipase C is present in most cell types, and most of its activity is cytosolic, although it has been found in various membranes preparations.[26] Several soluble enzyme activities have been identified in a variety of tissues and appear to differ in molecular mass, pH optima, and calcium dependence (see ref. 26 for review).[27,28] Hoffman and Majerus purified a 65-kd protein called PLC-I from sheep seminal vesicles; it had a pH optimum of 7.0 when assayed in the presence of deoxycholate.[29] A second distinct enzyme from the same source, designated PLC-II with a molecular weight of 85 kd, was later isolated. A soluble PI-specific phospholipase C was purified from bovine brain. This enzyme had a molecular weight of 88 kd, required calcium for maximum activity, and had an optimum pH of 6.5.[30] This enzyme cleaved all phosphoinositides; however, the hydrolytic rate of

polyphosphoinositides (PIP2 and PIP) was higher than that of PI. Other PI-specific PLC enzymes have also been purified from bovine brain.[26] Wolf and Gross partially purified a phospholipase C from canine myocardium that was active at a neutral pH.[31] The enzyme was specific for phosphatidylcholine and did not hydrolyse PI or spingomyelin. The results of this study are interesting since experiments with phorbal esters by Kolesnick and Paley have shown a cyclic degradation and resynthesis of PC similar to the PI cycle, described later.[28] These observations have led to the concept that agonist-induced phosphatidylcholine breakdown by a phosphatidylcholine-specific phospholipase C can also play an important role in cellular activation (for review see ref. 32).

Phospholipase C is a highly regulated enzyme. Activation occurs in response to receptor-regulated signals.[32] Phospholipase C apparently utilizes intermediary guanine nucleotide proteins, the so-called G proteins, in a manner analogous to the adenylyl cyclase system.[33] Phospholipase C activity may also be controlled by the physicochemical state of the membrane. Using ^{32}P-labeled PI, Irvine et al. demonstrated that the activity of the soluble, calcium-dependent PI-specific phosphodiesterase was inhibited by KCl at a pH of 7.0 in the presence of magnesium, positively charged proteins such as histones, and phospholipids containing a choline head group.[34] Choline phospholipids caused a 90% inhibition at an equimolar ratio to PI. Other mechanisms of regulation related to cellular signal transduction will be discussed later.

Phospholipase C is calcium-sensitive when calcium-concentration is maintained in the physiologic range. Insight into the role of G proteins in phospholipase activation has been further extended by observations that a G protein can stimulate phospholipase C in isolated plasma membranes of neutrophils. The presence of G protein reduced the calcium requirement for this enzyme activity from superphysiological to normal physiological intracellular concentration.[35] It has been proposed that the cytoplasmic calcium concentration as well as the phosphoinositide composition of the membrane may influence the relative rates of hydrolysis of the various phosphoinositides.

Lysophospholipases. Lysophospholipases are widely distributed in the cell.[36] Phospholipases A_1 and A_2 produce isomeric lysophosphoglycerides. In view of their detergent properties, lysophosphoglycerides can produce membrane destabilization. Thus, to facilitate removal two distinct metabolic pathways are available:

1. Acylation of lysophosphoglycerides occur in many subcellular membranes by Acyl-CoA: lysophosphoglyceride acyl transferase. This enzyme has a high concentration in the endoplasmic reticulum.
2. Alternatively, lysophospholipases may catabolize lysophospholipids.

Several lysophospholipases have been reported. De Jong et al. described lysophospholipase I and II in beef liver.[37] Lysophospholipase I is a soluble enzyme, while lysophospholipase II is membrane-bound and is found in the microsomal fraction. The soluble lysophospholipase I in the cytosol can also act on membrane-bound lysophospholipids and, therefore, contribute to the catabolism of membrane-bound substrates. However, this enzyme may act preferentially on the lysophospholipids in the outer surface of artificial membrane vesicles.

Cell Organelle Membranes and Phospholipases

The Plasma Membrane. Witt et al. described the purification of acyl hydrolase from plasma membranes of *Bacillus cereus*.[38] Franson et al. also described phospholipase A activity in highly enriched preparations of cardiac sarcolemma from hamster and dog.[39] Hamster cardiac myocytes contained phospholipase A_1 with optimal activity at pH 6.0 and 9.0 in the presence of 5 mM calcium; the activity was inhibited by EDTA. The membrane also contained a phospholipase A_2 with a pH optimum of 7.0 sensitive to calcium at 5 mM. Other cells such as hepatocytes and adipocytes showed similar phospholipase A_1 and A_2 activities.[40,41] It was shown that the phospholipase A_1 in the liver plasma membrane could be easily solubilized in the presence of heparin, suggesting that the serum phospholipase activity is chiefly of hepatic origin.[42,43] Recent data indicate that the phospholipase A_1 in the liver plasma membrane is located in the outer leaflet of the membrane, while phospholipase A_2 appears to be located in the inner leaflet, thus acting on endogenous substrates. The plasma membrane phospholipase A_1 appears to act mainly as a transacylase if an acyl receptor is available.[44]

Mitochondria. Mitochondria require phospholipid and ubiquinone to function.[45-48] The succinate-to-cytochrome C-mediated electron transport has been measured by Yu et al.[47] In these studies, when phospholipid and ubiquinone were removed, the membrane electron transport collapsed. Mitochondria contain a calcium-dependent phospholipase A_2 with an optimum pH of 9.5 (Figure 8-1). The enzymatic activity appears to be present both in the inner and outer membranes, although it is unclear if the same protein is present in both membranes. The purified protein causes the highest level of hydrolysis of phosphatidylethanolamine at a pH of 9.5, but the same enzyme hydrolyses phosphatidylserine optimally at a pH of 7.4.

Under conditions of ischemia, Smith et al. demonstrated the preferential loss of cardiolipin as an accumulative phenomenon that peaked at the time when irreversible cell injury and mitochondrial dysfunction ensued.[49] Riley and Pfeiffer reported that a number of metabolites and toxins that cause calcium release from mitochondria do so by altering the permeability of the inner membrane through disturbing the deacylation and reacylation cycle.[50] Brockenmeier et al. proposed that this disturbance causes accumulation of fatty acids and lysophospholipids in the hydrophobic environments of the mitochondrial membrane.[51]

Serum lipid–lowering drugs such as WY-14643 and clofibrate inhibit mitochondrial acyl–CoA:Lysophospholipid acyl transferase and cause increased permeability of mitochondria to calcium. It has been suggested that the change in permeability is due to lysophospholipid. Nachbaur et al. described the formation of phosphatidic acid through acylation of glycerophosphate in microsomes and lysophosphatidic acid in the outer membrane of mitochondria.[25] Lysophosphatidic acid, however, accumulated in mitochondria and was not processed further, due to a lack of acylating enzyme in the outer membrane. They concluded that mitochondria appear rather passive in renewal of phospholipids, both attacking exogenous substrates or re-acylating major structural components of phosphatidylcholine and phosphatidylethanolamine. Brdiczka and Barnard identified a calcium precipitable, soluble lipoprotein subfraction from brown fat and liver mitochondria representing mitochondrial matrix granules.[49]

Microsomes. Microsomal membranes (or endoplasmic reticulum) contain enzymes for the *de novo* synthesis of phospholipids and cholesterol assembly for other membranes in the cell.[24] A phospholipid modifying function has also been associated with the acylation–deacylation cycle in microsomes. For the complete cycle, microsomes contain phospholipase A_1, A_2, and lysophospholipase. The hydrolytic activity of the phospholipases has two pH optima: one is the calcium-dependent phospholipase A_1 at pH 7.5, and the other is the calcium-dependent phospholipase A_2 with optimum activity at pH 9.5.[52] Phospholipase A_1, stimulated by Triton X-100, was not found in liver hepatoma, although it was found in normal and regenerating liver. The activity of lysophospholipase is inhibited by calcium and Triton X-100.

Lysosomes. The detection and identification of lysosomal phospholipases has been difficult, since pure organelles are not easily isolated. Ruth et al. used specific isolation techniques that guaranteed lysosomal samples of high purity to investigate and to characterize lysosomal phospholipases.[53] It was shown that there was a temperature-dependent loss of latency of lysosomal enzymes.[54,55] (The term *latency*, in relation to lysosomal enzymes, describes the situation that arises when enzymes are separated from the substrate by the lysosomal membrane and only further destruction of membranes releases the hydrolytic enzymes of lysosomes to act on their substrates; for further definition of latency, see Chapter 6. Lysosomal phospholipids show a dramatic elevation in their content of lysophospholipids and fatty acids; at the same time, the latency of lysosomal enzymes is lost.[56] The soluble phospholipase from lysosome is inhibited by albumin, heparin, and protamine. Defatted albumin inhibits loss of latency and fatty acid release.

Subcellular fractionation studies on resting and stimulated human neutrophils revealed phospholipase A_2 activity at pH of 5.5 and 7.5.[57] The acidic phospholipase activity was contained in azurphilic granules. The enzyme was soluble and could be released upon stimulation with calcium ionophore A23187.

Cytosol. Soluble phospholipase activity has been detected in the cytosol fraction in liver homogenates. It is suspected that both phospholipase A_1 and A_2 with neutral pH are present. In addition, an acidic phospholipase has been described, with a pH optimum of 3.6.

Golgi Apparatus. The Golgi membranes in the liver contain predominantly phospholipase A_1 activity. The Golgi complex also possesses transacylating activity to acylate lysophospholipid derivatives.

Phospholipases, Prostaglandins, and Phosphoinositol

Biosynthesis of Prostaglandins. Prostaglandins are 20-carbon fatty acid derivatives that act as local hormones. A wide variety of functions have been ascribed to prostaglandins.[24] These include contraction of smooth muscle cells, aggregation of platelets, and potentiation of inflammatory response. Formation of prostaglandins, prostacylins, thromboxanes, and/or leukotrienes requires free fatty acids such as arachidonic acid. Arachidonic acid, however, does not exist in free form but rather is esterified primarily in the second position in cellular phospholipids.[58] Generation of free arachidonic acid is an enzyme-mediated process that follows the stimulation of certain cell surface receptors or the formation of complement channels in nucleated cells.

Prostaglandins are short-lived compounds and are not stored in cells to any appreciable amount.[59] Thus, it is believed that any increase in prostaglandin level comes about by the active release of precursor arachidonic acid from membrane phospholipids. Since tissue phospholipids are the richest source of arachidonic acid, mechanisms must exist whereby phospholipases initiate the biosynthesis of prostaglandins. For example, phospholipase A_2 directly removes arachidonic acid from the second position. The lysophospholipid formed in the process of prostaglandin synthesis may either be degraded further by lysophospholipases to form free arachidonic acid and glycerol-3-phosphate or it may form the building block for the resynthesis of lipoproteins (deacylation–reacylation cycle). The sequential action of PI-specific phospholipase C and diacylglycerol or monoacylglycerol lipases also produce free arachidonic acid.

The Phosphatidylinositol Cycle. The breakdown of PI follows selective stimulation of cells by a range of stimuli. This breakdown and resynthesis has been named *the PI cycle.* The release of arachidonic acid can be linked to this cycle. In platelets, the interaction of thrombin with specific receptors induces the degradation of phosphoinositides by phospholipase C.[59] Phosphorylation of 1,2 diacylglycerol (DAG) by DAG kinase creates phosphatidic acid, a key substance in this primary response to stimulation. Calcium and cAMP are the main regulators of PI cycle. The redirection of phosphatidate to PI is stimulated by cAMP through the enzyme CTP-phosphatidatecitidyl transferase to form CDP–DAG, the substrate for CDP-1,2-diacylglycerol-inositol phosphatidyl transferase. Calcium and lysophosphatidate inhibit redirection of phosphatidic acid to PI and cause release of arachidonic acid.

Phosphatidic Acid and Release of Arachidonate. In experiments in which oxidation of arachidonic acid was inhibited, it was shown that formation of phosphatidic acid precedes the release of arachidonic acid from platelets. It was also shown that diacylglycerol and phosphatidate stimulate phospholipase A_2 activity.[58,59]

MEMBRANE-MEDIATED CELL INJURY

The membrane theory of cell injury states that the primary targets of acute lethal cell injury are the cellular membranes. These membranes hold the key to cell viability because of their association with vital cell functions.

The Nature of Membrane Injury

Alterations of cell membranes leading to cell death may proceed gradually. The available information on the progression of cell injury suggests that these alterations, at least initially, are subtle and proceed in a stepwise fashion. Membrane alterations predominantly inflict changes in lipids but may also cause changes in many membrane-associated proteins either directly or though alterations of the lipid environment, as discussed in the section "The Role of Phospholipases and Phospholipids in Membrane Turnover and Cell Integrity."

One example described by Araki et al. shows the inhibition of mitochondrial 2-oxoglutarate dehydrogenase complex following freeze–thaw injury.[60] The activity could be reconstituted by the addition of phospholipids. Aloj demonstrated that when membrane lipids in the isolated plasma membranes are modulated, the hormone receptor interactions are altered, possibly by associated changes of phospholipids and membrane-associated microtubules.[61] The guanylyl and adenylyl cyclase systems, which are intimately related to the interior messenger system, are also inhibited by alterations of the phospholipids. Cyclic AMP and GMP will interact with intermediate filaments and tubulin and thus influence cell shape, mobility, and division.[62]

Mechanisms of Membrane Damage by Products of Membrane Phospholipase

The Effect of Products of Lipid Hydrolysis on Membranes

Troyer et al. reported a rapid accumulation (up to 27-fold) of unesterified fatty acids, particularly arachidonic acid, and a 17-fold increase in lysophospholipids following exposure of pig kidney cells to $HgCl_2$.[9] These changes were accompanied by a 26-fold decline of phosphatidylethanolamine. Radiolabel studies additionally disclosed the accumulation of lysophosphatidylcholine and a decrease of phosphatidylcholine. These biochemical changes were accompanied by the formation of blebs on the cell surface. Irreversible cell injury, indicated by flocculent densities in the matrix of mitochondria, ensued soon thereafter. Addition of fatty acids such as oleic acid, arachidonic acid, or linoleic acid caused the same bleb formation, while addition of lysophospholipids such as lysophosphatidyleth-

anolamine or lysophosphatidylcholine was accompanied by cell death but not by blebbing.

High Levels of Circulating Fatty Acids

High levels of circulating fatty acids have been observed in patients following myocardial infarction.[63–67]. This increase in fatty acids has been linked to the occurrence of arrhythmia. The importance of these observations may pertain to a general alteration in membrane permeability induced by free fatty acids. There is also an increased rate of lipid synthesis in ischemia. Lipid droplets have been seen in the tissue surrounding myocardial infarction between 1 and 5 hours. This may be an expressing of a depressed fatty acid catabolism. For example, free fatty acids may inhibit the citric acid cycle via inhibition of pyruvate dehydrogenase by NADH. Thus, NADH accumulates in the ischemic heart tissue. Fatty acid oxidation itself is inhibited for several reasons involving the citric acid cycle, but also the availability of CoA. Last, there may be metabolic block because of accumulation of long chain fatty acids, beta-oxidation being the rate-limiting step.

Production of Lysophosphatides

Accumulated lyso compounds may affect certain selective functions in membranes. For example, Coraboeuf et al. showed that lysophosphatides disturb electrolyte regulation by the plasma membrane and therefore induce arrhythmias.[67] Taylor et al. described early changes of phospholipids in ischemic myocardium and demonstrated a correlation with the arrhythmogenic effect of increased lysophosphatides.[68]

Such lipolytic products have an arrhythmogenic effect on cultured heart cells as well as on intact hearts.[69] Lysophosphotidylcholine has important functions in the synthesis of subspecies of phospholipids and in regulation of membrane-bound enzymes. Membrane fusion is influenced by lysophosphatidylcholine.[70] Adenyl cyclase is inhibited by lysophosphatidylcholine and guanylate cyclase is stimulated in the heart, fibroblasts, and neuroblastoma cells.[71-73]. Lysophosphatidylcholine stimulates alkaline phosphatase in Hela cells.[74] Wenzel and Innis tested products that were produced by activation of phospholipase A_2 on cultured heart cells and found a marked effect on the rhythm of heart cells, which could be inhibited by Quinacrine, a phospholipase A_2 inhibitor.[69] Lysophosphatidylcholine was added to prevent reoxygenation-induced arrhythmia.

FACTORS THAT CAN PROMOTE MEMBRANE-MEDIATED CELL INJURY

Complement Injury and Phospholipases

Channel Formation by Complement and Its Consequences

The terminal complement proteins, C5–C9, are cytolytic against bacteria, fungi, protozoa, viruses, and cells of higher organisms by virtue of their ability to form transmembrane channels in the target membrane. C5b-9 channels range in size from 0.7 to 10 nm in diameter, whereas C5b-8 channels are smaller in size and are unstable.[75] They are assembled following sequential activation of C5–C9, resulting in transformation of the hydrophobic complement proteins into amphiphilic complexes and insertion of complement peptides into the target membrane.[75,76]

In addition to their cytolytic properties, there is growing evidence that the terminal complement complexes can modulate physiologic and metabolic target cell behavior without causing cell death. The observed phenomena include (1) enhancement of membrane lipid methylation, (2) superoxide production, (3) elimination of terminal complement complexes, (4) activation of platelet prothrombinase activity, and (5) release of interleukin-1 from glomerular me-

sangial cells.[77-82] In addition, C5b-8 and C5b-9 stimulate membrane lipid hydrolysis in inflammatory cells, including macrophages, monocytes, polymorphonuclear leukocytes (PMNs), platelets, and primary cells such as oligodendrocytes and glomerular mesangial cells.[82-90] The latter phenomenon represents a role for C5b-8 and C5b-9 in stimulation of arachidonic acid release and production of its metabolic conversion products; this is in addition to the previously described effects of complement cleavage products C_{3a}, C_{3b}, C_{3bi} and C_{5a} in this function.[91-94] The effect of these peptides, however, appears to be mediated through receptor–ligand interaction, since they stimulate mediator production from cells bearing specific receptors for these peptides. In contrast, C5b-9 do not interact with specific receptors on cells and can attack virtually any target cell population.

Calcium and Complement

Measurements of intracellular calcium have demonstrated that Ca^{2+} flux is initiated through noncytolytic C5b-8 and C5b-9 channels that appear, in part, to stimulate membrane-associated cell functions.[95,96] Seeger et al. also demonstrated that noncytolytic C5b-8 complexes induce a significant uptake of $^{45}Ca^{2+}$ as early as 30 seconds in human PMNs.[88] The calcium level showed a steady increase over a 30-minute interval. Such alteration of ion entry into cells can modulate the activity of calcium-dependent phospholipases that can lead to arachidonic acid release and production of its oxygenated derivatives.

Studies investigating the role of extracellular Ca^{2+} in arachidonic acid release from rat PMNs showed that the presence of EGTA in the medium drastically reduced the release of arachidonic acid and its cyclooxygenase products.[85]. The presence of EGTA also abolished leukotriene B_4 (LTB$_4$) production by C5-8 and C5b-9 from human and rat PMNs, respectively.[85,88] These findings support the suggestion that 5-lipooxygenase is calcium dependent.[97] Recent studies by Shirazi et al. have also demonstrated that an absolute requirement exists for extracellular calcium in arachidonic acid release by C5b-9 from oligodendrocytes $\times$ C$_6$ glioma cell hybrids.[90]

Activation of Phospholipases by Complement

Several studies have suggested that phospholipase A$_2$ has a role in arachidonic acid release and mediator production by C5b-8 and C5b-9. Betz and Hansch showed that bromophenacyl bromide, chlorpromazine, and procaine inhibited arachidonic acid release by C5b-9 in Ehrlich ascites tumor cells.[98] Mepacrine, in a dose-dependent manner, could abolish LTB$_4$ production by C5b-8 in human PMNs.[88] Recent studies by Shirazi et al. have shown that arachidonic acid release by C5b-9 from C6 glioma $\times$ oligodendrocyte cell hybrids could also be abolished by meparine and bromophenacyl bromide. In addition, they have demonstrated activation of PI-specific phospholipase C by C5b-9 through direct measurements of various inositol phosphate species.[90]

Suppression of Reacylation by Complement

It has been proposed that stimuli for arachidonic acid release may both modulate phospholipase activity and suppress reacylation of free arachidonic acid by inhibiting acyltransferase activity.[99] Kroner observed arachidonic acid generation from bone marrow–derived macrophages when acylation was inhibited by various agents, including the Ca^{2+} ionophore A23187.[100] Hansch et al. showed that lyso-acyltransferase was inhibited by C5b-9 as well as nystatin, a channel former, while at the same time prostaglandin E$_2$ and thromboxane B$_2$ production was increased in human platelets.[87] This phenomenon might illustrate an alternative mechanism for C5b-8 and C5b-9 to control total cellular levels of free arachi-

donic acid. In renal glomerular mesangial cells, C5b-9 stimulated the incorporation and release of preincorporated arachidonic acid from various phospholipids, indicating that both acylation and deacylation are operative.[82]

Free Radicals and Phospholipases

Definitions

Free radicals are substances with an unbalanced electron, such as superoxide, hydrogen peroxide, hydroxyl radicals, hypochlorous acid, N-Chloramine, halides (such as iodine), hypobromous acid, and hypoxanthine. Cells are equipped with powerful antioxidant systems. Such systems are superoxide dismutase, catalase, and glutathione peroxidase. Other antioxidant substances are alpha-tocopherol, beta-carotene, thiols, thioethers, and ascorbic acid.

Generation of Free Radicals

Generation of free radicals has been considered by some authors to initiate membrane attack by phospholipases.[101–104] An association between lipid peroxidation and enhanced phospholipase A activity has indeed been demonstrated in various membranes such as rat liver mitochondria, hepatic lysosomes, and rat hepatic microsomes. Sevanian et al. postulated that fatty acyl-peroxides are the preferred substrate for phospholipases since there is increased phospholipase activity following exposure to membranes injured by peroxidation.[102] Lipid peroxidation consists of distinct phases of initiation and propagation. Using microsomal membranes, Beckman et al. showed that a correlation exists between lipid peroxidation and phospholipase A activation since a variety of promoters of lipid peroxidation act directly or indirectly as stimuli of phospholipase activation beyond the initiation.[104]

Mechanism of Calcium Potentiation of Phospholipases and Free Radical Injury

Malis and Bonventre studied the augmented effects of ischemic and toxic injury as a result of accumulation of cytosolic calcium and free radicals.[102] The authors proposed that calcium and free radicals are particularly toxic to mitochondrial membranes, damaging ATP synthesis and diminishing energy-dependent mechanisms. The deficiency in energy production could lead to cell death. In this model, oxygen-free radicals were generated by hypoxanthine and xanthine oxidase in the postischemic period *in vitro*. With site I–dependent substrates such as pyruvate and malate, calcium pretreatment followed by exposure to oxygen free radical resulted in inhibition of electron transport chain functions and complete uncoupling of oxidative phosphorylation. The effects could be mitigated by dibucaine, a phospholipase A_2 inhibitor. The electron transport chain defects mainly affected the NADH CoQ-reductase. With site II substrates such as succinate, the electron transport chain effects were not evident. Respiration even remained partially coupled. Calcium and free radicals combined caused 55% depression of ATPase activity and 65% depression of adenine nucleotide translocase. Free radicals alone had only marginal effects on the ATPase and no effects on adenine nucleotide translocase. Dibucaine, again, could prevent these changes. The authors postulated that calcium-mediated phospholipase A_2 activation was the sole promoter of the damage, while free radicals may have played only an enhancing role.

The Relationship between Free Radicals and Phospholipase

In a great number of experimental situations, the suggestion exists that the cell injury induced by free radical and by phospholipases are synergistic.

Microsomes. The involvement of phospholipases A in microsomal lipid peroxidation was investigated by Beckman et al.[104] Two types of inhibitors were used, one to inhibit phospholipase A — chloropromazine and mepacrine; the other to inhibit peroxidation — bromophenacyl bromide. Lipid peroxidation response to different peroxidation systems was completely eliminated by exposure of membrane particles to bromophenacyl bromide in the presence of glutathione.

Plasma Membrane. Jain described the reaction of red blood cell membranes to peroxidation injury and their protection by vitamin E.[105] In red blood cell membranes, peroxidation products stimulated catabolism of phospholipids in the presence of exogenously added bee-venom phospholipase A_2. This resulted in a shift of inner-leaflet phospholipids, phosphatidylserine and phosphatidylethanolamine, to the outer leaflet. This effect could be prevented by vitamin E.

The Role of Glutathione, Glutathione Transferase, and Phospholipase

Tan et al. described the inhibition of microsomal lipid peroxidation by glutathione, glutathione transferase B, and arachidonic acid in connection with phospholipase A_2 (see Chapter 7).[106]

Studies on Lipid Bilayers

Recently, Sevanian et al. examined the relationship between the release of membrane-derived lipid peroxidation products and phospholipases.[102] In these experiments, artificial membrane systems consisting of phosphatidylcholine liposomes with phospholipase A_2 served as test systems and were mixed with microsomes. Peroxidation was initiated in microsomes using iron ascorbate. It was shown that lysophosphatidylcholine was released after the initiation of lipid peroxidation. The process could be inhibited by using bromophenylacyl bromide and mepacrine; both these agents are described as inhibitors of phospholipase A_2.

Lipoxygenase-dependent oxygen consumption and peroxide transfer were dependent on phospholipases and were also inhibited by phospholipase inhibitors. This system displayed the intermembrane transfer of free radicals and a membrane repairing effect of phospholipase A_2. Tan et al., using a similar artificial membrane system, concluded that inhibition of microsomal lipid peroxidation *in vitro* requires the action of phospholipase A_2, which releases acylhydroperoxide from peroxidized phospholipids.[106] Glutathione peroxidases are also needed to reduce acylperoxides. Lipid hydroxyperoxides formed by subjecting phosphatidylcholine liposomes to iron ascorbate–induced peroxidation were effectively reduced by glutathione peroxidase when phospholipase A_2 was present. Low levels of glutathione peroxidase activity were observed in the absence of phospholipase A_2. Thus, by cleaving fatty acids from phospholipids and interacting with a detoxifying system, phospholipases may reduce lipid peroxidation products.

Studies on Cell Organelle Membranes

Such interaction, however, may not apply to cellular membranes *in situ* or after isolation of organelles. Malis and Benventre examined the combined effects of calcium and free radicals on membranes in particulate mitochondrial membranes of the kidney.[101] Free radicals were introduced by xanthine and hypoxanthine oxidase. Calcium exposure was carried out and respiration was stimulated using site I substrates. The exposure to free radicals resulted in inhibition of electron transport and uncoupling of mitochondria. These effects could be inhibited by dibucaine, a phospholipase A_2 inhibitor. The inhibitor prevented membrane and cell damage only in part. It could not stop the selective alteration of NADH CoQ reductase and only partially preserved mitochondrial ATPase, adenine nucleotide translocase, and calcium sequestration. Miyazaki et al. studied the reperfusion injury after 15 minutes occlusion of the left anterior descending coronary artery, which was re-

flected in mitochondrial function.[107] The changes in mitochondrial functions in the 30% of arrhythmic animals and in isolated mitochondria suggested that phospholipase C activation inflicted the damage on mitochondria.

CHECKS AND BALANCES: ANTIPHOSPHOLIPASE ACTION AND INHIBITORS

Phospholipid content, turnover, modification, and synthesis are highly controlled processes in the cell. Membrane organization, admixture of sphingomyelin with cholesterol, the orientation of phospholipids, and the composition of fatty acids greatly determine cell behavior and require regulation. One component of this regulatory process is the stimulating agents that enhance catalytic action, particularly calcium. Another component is inhibitors that balance such catalytic actions.

Endogenous Phospholipase Inhibitors

DNA-directed Synthesis of Phospholipase A_2 Inhibitor

Van Den Bosch proposed a tentative scheme for activation and inhibition of naturally occurring phospholipases.[108] In his scheme, the nuclear DNA directs the synthesis of an inhibitor protein. Such inhibitor synthesis can be mediated by corticosteroid via binding to a receptor. Removal of the inhibitor also appears to be regulated. For example, rabbit aorta contracting substance releasing factor (RCF–RF), a low peptide found in immunologically shocked lung effluent, may stimulate phospholipase A_2 activity by dissociating the inhibitor from the active enzyme.

Lipocortin and Calpactins

Lipocortin refers to a group of proteins whose synthesis and release are stimulated by glucocorticosteroids and that probably mediate the antiinflammatory function of steroids.[109-116] The factor was first described in macrophages, where it was called *macrocortin*. It has a molecular weight of 15 kd. It could inhibit phospholipase A_2 activity when labeled phospholipid was used as substrate.[109] Hirata et al. independently demonstrated the presence of a phospholipase A_2 inhibitory protein with a molecular weight of 40 kd in the particulate fraction of steroid-treated neutrophils called *lipomodulin*. Another factor was also synthesized by rat renomedullary interstitial cells that produce a large amount of prostaglandins following steroid treatment.[111] This factor, referred to as *renocortin*, proved to be a mixture of two proteins, one having a molecular weight of 15 kd and the other a molecular weight of 30 kd. The proteins just described were ultimately proved to be chemically and functionally identical, and the apparent discrepancy in molecular weight was found to be caused by proteolysis.[112]

Significantly, Hirata observed that lipocortin is incapable of binding to phospholipase A_2 when it is phosphorylated. Subsequent studies showed that the antiphospholipase A_2 activity of lipocortin is regulated by protein kinase C–dependent phosphorylation of a 40-kd protein, as shown in human thymocytes.[113] Studies in platelets have suggested that phosphorylation of a 40-kd protein by protein kinase C, associated with the release of serotonin and lysosomal enzymes, may be involved in arachidonic acid release. This 40-kd protein possessed antiphospholipase A_2 activity and cross-reacted with specific antilipocortin antibodies.[114] Lipocortin is identified as a 35- to 38-kd protein in human placenta and may occur in at least two forms: lipocortin I and II. Antiphospholipase A_2 activity of lipocortins is mediated through modification of substrates, as indicated by recent reports.[115] Lipocortins are most likely similar, if not identical, to the factor described by Van den Bosch.[108] The two proteins have been sequenced and cloned with the surprising result that the cDNA for calpactin II, a calcium-binding protein associated with the cytoskeleton, is at least 50% or more homologous to lipocortin I.[116] It has been demonstrated further that the cDNA for lipocortin I is identical to calpactin II. Calpactins are membrane cytoskeletal proteins. They serve as major substrates for tyrosine protein kinase, a viral and growth factor receptor.

Pharmacological Inhibition of Phospholipases

Pharmacological substances that have been shown to have inhibitory effects on phospholipases include: bromophenacyl bromide, chloropromazine, mepacrine, dibucaine, and quinacrine.[117]

Other agents described are trifluoroperazine and N-(6-aminohexyl)-5-chloro-1-naphtalene sulfonamide (W-7). Trifluoroperazine abolishes soluble phospholipase A_2 and particulate phospholipase C antivities. N-(6-aminohexyl)-5-chloro-a-naphtalene sulfonamide, or W-7, is a calmodulin antagonist.[118] It is thought that it interferes with the calcium and calmodulin-induced release of arachidonic acid and the subsequent oxidation to prostaglandins. Simultaneously, it reduces the decline in labeled phospholipids, but does not suppress labeled diacylglyceride and inositol accumulation.

SUMMARY: THE THEORY OF MEMBRANE-MEDIATED CELL INJURY

The study of cell injury in subcellular systems has supported fairly convincingly the hypothesis that early cell injury involves alteration of cell membranes, both the plasma membrane and organelle membranes.[1-16,119-126] The plasma membrane, which is responsible for the disparity of the interior milieu and the extracellular environment, plays a pivotal role in cell injury. The sequence of the plasma membrane changes in ischemic injury is less clear than the sequence of such injuries as complement injury or mercurial poisoning. The effect of membrane-mediated cell injury in ischemia as a result of autolytic processes, for example, has fascinating aspects. Thus, abundant literature exists showing electron micrographs of disrupted membranes of heart.[121] (See Figures 8-1 through 8-3.) Other, more basic, studies have shown inactivation of enzymes, particularly those that require phospholipids. This has been accepted as an indicator of membrane lytic activity in the course of cell injury. The altered movements of receptors or the appearance of lipid domains in the membrane have also been shown to be the consequence of ischemic injury and an indicator of altered membrane structure. Admission of horseradish peroxidase molecules, vital dye uptake, and leakage of cytoplasmic enzymes such as lactate dehydrogenase, CPK, and SGOT, among others, from the cytosol essentially describe the same disturbed functions.

In spite of such elaborate knowledge, the mechanisms responsible for plasma membrane–mediated injury have remained speculative at best. Table 8-3 shows the possibilities that have been disputed in the recent literature.

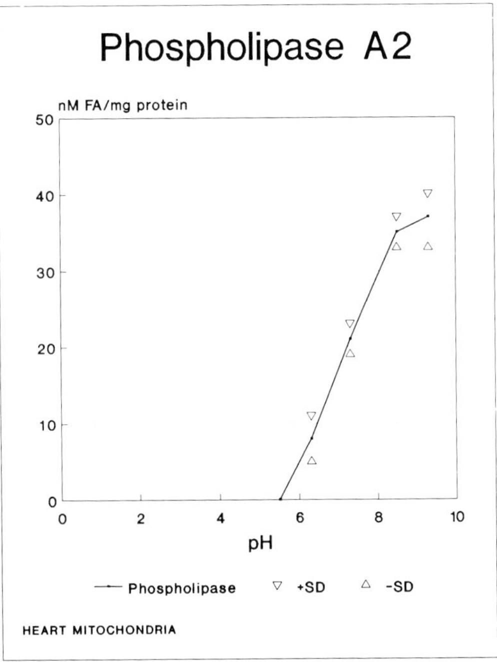

Figure 8-2. Graph showing mitochondrial-associated phospholipase A_2.

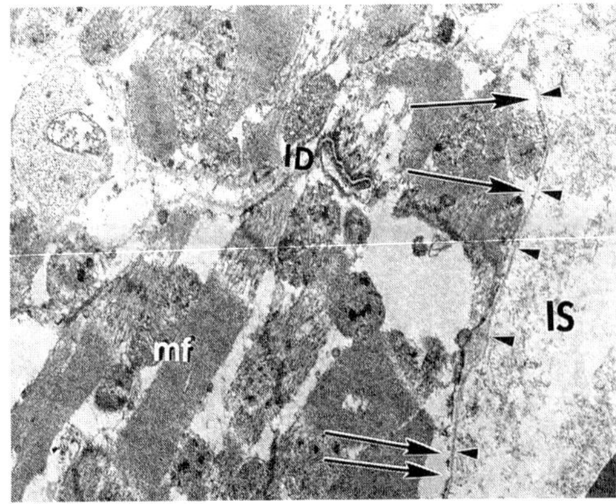

Figure 8-1. Electron micrograph showing heart cells that have undergone ischemic injury. Note the interruption of the plasma membrane.

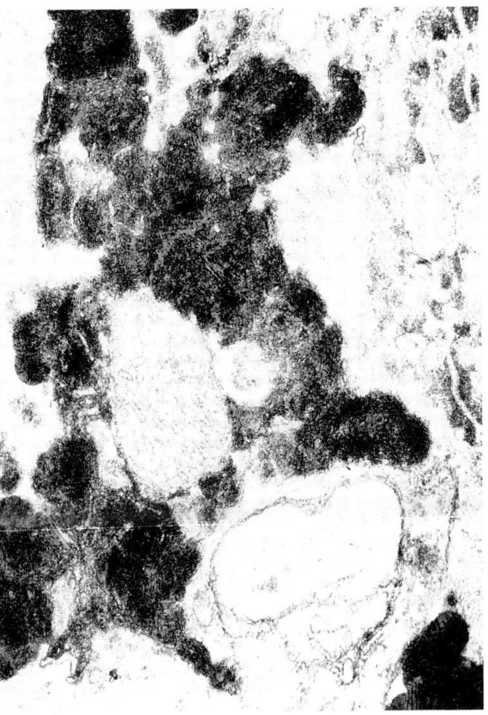

Figure 8-3. Hypokalemia. In hypokalemia, phospholipase is inhibited with special procedures (glutaraldehyde, tannic acid fixation). Accumulation of fingerprint pattern can be detected in cardiac myocytes.

TABLE 8-3
Proposed Mechanism of Membrane Alteration

1. Increased disorder by insertion of fatty acids into the lipid domain of the membrane.
2. Increased membrane disorder due to insertion of lysophospholipids.
3. Disorder of the hydrophobic protein–membrane interaction.
4. Unmixing of membranes and increase of the cholesterol domain.
5. Alteration of enzymes related to transmethylation.

Increased Disorder by Insertion of Fatty Acids into the Lipid Domain of the Membrane

Early attempts to identify the role of phospholipases in cell injury have focused on the effect of free fatty acids. For example, it could be shown that fatty acids were arrhythmogenic, supposedly on the basis of a partial collapse of the electrolyte gradient. The same effect could be achieved in single heart cells, in isolated perfused hearts, and in whole animals. These changes may be aggrevated by other disturbances of cell equilibrium such as anoxia. Extrapolating from the data with exogenously added lipids, it was postulated that the generation of intrinsic fatty acids by phospholipases from membrane phospholipids had a similar devastating effect. Other lipids such as esterified cholesterol were also thought to contribute to the membrane alteration. There are several problems, however, with these concepts:

1. One may ask if such floating of free fatty acids in a hydrophobic environment is really possible. It may, for example, be that membrane-associated enzymes such as lysophospholipases and acyltransferases immediately transfer the fatty acids and thus successfully prevent the existence of free fatty acids in the membranes, unless the enzyme function is otherwise inhibited.
2. One may also ask in what form this postulated disturbance takes place, and if channel formation by free fatty acids alone is really possible.
3. Alternatively, could these free fatty acids induce micelle formation?

Increased Membrane Disorder Due to Insertion of Lysophospholipids

Experimental systems *in vitro* have shown that the generation and presence of lysophosphatides in the phospholipid bilayer results in the breakdown of membrane gradients. An example of these alterations is the following experiment: Riley and Pfeiffer inhibited the acylation and reacylation cycle by hypolipidemic drugs such as clofibrate and WY-14643 and achieved a rapid release of calcium from mitochondria.[50] This was apparently due to the increase of membrane permeability caused by an overabundance of lysophospholipids.

Disorder of Hydrophobic Protein–Membrane Interaction

Relation between Phospholipase Action and Release of NADH Dehydrogenase

Externally added phospholipase A from *Naja naja* digests membrane-bound phospholipids such as phosphatidylethanolamine, lecithin, and cardiolipin. Awashti et al. demonstrated that the release of NADH dehydrogenase from the inner membrane of bovine cardiac mitochondria is associated with the hydrolysis of cardiolipin and alteration of oxidative phosphorylation.[120]

Transmethylation of Plasma Membranes

Transmethylation is the enzymatic transfer of methyl-groups to phosphatidylethanolamine in the inner leaflet of the membrane bilayer, thus producing phosphatidylcholine. The phospholipid may then flip-flop to the outer leaflet. This transmethylation process has been postulated to serve as a signal in chemotaxis. Stimulation of transmethylation has also been observed in cells bearing C5b-9 complement complexes, and has been proposed to have a protective effect, preventing cell injury by complement. In these experiments, inhibition of transmethylation could enhance cell killing by complement.[77]

Unmixing of the Membranes

The Generation of Lipid Domains

More recently, it has been proposed that accelerated lipid degradation in toxic and hypoxic injury may lead to the generation of "lipid domains" of different phospholipid and lipid makeup. An example is ischemic liver, where the molecular order of the microsomal membrane was changed, resulting in an increase in the ratio of cholesterol to phospholipids.[124] Similar changes were described in the plasma membrane. Domains of different molecular structures were observed by electron spin resonance, using lipid-soluble spin labels that were introduced into the plasma membranes and microsomal membranes.[122] Alternatively, ^{31}P-nuclear magnetic resonance of ischemic microsomal membranes were also performed and showed similar data.[123]

Phospholipid Reorganization and Bilayer Destabilization

Freeze fracture studies by Post and co-workers showed ultrastructural changes of sarcolemmal membranes following ischemia and reperfusion with calcium containing solutions.[125] These studies showed remarkable aggregation of the sarcolemmal intramembranous particles after ischemia. This aggregation was enhanced upon reperfusion. In addition, disruption of sarcolemma occurred with and after reperfusion, a finding that is also confirmed by Naylor et al.[126] Post and co-workers suggested that these changes represent lateral phase separa-

tion of phospholipids and destabilization of the lipid bilayers.[125] It has been proposed that increased intracellular pH and calcium concentrations during reperfusion may also enhance such changes.

CONCLUSIONS

There is sufficient evidence from inhibition studies, direct observations, and detailed explorations of cell organelles and organelle-related enzymes that significant alterations of internal and external phospholipid membranes occur in response to toxic, ischemic, and physical injuries. Calcium, pH, and peroxidation all play distinct parts in these processes. Catabolism of phospholipids may produce the irreversible cell injury, and it is very likely that the irreversibility is associated with the dissolution of functioning membranes (see Figure 3).

REFERENCES

1. Chien KR, Sen A, Reynolds R, et al. Release of arachidonate from membrane phospholipids in cultured neonatal rat myocardial cells during adenosine triphosphate depletion: correlation with progressive injury. *J Clin Invest.* 1985; 75:1770.
2. Chien KR, Willerson J, Buja LM. Phospholipid alterations and membrane injury during myocardial ischemia. *Adv Myocardiol.* 1985;5:347.
3. Chien KR, Abrams J, Pfau RG, Farber JL. Prevention by chloropromazine of ischemic liver cell death. *Am J Pathol.* 1977;88:539.
4. Chien KR, Abrams, Serroni A, Martin JT, Farber JL. Accelerated phospholipid turnover and associated membrane dysfunction in irreversible ischemic liver cell injury. *J Biol Chem.* 1978; 253:4809.
5. Chien KR, Pfau RG, Farber JL. Ischemic myocardial cell injury: prevention by chlorpromazine of the accelerated phospholipid degradation and the associated membrane dysfunction. *Am J Pathol.* 1979;97:505.
6. Chien KR, Sherman SC, Mittnacht S Jr, Farber JL. Microsomal membrane structure and function subsequent to calcium activation of an endogenous phospholipase. *Arch Biochem Biophys.* 1980;205:614.
7. Smith MW, Collan Y, Kahng MW, Trump BF. Changes in mitochondrial lipids of rat kidney during ischemia. *Biochem Biophys Arch.* 1980; 618:192.

8. Chien KR, Farber JL. Microsomal membrane dysfunction in ischemic rat liver cell. *Arch Biochem Biophys.* 1977;180:191.

9. Troyer DA, Kreisberg JI, Venkatachaloam MA. Lipid alteration in LLC-PK1 cells exposed to mercuric chloride. *Kidney Int.* 1986;29:530.

10. Curtis MT, Gilfor D, Farber JL. Lipid peroxidation increases the molecular order of microsomal membranes. *Arch Biochem Biophys.* 1984; 235:644.

11. El-Mofty SK, Scrutton MC, Serroni A, Nicolini C, Farber JL. Early reversible plasma membrane injury in galactosamine-induced liver cell death. *Am J Pathol.* 1975;79:579.

12. Farber JL, El-Mofty SK. The biochemical pathology of liver cell necrosis. *Am J Pathol.* 1975; 81:237.

13. Farber JL, El-Mofty SK, Schanne FAX, Aleo JJ Jr, Serroni A. Intracellular calcium homeostasis in galactosamine-intoxicated rat liver cells: active sequestration of calcium by microsomes and mitochondria. *Arch Biochem Biophys.* 1977; 178:617.

14. Farber JL, Martin JT, Chien KR. Irreversible ischemic cell injury: prevention by chlorpromazine of the aggregation of the intramembranous particles of rat liver plasma membrane. *Am J Pathol.* 1978;92:713.

15. Farber JL. Biology of disease: membrane injury and calcium homeostasis in the pathogenesis of coagulative necrosis. *Lab Invest.* 1982;47:114.

16. Katz AM, Messineo FC. Lipid–membrane interactions in the pathogenesis of ischemic damage in the myocardium. *Circ Res.* 1981;48:1.

17. Van Deenen LL, de Gier J, Demel RA. Lipid and lipid protein interaction in model systems and membranes. *Ann NY Acad Sci.* 1975;264:124.

18. Low MG, Limbrick AR, Finnean JB. Phospholipase C *Bacillus cereus* acts only at the inner surface of the erythrocyte membrane. *FEBS Lett.* 1973;34:1.

19. Dabbein-Sala F, Pitotti A, Bruni A. Regulation of phospholipid–ATPase complex interaction by the adenine nucleotide carrier. *Biochim Biophys Acta.* 1981;637:400–407.

20. Helmreich EJM, Elson EI. Mobility of proteins and lipids in membranes. *Adv Cyc Nucleo Prot Phosph Res.* 1984;18:1.

21. Schlessinger J. The mechanism and role of hormone-induced clustering of membrane receptors. *Trends Biochem Sci.* 1980;5:210.

22. Cherry RJ, Burkli A, Busslinger M, Schneider G, Parish GR. Rotational diffusion of band 3 protein in human erythrocyte membrane. *Biochim Biophys Acta.* 1979;559:289.

23. Cherry RJ, Schneider G. A spectroscopic technique for measuring slow rotational diffusion of bacteriorhodopsin in lipid membranes. *Biochemistry.* 1976;15:3657.

24. Alberts B, Bray D, Lewis J, Raff M, Roberts K, Watson JD. *Molecular Biology of the Cell.* New York: Garland; 1984.

25. Nachbaur J, Colbeau A, Vignais PM. Distribution of membrane confined phospholipases A in the rat hepatocyte. *Biochim Biophys Acta.* 1972;274:426.

26. Rhee SG, Suh PG, Ryn SH, Lee SY. Studies of inositol phospholipid-specific phospholipase C. *Science.* 1989;244:546.

27. Low MG, Weglicki WB. Resolution of myocardial phospholipase C into several forms with distinct properties. *Biochem J.* 1983;215:325.

28. Kolesnick RN, Paley AE. 1,2-diacylglycerols and phorbal esters stimulate phosphatidylcholine metabolism in GH3 pituitary cells. *J Biol Chem.* 1987;262:9204.

29. Hoffman SL, Majerus PW. Identification of properties of two distinct phosphatidylinositol-specific phospholipase C enzymes from sheep seminal vesicle glands. *J Biol Chem.* 1982;257:6461.

30. Rebecchi MJ, Rosen OM. Purification of phosphoinositol-specific phospholipase C from bovine brain. *J Biol Chem.* 1987;262:12526.

31. Wolf RA, Gross RW. Identification of neutral active phospholipase C which hydrolyzes choline glycerophospholipids and plasmalogen selective phospholipase A₂ in canine myocardium. J Biol Chem. 1985;260:7295.

32. Exton JH. Mechanism of action of calcium-mobilizing agonists: some variations on a young theme. *FASEB J.* 1988;48:2670.

33. Putney J. Formation and action of calcium mobilizing messenger inositol 1,4,5-triphosphate: editorial review. *Am J Physiol.* 1987;252:G149.

34. Irvine RF, Hemington N, Dawson RMC. The calcium-dependent phosphatidylinositol phosphodiesterase of rat brain. *Eur J Biochem.* 1979;99:525.

35. Smith CM, Cox CC, Synderman R. Receptor coupled activation of phosphoinositol-specific phospholipase C by an N protein. *Science.* 1986;232:97.

36. Van den Bosch H, Van den Besselaar AMHP. Intracellular formation and removal of lysophospholipids. *Adv Prostaglandin Thromboxane Res.* 1978:69–75.

37. De Jong JGN, Van den Bosch H, Rijken D, Van Deenen LLM. Studies on lysophospholipase, III: the complete purification of two proteins with lysophospholipase activity from beef liver. *Biochim Biophys Acta.* 1974;369:50.

38. Witt W, Bruller HJ, Falker G, Fuhrmann GF. Purification and properties of a phospholipid acyl hydrolase from plasma membranes of *Saccharomyces cerevisiae. Biochim Biophys Acta.* 1982; 711:403.

39. Franson RC, Pang DC, Towle DW, Weglicki WB. Phospholipase A activity of highly enriched preparations of cardiac sarcolemma from hamster and dog. *J Mol Cell Cardiol.* 1978;10:921.

40. Waite M, Sisson P. Studies on the substrate specificity of phospholipase A₁ of the plasma membrane of rat tissue. *J Biol Chem.* 1974; 249:6401.

41. Bereziat G, Wolf C, Colard O, Polonovski J. Phospholipases of plasmic membranes of adipose tissue: possible intermediates for insulin action. *Adv Exp Med Biol.* 1978;101:191–199.

42. Waite M, Sisson P. Solubilization by heparin of the phospholipase A₁ from the plasma membrane of rat liver. *J Biol Chem.* 1973;248:7201.

43. Zieve FJ, Zieve L. Post heparin phospholipase

and post heparin lipase have different tissue origins. *Biochem Biophys Res Comm*. 1972;47:1480.
44. Waite M, Sisson P. Utilization of neutral glycerides and phosphoethanolamine by phospholipase A_1 of the plasma membrane of rat liver. *J Biol Chem*. 1973;248:7985.
45. Moller F, Wilson JE. The influence of specific phospholipids on the interaction of hexokinase with the outer mitochondrial membrane. J Neurochem. 1983;41:1109.
46. Vignais PV, Douce R, Lanquir GJ. Binding of radioactively labeled carboxyatractyloside, atractyloside and bongkrekic acid to the ADP translocator of potatoe mitochondria. *Biochim Biophys Acta*. 1976;440:688.
47. Yu Ca, Yu L, King TE. Reconstitution of succinate-Q reductase. *Biochem Biophys Res Commun*. 1978;79:939.
48. Brdiczka D, Barnard T. Mitochondrial matrix granules in soft tissue, II: isolation and initial characterization of a calcium precipitable, soluble lipoprotein subfraction from brown fat and liver mitochondria. *Exp Cell Res*. 1980;126:127.
49. Smith MW, Collan Y, Kahng MW, Trump BF. Changes in mitochondrial lipids of rat kidney during ischemia. *Biochim Biophys Acta*. 1980; 618:192.
50. Riley WW, Pfeiffer DR. The effect of Ca^{2+} and Acyl-CoA: phospholipid-acyltransferase inhibitors on permeability properties of the liver mitochondrial inner membrane. *J Biol Chem*. 1986; 261:14018.
51. Brockenmeier KM, Schmid PC, Schmid HHO, Pfeiffer DR. Effects of phospholipase A_2 inhibitors on rutherium red–induced Ca^{2+} release from mitochondria. *J Biol Chem*. 1985;260:105–113.
52. Lumb RH, Allen KF. Properties of microsomal phospholipases in rat liver and hepatoma. *Biochim Biophys Acta*. 1976;450:175.
53. Ruth RC, Kennett FF, Weglicki WB. A new technique for isolation of particulate lysosome activity from canine and rat myocardium. *J Mol Cell Cardiol*. 1978;10:739.
54. Ruth RC, Weglicki WB. The temperature-dependence of the loss of latency of lysosomal enzymes. *Biochem J*. 1978;172:163.
55. Weglicki WB, Owens K, Ruth RC. Activity of endogenous myocardial lipases during incubation at acid pH. *Cardiovasc Res*. 1974;8:837.
56. Weglicki WB, Ruth RC, Owens K, Griffin HD, Waite BM. Changes in lipid composition of Triton filled lysosomes during lysis: association with activation of acid-active lipases and phospholipases. *Biochim Biophys Acta*. 1974;337:145–152.
57. Balsinde J, Diez E, Schuller A, Mollinedo F. Phospholipase A activity in resting and activated human neutrophils: substrate specificity, pH dependence, and subcellular localization. *J Biol Chem*. 1988;263:1929.
58. Lapetina EG. Regulation of arachidonic acid production: role of phospholipase C and A_2. *Trends in Pharm Sci*. 1982;3:115.
59. Lapetina EG, Cuatrecasas P. Rapid inactivation of cyclooxygenase activity after stimulation of intact platelets. *Proc Natl Acad Sci USA*. 1979; 76:121.
60. Araki T. Inactivation of mitochondrial 2-oxyglutarate dehydrogenase complex as a result of phospholipid degradation induced by freeze-thawing. *Biochim Biophys Acta*. 1977;496:532.
61. Aloj SM. Membrane lipids and modulation of hormone receptor expression. *Horiz Biochem Biophys*. 1982;6:83.
62. Kameyama Y, Kudo S, Ohki K, Nozawa Y. Differential inhibitory effect by phospholipase A_2 on guanylate and adenylate cyclase in Tetrahymena plasma membrane. Jpn J Exp Med. 1982;52:183.
63. Vetter NJ, Strange RC, Adams W. Initial metabolic and normal response to acute myocardial infarction. *Lancet*. 1974;1:184.
64. Opie LH. Metabolic response during impending myocardial infarction, I: relevance of studies of glucose and fatty acid metabolism in animals. *Circulation*. 1972;45:483.
65. Opie LH, Lubbe WF. Are free fatty acids arrhythmogenic? *J Mol Cell Cardiol*. 1975;7:155. (letter)
66. Opie LH. Free fatty acids and acute myocardial infarction. *Lancet*. 1974;2:903.
67. Coraboeuf E, Dernbaix E, Hoerter J. Control of ionic permeabilities in normal and ischemic hearts. *Circ Res*. 1974;38(suppl 1):92.
68. Taylor IM, Shaikh NA, Downar E. Ultrastructural changes of ischemic injury due to coronary artery occlusion in the porcine heart. *J Mol Cell Cardiol*. 1984;16:79.
69. Wenzel DG, Innis JD. Arrhythmogenic and antiarrhythmogenic effects of lipolytic factors on cultured rat heart cells. *Res Commun Chem Pathol Pharmacol*. 1983;41:383.
70. Lucy JA. The fusion of biological membranes. *Nature*. 1970;227:815.
71. Shier WT, Trotter JT. Stimulation of cell surface phospholipase A_2 and prostaglandin synthesis in 3T3 mouse fibroblast by phallolysin, a toxin from Amanita phalloides. *Biochim Biophys Acta*. 1980;619:235.
72. Shier WT, Baldwin JH, Nilsen-Hamilton M, Hamilton R, Thanassi NM. Regulation of guanylate and adenylate cyclase activities by lysolecithin. *Proc Natl Acad Sci USA*. 1976;73:1586.
73. Zwiller J, Ciesielski-Treska J, Mandel P. Effect of lysolecithin on guanylate and adenylate cyclase activities in neuroblastoma cells in culture. *FEBS Lett*. 1976;69:286.
74. Hung SC, Melnykovych G. Alkaline phosphatase in Hela cells: stimulation by phospholipase and lysophosphatidylcholine. *Biochim Biophys Acta*. 1976;42:409.
75. Mayer MM, Michaelis DW, Ramm LE, Whitlow MB, Willoughby JB, Shin ML. Membrane damage by complement. *Crit Rev Immunol*. 1981; 2:133.
76. Muller-Eberhard HJ. The membrane attack complex. *Springer Sem Immunopathol*. 1984;7:93.
77. Hansch GM, Betz M, Shin ML. Cytolysis of nucleated cells by complement: inhibition of membrane-transmethylation enhances cell death by C5b-9. *J Immunol*. 1984;132:1440.
78. Roberts APA, Morgan BP, Campbell AK. 2-chloroadenosine inhibits complement-induced reactive metabolite reduction and recovery of human polymorphonuclear leucocyte attacked by

complement. *Biochem Biophys Res Comm.* 1985;126:692.

79. Carney DF, Koski CL, Shin ML. Elimination of terminal complement intermediates from the plasma membrane of nucleated cells: the rate of disappearance differs for cells carrying C5b-7 or C5b-8 or a mixture of C5b-8 with a limited number of C5b-9. *J Immunol.* 1985;134:1804.

80. Morgan BP, Dankert JR, Esser AF. Recovery of human neutrophils from complement attack: removal of the membrane attack complex by endocytosis and exocytosis. *J Immunol.* 1986;138:246.

81. Wiedmer T, Esmon CT, Sims PJ. On the mechanism by which complement proteins C5b-9 increase platelet prothrombinase activity. *J Biol Chem.* 1986;261:14587.

82. Lovett DH, Hansch GM, Goppelt M, Resch K, Gemsa D. Activation of glomerular mesengial cells by the terminal membrane attack complex of complement. J Immunol. 1987;138:2473.

83. Imagawa DK, Osfichin NE, Paznekas WA, Shin ML, Mayer MM. Consequences of cell membrane attack by complement: release of arachidonate and formation of inflammatory derivatives. *Proc Natl Acad Sci USA.* 1983;80:6647.

84. Imagawa DK, Osfchin NE, Ramm LE, Hammer CH, Shin ML, Mayer MM. Release of arachidonic acid and formation of oxygenated derivatives after complement attack on macrophages: role of channel formation. *J Immunol.* 1986;136:4637.

85. Imagawa DK, Barbour SE, Morgan PB, Wright TM, Shin HS, Ramm LE. The role of C-9 and calcium in the generation of arachiodonic acid and its metabolites from rat polymorph nuclear leukocytes. *Mol Immunol.* 1987;24:1263.

86. Hansch GM, Seitz M, Martinotti G, Betz M, Rauterburg EW, Gemsa D. Macrophages release arachidonic acid, prostaglandin E_2 and thromboxane in response to late complement components. *J Immunol.* 1984;133:2145.

87. Hansch GM, Gemsa D, Resch K. Induction of prostanoid synthesis in human platelets by the late components of C5b-9 and channel forming antibiotic nystatin: inhibition of reacylation of liberated arachidonic acid. *J Immunol.* 1985;135:1320.

88. Seeger W, Suttrop N, Hellwig A, Bahki S. Noncytolytic terminal complement complex may serve on calcium gates to elicit leukotriene B_4 generation in human polymorphonuclear leukocytes. *J Immunol.* 1986;137:1286.

89. Shirazi Y, Imagawa DK, Shin ML. Release of leukotriene B_4 from sublethally injured oligodendrocytes by terminal complement complexes. *J Neurochem.* 1987;48:271.

90. Shirazi Y, McMorris FA, Shin ML. Arachidonic acid mobilation and phosphoinositide turnover complement complex C5b-9 in rat oligodendrocyte X C_6 glioma cell hybrids. *NY Acad Sci.* 1988;540:381.

91. Hartung HP, Bitter-Suermann D, Hadding U. Induction of thromboxane release from macrophages by anaphylatoxic peptide C3a of complement and synthetic hexapeptide C3a 72-77. *J Immunol.* 1983;130:1345.

92. Hartung HP, Hadding U, Bitter-Suermann D, Gemsa D. Stimulation of prostaglandin E and thromboxane synthesis in macrophages by purified C3b. *J Immunol.* 1983;130:2861.

93. Regal JF. C5a-induced aortic contraction: effect of an antihistamine and inhibitors of arachidonic metabolism. *J Pharmacol Exp Ther.* 1982;222:102.

94. Rutherford B, Schenkein HA. C3 cleavage products stimulate release of prostaglandins by human mononuclear phagocytes *in vitro. J Immunol.* 1983;130:874.

95. Carney DF, Hammer CH, Shin ML. Elimination of terminal complement complexes in the plasma membrane of nucleated cells: effect of extracellular Ca^{2+} and association with cellular Ca^{2+}. *J Immunol.* 1986;137:263.

96. Morgan BP, Campbell AK. A recovery of human polymorphonuclear leucocytes from sublytic complement attack is mediated by changes in intracellular free calcium. *Biochem J.* 1985;231:205.

97. Jakschick BA, Sun FF, Lee L, Steinhoff MM. Calcium stimulation of normal lipoxygenase. *Biochem Biophys Res Comm.* 1980;95:103.

98. Betz M, Hansch G. Release of arachidonic acid: a new function of the complement components. *Immunobiology.* 1984;166:473.

99. Irvine RF. How is the level of free arachidonic acid controlled in mammalian cells? *Biochem J.* 1982;204:3.

100. Kroner EE, Peskar BA, Fischer H, Farber E. Control of arachidonic acid accumulation in bone marrow derived macrophages by acyl-transferases. *J Biol Chem.* 1981;256:3690.

101. Malis CD, Bonventre JV. Mechanism of calcium potentiation of oxygen free radical injury to renal mitochondria. *J Biol Chem.* 1986;261:14201.

102. Sevenian A, Muakkassah-Kelly SF, Montestruque S. The influence of phospholipase A_2 and glutathione peroxidase on the elimination of membrane lipid peroxides. *Arch Biochem Biophys.* 1983;223:441.

103. Weglicki WB, Dickens BF, Mak IT. Enhanced phospholipid degradation and lysophospholipid production due to free radicals. *Biochem Biophys Res Commun.* 1984;124:229.

104. Beckman JK, Borowitz SM, Burr IM. The role of phospholipase A activity in liver microsomal lipid peroxidation. *J Biol Chem.* 1987;262:1479.

105. Jain SK. Vitamin E and stabilization of membrane lipid organization in red blood cells with peroxidative damage. *Biomed Biochim Acta.* 1983;42:S43.

106. Tan KH, Meyer DJ, Belin J, Ketterer B. Inhibition of microsomal lipid peroxidation by glutathione and glutathione transferase B and AA. Role of endogenous phospholipase A_2. *Biochem J.* 1984;220:243.

107. Miyazaki Y, Nagai S, Ogawa K, Satake T, Sugiyama S, Ozawa T. The role of phospholipase in mitochondrial dysfunction after coronary reperfusion in the canine myocardium. *Jpn Circ J.* 1984;48:498.

108. Van Den Bosch H. Intracellular phospholipase A. *Biochim et Biophys Acta.* 1980;604:191.

109. Blackwell GJ, Carnuccio R, DiRosa M, Flower RS, Parente L, Persico P. Macrocortin: a polypeptide causing the anti-phospholipase effect of glucocorticoids. *Nature.* 1980;287:147.
110. Hirata F. The regulation of lipomodulin, a phospholipase inhibitory protein in rabbit neutrophils by phosphorylation. *J Biol Chem.* 1981;256:7730.
111. Cloix JF, Colard O, Rothhut B, Russo-Marie F. Characterization and partial purification of "renocortin": two polypeptides formed in renal cells causing the antiphospholipase-like action of glucocorticoids. *Br J Pharmacol.* 1983;79:313.
112. DiRosa M, Flower RS, Hirata F, Parente L, Russo-Marie F. Nomenclature announcement: antiphospholipase proteins. *Prostaglandins.* 1984; 28:441.
113. Hirata F, Matsuda K, Notsu Y, Hattori I, Carmine R. Phosphorylation of tyrosine residue of lipomodulin in nitrogen-stimulated murine thymocytes. *Proc Natl Acad Sci USA.* 1984;81:4717.
114. Tongui L, Rothut B, Shaw AM, Fradin A, Vargaftig BB, Russo-Marie F. Platelet activation: a role for 40 k anti-phospholipase A_2 protein indistinguishable from lipocortin. *Nature.* 1986;321: 177.
115. Davidson FF, Dennis EA, Powell M, Glenny JR. Inhibition of phospholipase A_2 by "lipocortin" and "Calpactins": an effect of binding to substrate phospholipids. *J Biol Chem.* 1987;262: 1698–1705.
116. Huang KS, Wallner BP, Mattaliano RJ, et al. Two human 35kd inhibitors of phospholipase A_2 are related to substrates of pp 60 and of the epidermal growth factor receptor/kinase. *Cell.* 1986;46:191.
117. Chiariello M, Ambrosio G, Cappelli-Bigazzi M, Nevola E. Inhibition of ischemia-induced phospholipase activation by quinacine protects jeopardized myocardium in rats with coronary artery occlusion. *J Pharmacol Exp Res.* 1987;241:560.
118. Craven PA, De Rubertis FR. Ca^{2+}-calmodulin-dependent release of arachidonic acid for renal medullary prostaglandin synthesis: evidence of involvement of phospholipase A_2 and C. *J Biol Chem.* 1983;258:4814.
119. Otomeri T, Frauzen L, Lindwerke D, Tagesson K. Increase of phospholipase A_2 and decreased lysophospholipase activity: the small intestinal mucosa after ischemia and revascularization. *Gut.* 1987;XX:1445.
120. Awashti YC, Ruzicka FJ, Crane FL. The relation between phospholipase action and release of NADH dehydrogenase from mitochondrial membrane. *Biochim et Biophys Acta.* 1970; 203:233.
121. Young EE, Farber JL. Accelerated phospholipid degradation in anoxic rat hepatocytes. *Arch Biochem Biophys.* 1983;211:312.
122. Taraschi TF, Petrovich DR, Farber JL. Spin-labeled phospholipids reveal domains of differing molecular membranes of ischemic rat liver. *Biochim Biophys Acta.* (in press)
123. Taraschi TF, Petrovich DR, Farber JL. ^{31}P nuclear magnetic resonance study of ischemic microsomal membranes: demonstration of phospholipid domains. *Biochim Biophys Acta.* (in press)
124. Petrovich DR, Finkelstein SD, Waring AJ, Farber JL. Liver ischemia increases the molecular order of microsomal membranes by increasing the cholesterol to phospholipid ratio. *J Biol Chem.* 1984;159:13217.
125. Post JA, Ruigrok TJC, Vekleij AJ. Phospholipid reorganization and bilayer destabilization during myocardial ischemia and reperfusion: a hypothesis. *J Mol Cell Cardiol.* 1988;20(suppl II):107.
126. Naylor WG, Panagiotopoulos S, Elz JS, Daly MJ. Calcium-mediated damage during post ischemic reperfusion. *J Mol Cell Cardiol.* 1988;20:41.

9

Cardiac Adenine Nucleotide Metabolism in Cell Injury and Restitution

Heinz-Gerd Zimmer

INTRODUCTION

It is well established that high energy phosphates play an important role in the maintenance of structure, function, and metabolism of an organ. In this chapter, the major emphasis will be on the myocardium. In this organ, contraction, the ionic pumps, phosphorylation processes, and several biosynthetic pathways are all energy dependent. It is therefore mandatory that ATP, ADP, creatine phosphate, and inorganic phosphate—thought to be instrumental in the cell's ability to balance the rate of energy conversion and work output—are kept fairly constant during the cardiac cycle and over a wide physiological range of work.[1,2] Any major perturbation in oxygen demand and supply leads to a rapid decline of high energy phosphates in the myocardium as well as in other organs.[3-7] It is thus reasonable to consider adenine nucleotide metabolism in cell injury. Almost all cardioprotective strategies developed in the past were directed toward normalizing the content of high energy phosphates with the ultimate aim of improving heart function.

In very general terms, to determine whether a metabolic compound is involved in causing cardiac injury when its content is changed, several experimental approaches may be utilized:

1. The content of the particular compound can be correlated with the respective functional states in different experimental conditions. This approach has been applied in regard to the ATP level to characterize the loss of function early during myocardial ischemia.

2. In another approach, the respective metabolic compound may be altered by an experimental procedure and its level at the end of the intervention period can then be related to function during the recovery period. This has been done, for instance, with respect to lactate, which accumulates during myocardial ischemia.[8]

3. In the third experimental design, the synthesis of the compound in question can be modified and then function can be evaluated or the morphology can be examined. This approach has been utilized in our own long-term studies on intact small laboratory animals with respect to cardiac adenine nucleotides. Procedures that can be applied to experimental *in vivo* models are required. The advantage of this design is that

physiologically relevant results can be obtained and that those metabolic interventions that turn out to be cardioprotective may then be used therapeutically in humans.

In this chapter, the metabolic changes associated with cell injury will first be discussed to illustrate that adenine nucleotide metabolism is an important and critical segment of metabolic events that have been incriminated in cell injury. Since our own studies were all done in closed-chest small animals, the experimental procedures necessary for those studies will be described, and the models used will be characterized in functional and metabolic terms. To put the subject into perspective, the pharmacologic and metabolic interventions utilized for cardioprotection will be discussed, with particular emphasis on those procedures that interfere directly with cardiac adenine nucleotide metabolism. From all these metabolic interventions a substrate has emerged that is particularly suited for the *in vivo* stimulation of the *de novo* synthesis of cardiac adenine nucleotides. The metabolic, functional, and morphologic consequences of this approach will then be delineated for several pathophysiologic *in vivo* models. Finally, criteria will be given that have been developed for selection of a cardioprotective substrate. It will be shown that all the specifications are met by ribose, a substrate that has been found to have prominent cardioprotective potency.

METABOLIC CHANGES ASSOCIATED WITH CELL INJURY

Depletion of High Energy Phosphates

It is well known that there is a rapid deterioration in heart function subsequent to global ischemia in the isolated perfused heart and after regional ischemia in the heart *in situ*, induced by ligation of a coronary artery.[9-11] In an isolated rat heart preparation, a substantial decrease in myocardial ATP and creatine phosphate content occurred after the onset of anoxia before contractile failure was evident.[12] Several other studies on isolated rat or rabbit heart preparations demonstrated a good correlation between cardiac function and myocardial ATP content.[13-15] Also, in the *in situ* dog ventricle, changes in ATP concentration were seen to parallel changes in mechanical function of the nonischemic portion of the infarcted myocardium.[16] Interestingly, brief periods of ischemia in the dog heart resulted in a marked depression of regional heart function that lasted for hours and in a decrease of the cardiac ATP content that lasted even for some days both in the dog and in the rat hearts.[17-21] In this context, it is appropriate to mention that repetitive episodes of brief ischemic periods did not lead to a cumulative depletion of high energy phosphate compounds.[22] The total adenine nucleotide content declined significantly with the first occlusion of 12 minutes duration in both the epicardium and the endocardium, and there was no further significant change in total adenine nucleotide content during the subsequent occlusion periods. This result may indicate that susceptible ATP pools may be depleted during the first ischemic period, while the more resistant pools may remain intact during the subsequent ischemic periods.

It has previously been suggested that a compartmentalization of adenine nucleotides exists.[15-26] Apart from studies showing a good correlation between myocardial ATP content and function, there is also evidence of a close relationship between the marked depletion of high energy phosphates and the development of lethal injury in morphologic terms; that is, irreversible myocardial cell damage occurring in the acutely ischemic myocardium.[27] Although there is a good correlation between residual ATP on the one hand and mechanical function and morphology on the other hand, a cause-and-effect relationship has not been clearly established.[8] Alternatively, the decline in total tissue ATP to a critical level may not be the determining factor for the initiation of cell injury but rather the source from which residual tissue ATP is derived. It has been shown that ATP produced by glycolysis was better able to delay or prevent ischemic contracture than ATP produced by oxidative phosphorylation.[26]

Accumulation of Glycolytic Products

In a recent study, Neely and Grotyohann found that there was a good correlation between residual ATP and heart function only for isolated rat hearts in which high levels of lactate accumulated when 2.5 mmol Ca^{++} were present in the perfusion medium.[8] Depletion of glycogen and removal of lactate prior to the ischemic period resulted in a much better recovery of ventricular function. An excellent negative correlation was observed between tissue levels of lactate during ischemia and recovery of function with reperfusion. Also, when lactate was added to the perfusion medium prior to ischemia, ventricular function during the postischemic period deteriorated linearly with added perfusate lactate. From this study it appeared that ventricular function with reperfusion is inversely related to tissue lactate during ischemia, with no obvious correlation existing between residual ATP and functional recovery. So, high levels of lactate during ischemia seem to be associated with accelerated cellular damage independent of the ATP pool.

The role of metabolic products such as lactate in the development of ischemic damage has been appreciated for many years. Tennant and Wiggers in 1935 had recognized that a reduction in coronary flow has a profound effect on cardiac contractility, and the studies of Tennant already pointed to excess lactate as a possible factor in preventing contraction under anoxic conditions.[10,11] However, the mechanism for the

harmful effects of lactate accumulation is not known. The tissue damage associated with high tissue levels of lactate during ischemia may be mediated by changes in intracellular pH. In the isolated interventricular septum of the rabbit heart, the onset of acidosis preceded any decline in mechanical function during total ischemia.[28]

In another study, intracellular acidosis was estimated to account for from 40 to 50% of the depression in left ventricular developed pressure of the isolated perfused rat heart during the early phases of ischemia.[29] In the isolated perfused rat heart, the fall in intracellular and extracellular pH was found to be the principal determinant of the decline of pressure development during the early phase of myocardial ischemia.[30] In this context, the interference with the contractile process by the accumulation and retention of phosphate ions liberated from creatine phosphate and from ATP has to be among our considerations.[31] Increased cell phosphate levels may cause calcium to be sequestered and trapped in the sarcoplasmic reticulum and the mitochondria so that this cation is not available for participation in excitation–contraction coupling. In this way, the early "pump" failure of the ischemic heart may be explained.

Calcium Overload

The idea that calcium overload may be involved in causing cell injury was developed on the basis of experiments in which the effects of beta-adrenergic overstimulation of the heart were studied.[32] As is well known, sympathetic catecholamines increase cardiac contractility by enhancing Ca^{++}-dependent utilization of high energy phosphates in the contractile machinery. With high doses of isoproterenol, Ca^{++}-uptake and high energy phosphate breakdown become excessive, so that depletion of ATP and creatine phosphate results and focal myocardial cell lesions develop.[33] The situation is quite similar in the myocardium subsequent to ischemia. While no significant uptake of $^{45}Ca^{++}$ occurred after 60 minutes of ischemia produced by permanent occlusion of a coronary artery in the dog, 40 minutes of ischemia followed by 10 minutes of arterial reflow resulted in an 18-fold increase in Ca^{++}-uptake. Reversible myocardial injury induced by 10 minutes of ischemia was not followed by Ca^{++}-accumulation during a 20 minute reperfusion period.[34] The calcium antagonist verapamil reduced the net Ca^{++} gain in the rabbit interventricular septum when given before the ischemic period of 60 minutes duration and reperfusion for 30 minutes, but not when given upon, or 10 minutes prior to, reperfusion.[35]

Also, Ca^{++} ions have been shown to have deleterious effects upon reperfusion of the isolated rat heart after a Ca^{++}-free perfusion period, a phenomenon called *calcium paradox*.[36] It occurs as soon as Ca^{++} is readmitted to the heart and is characterized by an influx of Ca^{++} into the myocardial cell, exhaustion of high energy phosphates, the rapid onset of myocardial contracture, massive release of cell constituents and excessive ultrastructural damage. Interestingly, addition of free Ca^{++} ions to the perfusate during low flow ischemia in the isolated rat heart accelerated the onset of ventricular failure during the postischemic recovery period.[8] Thus, in this experimental condition also, Ca^{++} overload has a negative effect on heart function.

With respect to the mechanism, it has been shown that Ca^{++}-free solution produced a distinct separation of the membrane external lamina from the surface coat of the sarcolemma that was irreversible and correlated with increased cellular $^{45}Ca^{++}$ content and irreversible contracture upon reperfusion with normal Ca^{++}.[37] In addition, there is considerable disruption of the myofilaments, swelling of the mitochondria with formation of electron-dense particles, and disruption of the sarcolemma and intercalated discs. On the other hand, the calcium paradox may be associated with a defect in the ability of sarcoplasmic reticulum to regulate Ca^{++}.[38] The results of studies such as these emphasize that the sarcolemma is the critical structure.

Activation of Phospholipases

Phospholipids are integral constituents of the cell membrane. Changes in the barrier function of the sarcolemma to Ca^{++} ions may therefore be brought about by degradation of sarcolemmal phospholipids induced by phospholipases A_1 and A_2, which hydrolyze diacyl phospholipids to form lysophosphatides and free fatty acids.[39] Experimental myocardial ischemia in dogs resulted in a small but measurable decrease in total phospholipid content in the subendocardium. After 3 hours of ischemia, there was a 10% decrease in total phospholipid content, in particular in phosphatidyl choline and phosphatidyl ethanolamine. In regard to permeability disturbance, it was shown that sarcolemmal vesicles isolated from ischemic myocardium after 3 hours of ischemia had an increased Ca^{++} permeability and phosphatidyl choline depletion.[40] Thus, there was a relationship between alterations in phospholipid metabolism and the development of irreversible damage in the ischemic myocardium. *In vitro* studies on dog hearts, however, indicated that lysophospholipid accumulates in the ischemic myocardium, but that this accumulation occurs slowly and does not appear to be the dominant mechanism of plasma membrane fragmentation.[41]

Oxygen-derived Free Radicals

The superoxid free radical O_2^- has been implicated in several pathophysiological conditions : oxygen toxicity, inflammation, and ischemia-induced tissue injury.[42] In the myocardium, superoxide free radicals may be involved in reperfusion damage, the main source being the xanthine oxidase reaction. During ischemia, xanthine dehydrogenase has been demonstrated to convert to xanthine oxidase.[42,43] Also, hypoxanthine, the first substrate for this enzyme, originates quite rapidly during ischemia as a result of the adenine nucleotide degradation via adenosine and inosine.[6] When oxygen is supplied during reperfusion, there appears to be a burst of superoxide radical production. Using allopurinol, an inhibitor of xanthine oxidase, it was found in dogs that myocardial infarct size subsequent to 1 hour of total occlusion of the left anterior descending coronary artery followed by 4 hours of reperfusion was significantly reduced.[43] Also, superoxide dismutase and catalase, two specific free radical metabolizing enzymes, were capable of reducing infarct size in dogs.[43,44]

In another study, however, in which the circumflex coronary artery was occluded in dogs for 40 minutes followed by reperfusion for 4 days, allopurinol had no effect at all on infarct size.[45] Another source of superoxide free radicals may be the polymorphonuclear neutrophils. In this context it is interesting to mention that the antiinflammatory agent ibuprofen has been shown to have cardioprotective effects.[46] Also, the depletion of neutrophils reduced the extent of the leucocytic infiltrate in infarcted myocardium and was associated with a 43% reduction in the extent of irreversible myocardial injury.[47]

FUNCTIONAL AND METABOLIC CHARACTERIZATION OF EXPERIMENTAL *IN VIVO* MODELS OF MYOCARDIAL INJURY

Experimental Procedures

Continuous IV Infusion in Conscious Small Laboratory Animals

In order to administer drugs and metabolic substrates continuously in rats, a constant infusion model was applied. In ether anesthesia, a Steriflex catheter of 1 mm outer diameter (Vygon Company, Aachen) was implanted in the right jugular vein. It was then subcutaneously directed to the neck of the animal and connected to a 20-ml syringe that was mounted in an infusion pump (type 5003, Infors AG, Hofstetten bei Basel). The infusion rate in the animals (200–250 g body weight) was 5 ml/kg/hour. After surgery, the rats could move around freely in their individual cages.[48] The advantage of this infusion model is that it is inexpensive and versatile. The infusion rate can be changed according to the needs, and different agents can be administered in combination.

Measurement of Heart Function in Closed-Chest Rats

For measurements of left ventricular functional parameters, the ultraminiature catheter pressure transducer (model PR-249, Millar Instruments, Inc., Houston, Texas) was used. The sensor portion at the tip of the catheter is 5 mm in length and has an outside diameter of 0.9 mm, equivalent to 3 French. The catheter itself has an outside diameter of only 0.5 mm. The catheter was connected to a Millar transducer control unit (model TC-100) that was connected, with a HSE-Elektro-Manometer (Hugo Sachs Elektronik, March-Hugstetten). The maximal rate of rise in left ventricular pressure (LV dP/dt_{max}) was obtained with an electronic differentiation system (Physiol-Differentiator, Hugo Sachs Elektronik, March-Hugstetten). The recording system was a Gould Brush 2600 recorder. Calibration of pressure was made with a mercury pressure calibrator (type 367, Hugo Sachs Elektronik, March-Hugstetten).

The ultraminiature catheter pressure transducer was implanted in the right carotid artery using thiobutabarbital sodium (inactin[R]) anesthesia and was then advanced into the left ventricle. Heart rate and left ventricular pressure were directly obtained, as was LV dP/dt_{max}. After completion of these measurements, the catheter was withdrawn from the left ventricle and positioned in the aorta for the measurement of blood pressure. In addition, the pressure–rate product was calculated. The method of left heart catheterization in small laboratory animals is rapid, easy to perform, reliable and nondestructive.[49,50]

Measurement of Metabolic Parameters

Glucose-6-Phosphate Dehydrogenase and 6-Phosphogluconate Dehydrogenase. When the activity of glucose-6-phosphate dehydrogenase and 6-phosphogluconate dehydrogenase was to be measured, the respective organs were perfused free of blood. To do this in the heart, a cannula was placed in the ascending aorta after thoracotomy in ether anesthesia and then tightly fixed. The hearts were rapidly excised, and the coronary arteries were perfused via the cannula with an ice-cold KC1 solution 0.15 mol containing 8 ml of 0.02 mol $KHCO_3$/l) to remove blood and to stop beating. Perfusion of liver and kidney was done accordingly. Homogenization of the organs in the perfusion medium, centrifugation, dialysis of the supernatant, and measurement of the enzyme activities were done according to the methods of Glock and McLean.[51,52]

Availability of 5-Phosphoribosyl-1-Pyrophosphate. The available pool of 5-phosphoribosyl-1-pyrophosphate (PRPP) was assessed by measuring the incorporation of ^{14}C-adenine or ^{14}C-inosine into the adenine nucleotides after an *in vivo* exposure time of 15 minutes. PRPP is consumed in the reaction in which adenine or the originating hypoxanthine from inosine degradation are converted to AMP and IMP, respectively. The resulting total radioactivity of the adenine nucleotides serves as an indirect measure of the available PRPP pool.[53]

De novo Synthesis of Adenine Nucleotides. Rates of the *de novo* synthesis of adenine nucleotides were determined on the basis of the incorporation of 1-^{14}C-glycine into adenine nucleotides. The resulting radioactivity was related to the mean specific activity of the tissue glycine precursor pool. This procedure allowed researchers to obtain actual rates of adenine nucleotide *de novo* synthesis in various rat organs after an *in vivo* exposure time to 1-^{14}C-glycine of 60-minutes duration.[54] The content of adenine nucleotides was determined using the method of Gerlach et al. or by HPLC methods.[55]

Experimental *in Vivo* Models of Cardiac Injury in Rats

Overload Leading to Cardiac Hypertrophy

To induce cell injury, three models of cardiac overload were studied. In the first, pressure overload was induced by constriction of the abdominal aorta.[56] In hemodynamic terms, this hypertrophy model is characterized by a slight decrease in heart rate and a moderate increase in left ventricular systolic pressure and LV dP/dt_{max} during the first 3 days.[57] In the initial phase, that is, during the first 5 hours, there is a decline in the *de novo* synthesis of adenine nucleotides and in protein synthesis, as well as a decrease in the content of adenine nucleotides and creatine phosphate.[56,58] Later on, the biosynthetic processes are enhanced and precede the increase in heart weight.

The changes in left ventricular functional parameters occurring in the isoproterenol-induced cardiac hypertrophy are shown in Figure 9-1. Heart rate and LV dP/dt_{max} are increased initially and then return to the respective control levels. On the other hand, left ventricular systolic pressure is decreased.[57] In metabolic terms, there is an immediate stimulation of adenine nucleotide and protein synthesis as well as of other metabolic parameters.[58] Morphologically, the isoproterenol model is characterized by the occurrence of focal myocardial cell lesions.[33] A marked depression of all left ventricular hemodynamic parameters is induced when rats with aortic constriction

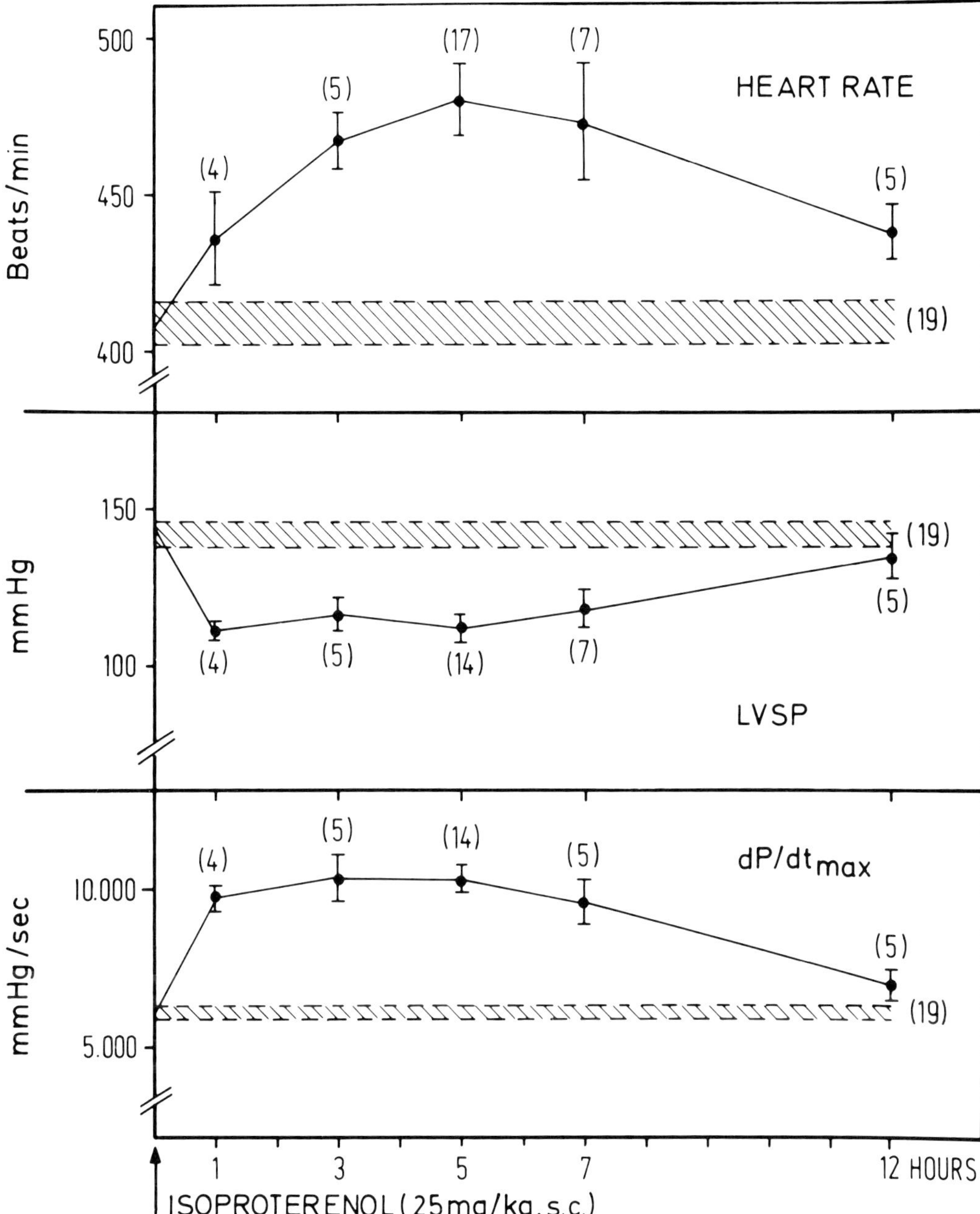

Figure 9-1. Effect of isoproterenol on heart rate, left ventricular systolic pressure (LVSP), and on the maximal rate of rise in left ventricular pressure (dP/dt_{max}) in rats. After the different periods of time following s.c. isoproterenol application, the Millar ultraminiature catheter pressure transducer (model PR-249) was inserted into the left ventricle in Inactin[R]-anesthesia. All isoproterenol-induced alterations were significantly different from controls except for LVSP after 12 hours. Mean values ± SEM; number of experiments in parentheses. (From Zimmer, 1984. By permission of Acadademic Press.[57]

are injected with a single s.c. dose of isoproterenol.[59]

Marked functional changes are also observed when rats are treated daily with triiodothyronine. In this cardiac hypertrophy model, heart rate and LV dP/dt_{max} are enhanced very early, and this enhancement is maintained for the duration of triiodothyronine treatment.[57] There is a variety of metabolic alterations associated with this hypertrophy model.[58]

Intermittent Asphyxic Periods

A suitable experimental model for short-term studies is the induction of asphyxia in anesthetized rats.[60] When artificial respiration is switched off repetitively for 1 minute each time with recovery periods of 3 minutes in between, heart rate, left ventricular systolic pressure and LV dP/dt_{max} decline very rapidly, while LVEDP becomes elevated (Figure 9-2). During the first two asphyxic periods, there is a small peak in LVSP and in LV dP/dt_{max}, possibly indicat-

ing the effect of catecholamines released during oxygen deficiency. The more asphyxic periods are applied, the deeper and longer in duration is the depression in ventricular function. In addition, arrhythmias develop that also become more pronounced as the number of asphyxic periods increases. Four such episodes of oxygen deficiency of 1-minute duration with an additional 30-second asphyxic period were sufficient to induce a depletion of ATP (Figure 9-2). Despite this decline in ATP, the functional parameters returned to the respective control very rapidly. This experimental model is limited in duration since the animals have to be kept in anesthesia and under artificial respiration. For long-term studies, another model is required.

Experimental Myocardial Infarction

The experimental myocardial infarction model is clearly designed for long-term studies when coronary artery ligation is done either permanently or transiently. The de-

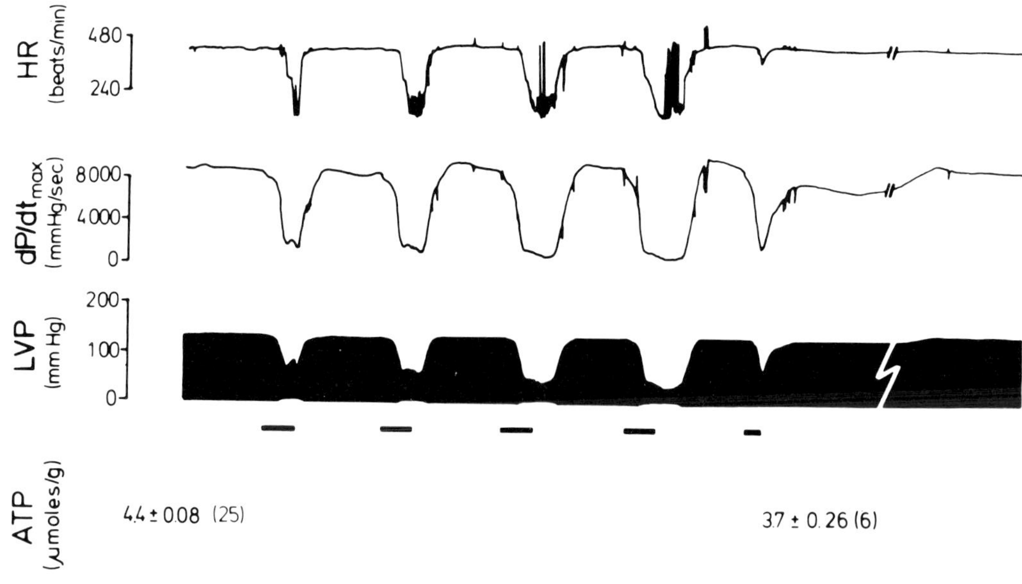

Figure 9-2. The effect of several periods of asphyxia (horizontal black bars) on heart rate (HR), dP/dt_{max}, and left ventricular pressure (LVP), as well as on the cardiac ATP level in rats anesthetized with ether and ventilated artificially using a Rodent respiration (Harvard Instruments). The ATP values are means ± SEM; number of experiments in parentheses.

scending branch of the left coronary artery in rats is ligated near the region between the origin of the pulmonary artery and the left auricle.[61] This leads to ischemia and ultimately to infarction of almost the entire left ventricular free wall and has marked hemodynamic effects on the left ventricle, as evidenced by a progressive decline in LVSP, LV dP/dt_{max}, and a pronounced elevation in LVEDP.[62]

In this situation, the heart is primarily dependent on the nonischemic portion. Accordingly, adaptive metabolic processes take place in this part of the myocardium, such as the increase in the activity of glucose-6-phosphate dehydrogenase and the stimulation of adenine nucleotide and protein synthesis.[62] When the left anterior descending (LAD) coronary artery is ligated for a brief period of time, the metabolic alterations as well as the influence of therapeutic and metabolic interventions can be studied in the previously ischemic part of the heart.[63]

PHARMACOLOGIC AND METABOLIC INTERVENTIONS FOR CARDIOPROTECTION

Pharmacologic Agents

In general, cardioprotection may be defined in two ways. First, all procedures that are suitable to protect the myocardium from risk factors such as high blood pressure, diabetes, and hyperlipidemia may be considered cardioprotective; that is, cardioprotection by prevention. On the other hand, in the case of imminent or apparent myocardial damage, one can attempt to limit the extent of cardiac injury or to achieve repair and restoration; that is, cardioprotection by restitution. In most acute clinical situations, the efforts are mainly aimed at restitution.

When myocardial injury is induced by ischemia, the most straightforward intervention is the attempt to restore coronary blood flow. During the last few years, this has been achieved by administration of intracoronary or intravenous streptokinase, which dissolves new thrombi; by mechanical transluminal recanalization of occluded coronary arteries; or by immediate coronary artery bypass grafting.[64-67] Another approach is aimed at reducing myocardial oxygen demand.[68,69] In this context, beta-receptor blockers have to be mentioned, since they decrease heart rate and contractility.[68] Also, calcium antagonists have been shown to exert beneficial effects.[32] Verapamil treatment resulted in significantly less necrosis, with minimal hemodynamic consequences after coronary artery occlusion in dogs. Higher doses were even more effective in limiting cardiac injury despite hypotension and heart block.[70] Moreover, verapamil and nifedipine preserved high energy phosphates to a certain extent both in the isolated perfused rat heart and in the dog heart *in vivo* when measured during the reperfusion period following 15 minutes of ischemia.[71,72]

Influence on Cardiac Carbohydrate and Lipid Metabolism

It is well established that in the ischemic myocardium, glycolysis is first stimulated, then decreased, and finally stops entirely.[73-75] Several approaches have therefore been developed to influence carbohydrate metabolism. Infusion of glucose–insulin–potassium is one possibility.[76] In experimental studies, a protective effect against ischemia and a reduction of myocardial necrosis has been shown.[77] In order for this intervention to be beneficial, the degree of perfusion must be adequate to prevent the accumulation of high tissue levels of lactate that inhibit glycolysis and prevent glycolytic stimulation by hyperglycemia and insulin.[78] In the early phase of acute myocardial infarction in humans, however, there was no clinically significant effect of this treatment with regard to ventricular arrhythmias.[79]

Application of pyruvate may be another potential approach.[80] However, intracoronary pyruvate failed to limit infarct size during a 90-minute reperfusion subsequent to 3 hours of circumflex coronary artery occlusion.[81] On the other hand, administration of dichloroacetate can be envisaged and has been shown to stimulate glycolysis by increasing the activities of phosphofructokinase and pyruvate dehydrogenase.[82]

With respect to fatty acid metabolism, myocardial lipolysis can be inhibited by beta-pyridyl-carbinol, since it has been shown that oxidation of free fatty acids increases the myocardial oxygen requirements.[83,84] Furthermore, carnitine may act by reversing the inhibition of adenine nucleotide translocase, which is induced by long chain fatty acyl-CoA esters and therefore attenuates the decline of ATP in the cytosol.[85-87]

Interventions in Myocardial Adenine Nucleotide Metabolism

In the discussion of cardiac adenine nucleotide metabolism, it is appropriate to consider first the degradation and synthesis of these high energy phosphates and then the possibilities of interfering with either of these processes (Figure 9-3). The degradation of ATP via ADP and AMP may proceed either via IMP or via adenosine. The pathway via adenosine seems to prevail in the myocardium.[88,89] During ischemia, adenine nucleotide breakdown in the cardiomyocyte proceeds only to inosine. Due to the localization of nucleoside phosphorylase and xanthine oxidase in microvascular endothelium, further degradation of inosine to hypoxanthine, xanthine, and uric acid

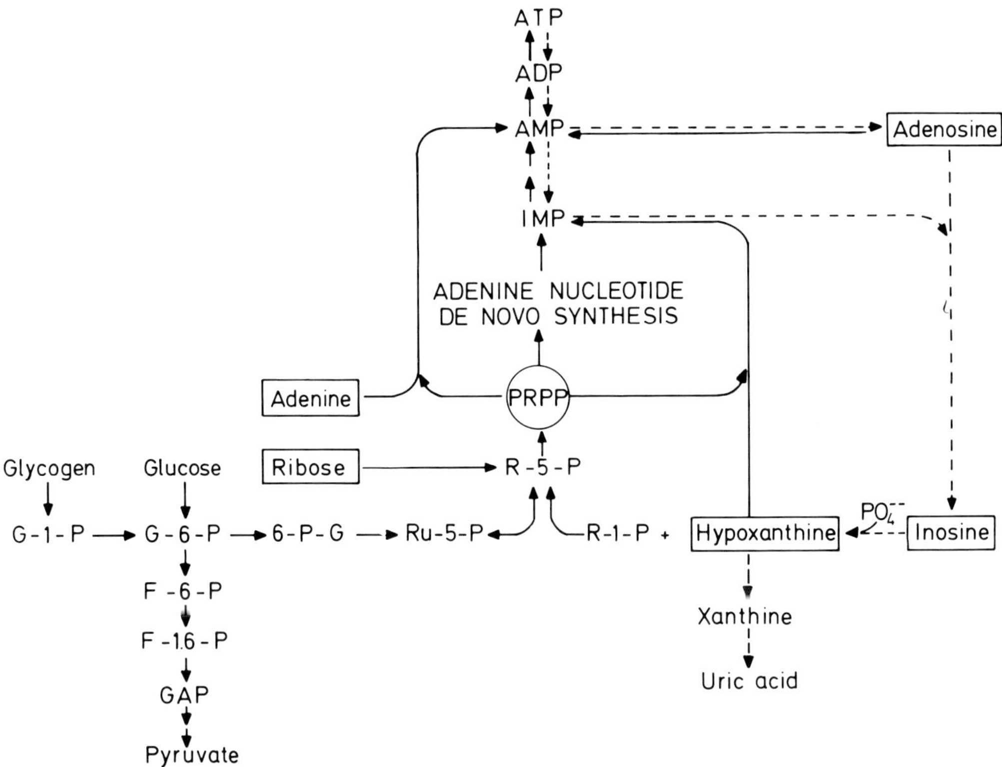

Figure 9-3. Pathways of adenine nucleotide degradation and synthesis in the myocardium. G-1-P: glucose-1-phosphate; G-6-P: glucose-6-phosphate; F-6-P: fructose-6-phosphate; F-1,6-P: fructose-1,6-phosphate; GAP: glycerinaldehydephosphate; 6-P-G: 6-phosphogluconate; Ru-5-P: ribulose-5-phosphate; R-1-P: ribose-1-phosphate; R-5-P: ribose-5-phosphate; PRPP: 5-phosphoribosyl-1-pyrophosphate.

occurs predominantly in the vascular space.[90,91]

There are two possibilities of restoring the cardiac adenine nucleotide pool once it has been depleted. The first is the utilization or reutilization of degradation products in the so-called salvage pathways. The most direct pathway is the conversion of adenosine to AMP by adenosine kinase. The other reactions need 5-phosphoribosyl-1-pyrophosphate, such as the reaction in which adenine is transformed to AMP by adenine phosphoribosyl-transferase. Inosine can be utilized only after its degradation to hypoxanthine. The latter is then converted to IMP by hypoxanthine phosphoribosyltransferase.[92]

In many experimental conditions, the degradation products to be utilized are washed out from the myocardium, for example, during the reperfusion period subsequent to ischemia, so that the "salvage pathways" will not become effective. In these situations, the heart is mainly dependent on the *de novo* formation of adenine nucleotides from small molecular precursor substances such as glycine, asparate, glutamine, folate, and carbon dioxide.[93] As is evident from Figure 9-3, the *de novo* synthesis (biosynthesis) of adenine nucleotides is entirely dependent on the available 5-phosphoribosyl-1-pyrophosphate. This is provided by ribose-5-phosphate which is derived either from ribose-1-phosphate or from ribulose-5-phosphate. The latter originates in the oxidative pentose phosphate pathway (see "The Oxidative Pentose Phosphate Pathway"), which is an alternative to glycolysis, the main route for glucose metabolism. There are some possibilities of interfering directly with cardiac adenine nucleotide metabolism.

Prevention of Purine Loss

One approach is the inhibition of adenosine transport by dipyridamole.[94,95] It has been reported that dipyridamole significantly increased coronary blood flow and improved global heart function during reperfusion following ischemia in open-chest

dogs.[96] Dipyridamole also reduced infarct size considerably in conscious dogs, most likely due to an increase in collateral blood flow.[97] Another approach is directed to inhibition of the catabolism of adenosine by blocking adenosine deaminase with erythro-9-(2-hydroxy-3-nonyl) adenine hydrochloride (EHNA).[98,99] In particular, the combined EHNA and adenosine treatment of dogs after a 20-minute normothermic ischemia resulted in recovery of ATP levels to 88% of the controls.[98]

The last approach to be mentioned here is the inhibition of xanthine oxidase by allopurinol, which was discussed in the context of oxygen-derived free radicals (see "Oxygen-derived Free Radicals"). It has not yet been determined conclusively whether the beneficial effect of allopurinol, when present, is due to prevention of superoxide radical production or to prevention of irreversible loss of hypoxanthine from the myocardium. It was demonstrated previously that allopurinol has profound effects in the ischemic myocardium and during hemorrhagic shock.[100–102]

Supply of Purine Compounds

An alternative possibility is to raise the adenine nucleotide level by increasing the supply of purine compounds, for example, during postischemic reperfusion. All substrates marked in Figure 9-3 appear to be suitable for this purpose. The most direct approach is the administration of adenosine, since it is immediately converted to AMP by adenosine kinase. Adenosine has been shown to increase the cardiac adenine nucleotide content, and this effect was observed also when the concomitant hypotension was prevented.[103–106]

The remaining compounds all require 5-phosphoribosyl-1-pyrophosphate. Adenine has been shown to have beneficial effects, especially when combined with ribose.[107,108] Inosine is first degraded to hypoxanthine, and this can then be converted to IMP.[92] In several models, inosine proved to affect the cardiac adenine nucleotide pool.[109–12] The acute hemodynamic effects of inosine in rats will be discussed later (see "Criteria for Selection of a Cardioprotective Substrate").

Stimulation of Adenine Nucleotide Biosynthesis

In view of the fact that the adenine nucleotide degradation products are released from the myocardial cell, washed out from the heart, and thus are not available for restitution unless supplied to the myocardium (see "Supply of Purine Compounds"), it is a logical approach to stimulate the *de novo* synthesis of cardiac adenine nucleotides. Two procedures have been developed in this regard. The first is the application of "AICAR" (5-amino-4-imidazolcarboxamide riboside).[113,114] This compound is an intermediate in the *de novo* synthesis pathway one step before the formation of IMP. It has been shown to stimulate myocardial adenine nucleotide biosynthesis; however, it is also converted to 5-amino-4-imidazolcarboxamide ribose triphosphate, which is not found in the heart muscle under normal conditions.[114]

On the other hand, ribose stimulates the formation of 5-phosphoribosyl-1-pyrophosphate, to which the individual components of the purine ring are successively added during the *de novo* formation. Thus, ribose interferes with the initial steps in purine nucleotide biosynthesis and induces a marked stimulation of this synthetic process.[53] Since pyridimine nucleotide synthesis is dependent on 5-phosphoribosyl-1-pyrophosphate, it is also enhanced by ribose.[115,116] The ribose-elicited stimulation of purine and pyrimidine nucleotide biosynthesis in the heart can best be explained when the oxidative pentose phosphate pathway is taken into consideration.

THE OXIDATIVE PENTOSE PHOSPHATE PATHWAY

Glucose-6-phosphate originating from glycogenolysis or from glucose taken up by the myocardial cell is metabolized mainly via glycolysis (Figure 9-4, right-hand side). An alternative possibility is metabolization via the oxidative pentose phosphate pathway (central portion in Figure 9-4), of which glucose-6-phosphate dehydrogenase (G-6-PDH) is the first and rate-limiting enzyme.[117] After a further step catalyzed by lactonase, 6-phosphogluconate is converted to ribulose-5-phosphate by 6-phosphogluconate dehydrogenase. From ribulose-5-phosphate originates ribose-5-phosphate, the immediate precursor for 5-phosphoribosyl-1-pyrophosphate. This substrate is essential both for pyrimidine nucleotide and purine nucleotide biosynthesis.

The oxidative pentose phosphate pathway has many important functions. It provides reducing equivalents in the form of NADPH. In this way, oxidized glutathione (GSSG) can be transformed to reduced glutathione (GSH). This system is of particular significance in erythrocytes. In patients with glucose-6-phosphate dehydrogenase deficiency, certain drugs lead to hemolytic anemia.[118] In the myocardium, an increase in GSSG may indicate oxidative stress such as occurs during postischemic reperfusion. Moreover, NADPH is needed for the synthesis of fatty acids and steroids, especially in the liver, mammary gland, and adrenal cortex. There are connections to glycolysis via the transaldolase and transketolase reactions that have not been investigated in full detail. The most important function of the oxidative pentose phosphate pathway in the present context is the generation of ribose-5-phosphate for the synthesis of 5-phosphoribosyl-1-pyrophosphate.

To assess the capacity of the oxidative pentose phosphate pathway in various organs of the rat, the activity of the first and rate-limiting enzyme, glucose-6-phosphate dehydrogenase, and the available pool of 5-phosphoribosyl-1-pyrophosphate were determined and compared with the rates of adenine nucleotide biosynthesis. Figure 9-5 shows that among the organs tested, the highest activity of glucose-6-phosphate dehydrogenase was found in the kidney and liver, followed the heart and skeletal muscle.

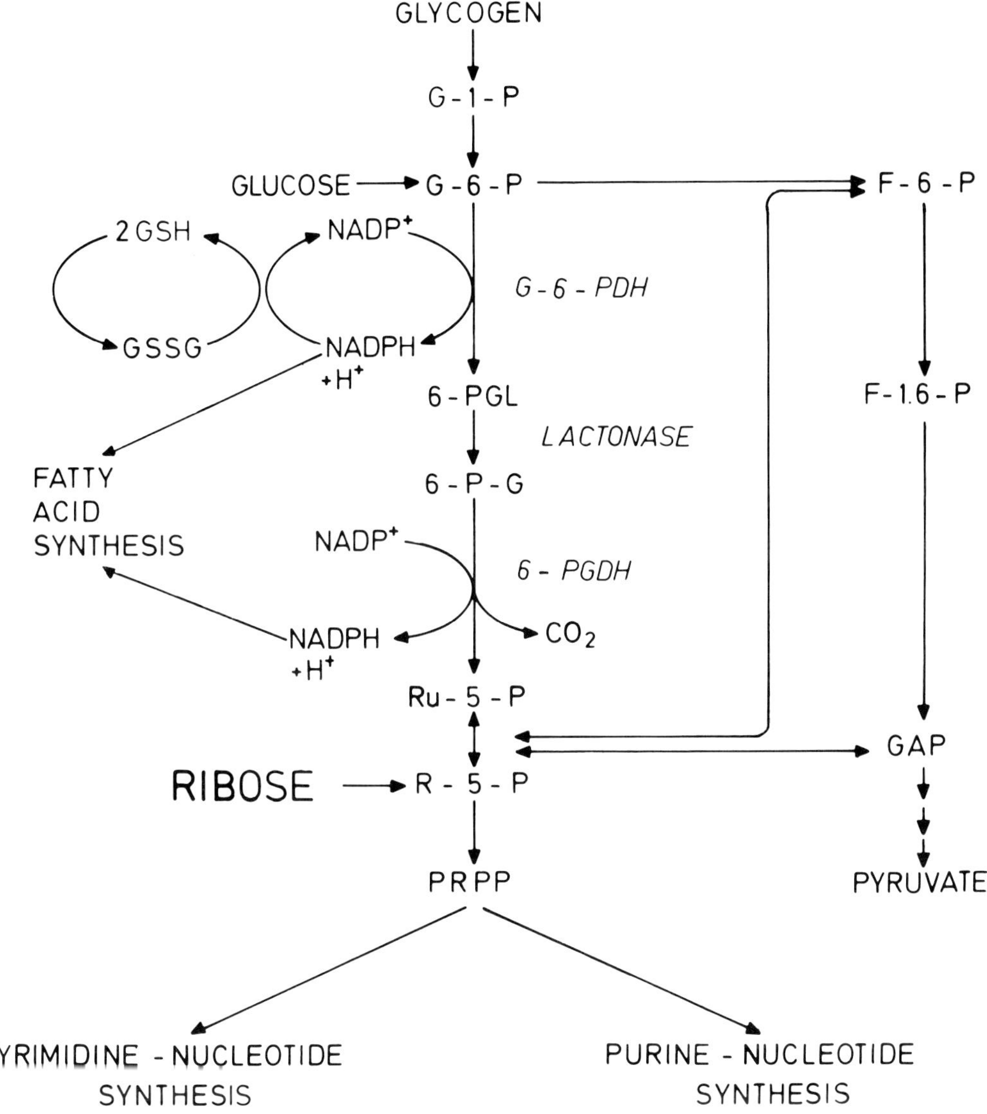

Figure 9-4. Schematic presentation of the oxidative pentose phosphate pathway. The connections of the nonoxidative segment via the transaldolase and transketolase reactions are indicated by arrows. Abbreviations as in Fig. 9-3. G-6-PDH: glucose-6-phosphate dehydrogenase; 6-P-GDH: 6-phosphogluconate dehydrogenase.

An essentially similar pattern was obtained in these organs regarding the available pool of 5-phosphoribosyl-1-pyrophosphate and the rate of adenine nucleotide biosynthesis (middle and bottom panel of Figure 9-5). From these results it appears that the capacity of the oxidative pentose phosphate pathway is very small in muscular organs. The consequence of this is that the pool of 5-phosphoribosyl-1-pyrophosphate is minute

and that adenine nucleotide biosynthesis is small in the heart and not measurable at all in the resting skeletal muscle.

There are essentially two possible ways to enhance the available pool of 5-phosphoribosyl-1-pyrophosphate and the synthesis of adenine nucleotides in muscular organs. In the myocardium, the first possibility is to increase the activity of glucose-6-phosphate dehydrogenase. In fact, it has been demon-

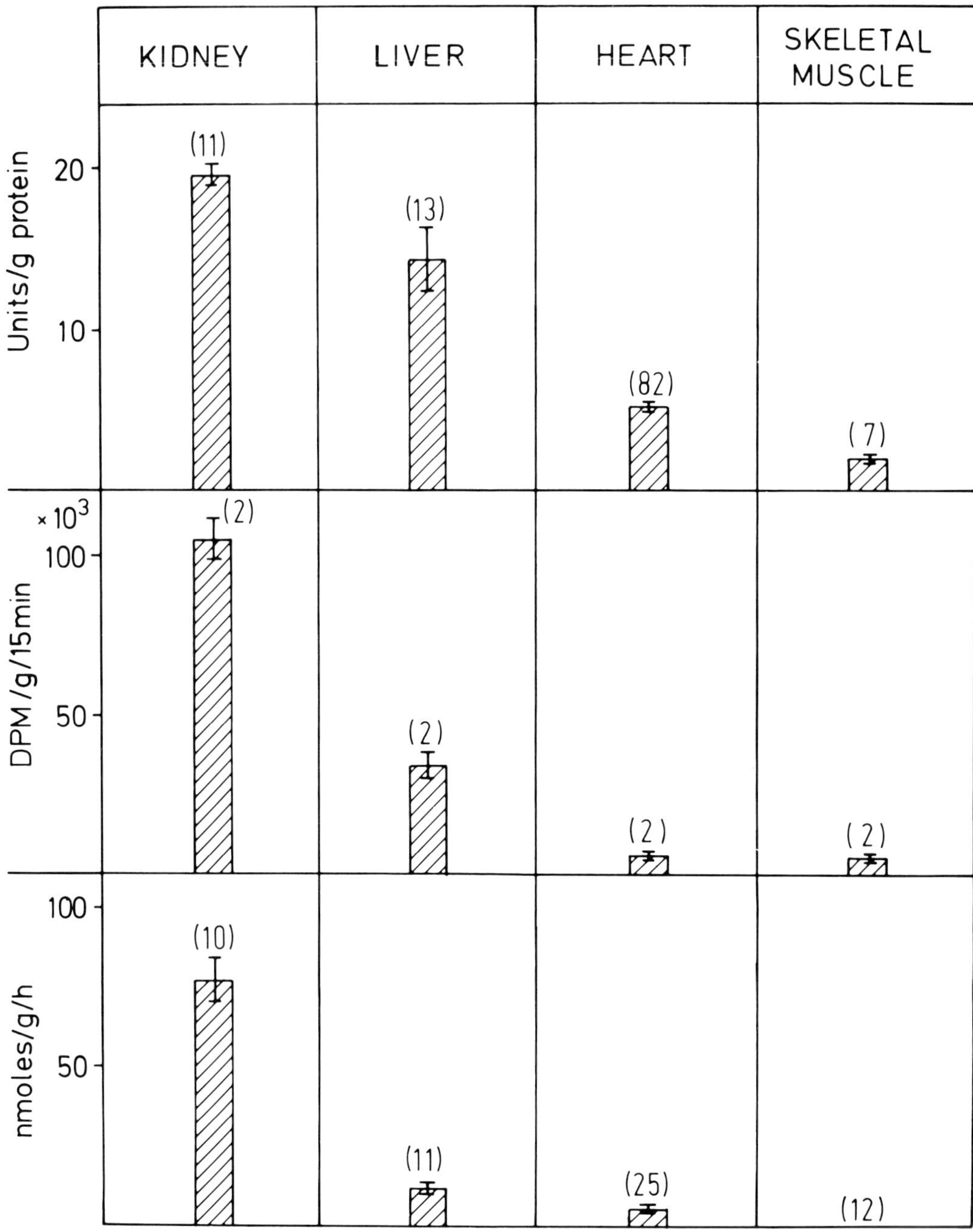

Figure 9-5. Activities of glucose-6-phosphate dehydrogenase (upper panel); the available pool of 5-phosphoribo-syl-1-pyrophosphate as evaluated by the incorporation of ^{14}C-inosine into adenine nucleotides (middle panel); and rates of adenine nucleotide biosynthesis (bottom panel) in various organs of the rat. Mean values ± SEM; number of experiments in parentheses.

strated that catecholamines such as isoproterenol and sympathomimetic agents such as dopamine and dobutamine stimulate glucose-6-phosphate dehydrogenase activity.[48] The mechanism by which this is brought about is the catecholamine-induced increase in the synthesis of the enzyme protein.[119] Since catecholamines stimulate heart rate and contractility and thus enhance the oxygen demand of the heart, it would be desirable to achieve a comparable metabolic stimulation without additional increase in heart function. This should be possible by bypassing the critical step in the oxidative pentose phosphate pathway by providing ribose (Figure 9-4). We have therefore examined what effect ribose may have on the available pool of 5-phosphoribosyl-1-pyrophosphate and on the rate of adenine nucleotide biosynthesis in the organs in which the capacity of the oxidative pentose phosphate pathway is so different (Figure 9-5).

Figure 9-6 shows the available pool of 5-phosphoribosyl-1-pyrophosphate in kidney, liver, heart, and skeletal muscle under normal resting conditions, and the effect of ribose as determined after 4 hours of continuous IV infusion in conscious, unrestrained rats. Ribose had no effect in kidney and liver. In heart and skeletal muscle, however, there was a threefold increase. A comparable picture emerged when the effect of ribose on the rates of adenine nucleotide biosynthesis was determined. In these experiments, ribose was administered as a single IV dose of 100 mg/kg, and the measurements were

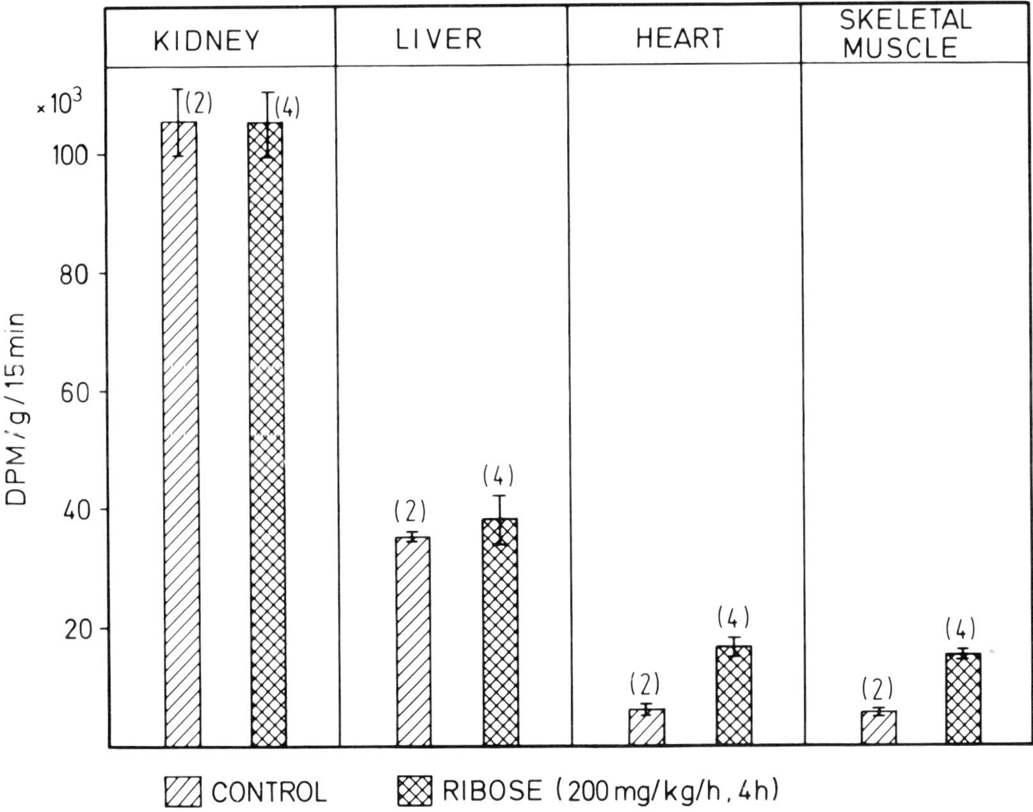

Figure 9-6. Available pool of 5-phosphoribosyl-1-pyrophosphate as assessed by the incorporation of ^{14}C-inosine into adenine nucleotides in various rat organs, and the effect of ribose 4 hours after continuous IV infusion. Data are means ± SEM; number of experiments in parentheses.

made after 60 minutes. Again, ribose had no influence in kidney and liver.

In the myocardium, there was a marked increase in adenine nucleotide biosynthesis, and in skeletal muscle, adenine nucleotide biosynthesis became measurable and was present in the same order of magnitude as in heart muscle (Figure 9-7). Thus, ribose has a stimulating effect only in muscular organs in which the activity of glucose-6-phosphate dehydrogenase is very low. This concept has recently been confirmed in studies on the isolated perfused rat heart and on mature rat cardiac myocytes.[120-122] This latter study demonstrated clearly that the metabolic influence of ribose is specific for muscle cells, and that endothelial or interstitial cells seem not to contribute to the results obtained in the heart *in vivo*. Thus, the effects of ribose are specific in two respects: (1) they are muscle-specific and (2) they are confined to pentoses and pentitols since glucose and fructose are without any influence on the essential metabolic parameters.[53]

Since ribose has such a marked stimulating effect already in the normal heart, it was of interest to examine whether it may enhance cardiac adenine nucleotide biosynthesis in pathophysiologic situations, and whether this would then result in appropriate changes in the adenine nucleotide content. The ultimate question, of course, concerns the possible relation between adenine nucleotide content and heart function. This will be examined in several *in vivo* models.

EFFECT OF RIBOSE ON CARDIAC ATP CONTENT AND FUNCTION IN SEVERAL PATHOPHYSIOLOGIC CONDITIONS

Overload Leading to Cardiac Hypertrophy

In all three models of cardiac hypertrophy, myocardial adenine nucleotide biosynthesis was enhanced in a typical time-dependent manner.[58] Table 9-1 shows the increased rates of adenine nucleotide synthesis for three different time periods: 3 days after aortic constriction; 1 day after application of a single s.c. dose of isoproterenol; and 2 days after triiodothyronine administration. When ribose was given as a single dose, there was a further stimulation of cardiac adenine nucleotide synthesis in every model examined.

Of particular interest is the isoproterenol model, since it is associated with pronounced hemodynamic alterations (see Figure 9-1). Whether the characteristic increase in LV dp/dt_{max} may be influenced by ribose was examined. As is shown in Figure 9-8, when ribose was administered at an infusion rate of 450 mg/kg/hour, the increase in LV dp/dt_{max} was further enhanced throughout almost the entire observation period. In an attempt to delineate which factor may be responsible for this ribose-elicited improvement in cardiac contractility, several metabolic parameters were determined 5 hours after isoproterenol administration; that is, at the peak of the hemodynamic alterations.

In Table 9-2, the further enhancement of contractility by ribose is indicated by percentage of control value. There was no further increase in the cAMP level by ribose. However, adenine nucleotide biosynthesis, which was higher after 5 hours than after 24 hours (Table 9-1), was further enhanced by ribose. As a consequence, the drop in ATP was markedly attenuated. It thus appears that ribose is capable of further enhancing an increased contractility, and this appears to be associated with the near-normalization of the ATP level.[50] In this context it may be mentioned that the complete normalization of the isoproterenol-induced ATP decline after 24 hours was characterized by the attenuation of the focal myocardial cell lesions that are typical for isoproterenol.[33,123] Now that we have shown this beneficial effect of ribose in functional and morphologic terms, the obvious question may be raised whether ribose is also able to normalize a reduced heart function.

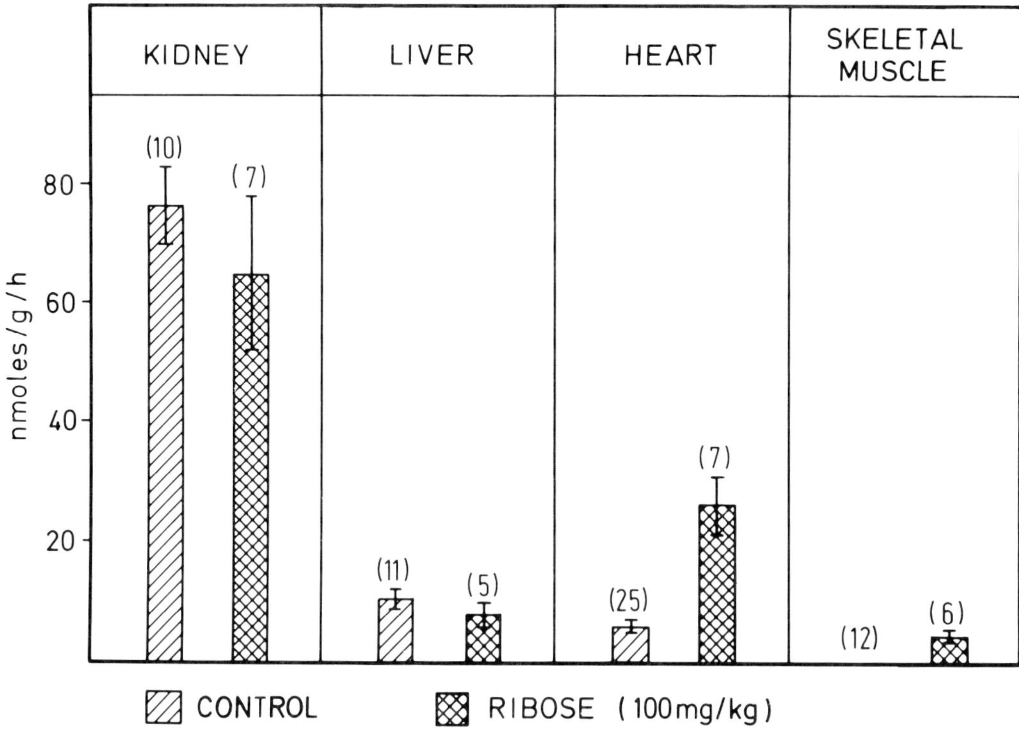

Figure 9-7. Effect of ribose on the biosynthesis of adenine nucleotides in several rat organs. Measurements were done 60 minutes after IV application of 1-^{14}C-glycine. Mean values $\pm$ SEM; number of experiments in parentheses.

To induce a decline in global heart function in rats, together with a depressed ATP level, two models of cardiac overload were combined: aortic constriction and isoproterenol application. Twenty-four hours after this combined intervention, the ATP content was significantly lower, and left ventricular systolic pressure as well as LV dp/dt_{max} were depressed (Figure 9-9). When ribose was given as a continuous IV infusion for 24

TABLE 9-1

Effect of Ribose (100 mg/kg, IV, 1-Hour Exposure Time) on the Rates of Adenine Nucleotide Biosynthesis (nmol/g/hour) in Control and Hypertrophying Rat Hearts[a]

	Without Ribose	With Ribose
Control	6.0 $\pm$ 0.7 (25)	26.7 $\pm$ 5.0 (7)
Aortic constriction (3 days)	16.4 $\pm$ 3.6 (9)	85.4 $\pm$ 19.8 (5)
Isoproterenol (1 day)	16.6 $\pm$ 0.7 (5)	199.6 $\pm$ 19.2 (4)
Triiodothyronine (2 days)	38.4 $\pm$ 2.8 (10)	181.0 $\pm$ 15.4 (5)

[a]The duration after initiation of hypertrophy is indicated in parentheses. Isoproterenol was administered as a single s.c. dose of 25 mg/kg. Triiodothyronine was applied every 24 hours in a s.c. dose of 0.2 mg/kg. Data are mean values $\pm$ SEM. Number of experiments in parentheses.

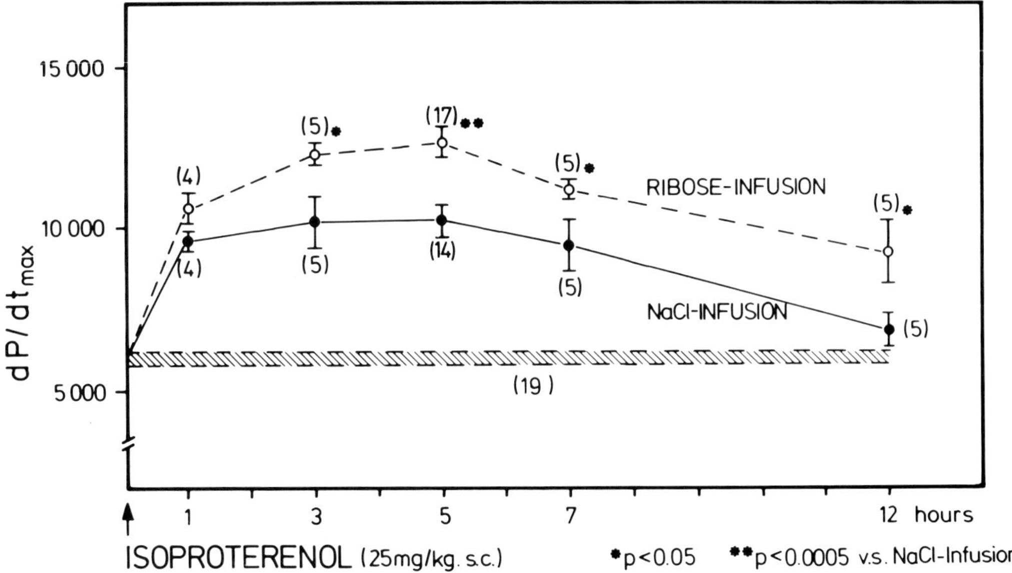

Figure 9-8. Effect of ribose administration (450 mg/kg/hour) for the periods of time indicated on the abcissa on the maximal rate of rise in left ventricular pressure (dp/dt_{max}). Mean values $\pm$ SEM; number of experiments in parentheses.

hours in a dose of 200 mg/kg/hour, both the ATP content and the left ventricular hemodynamic parameters were normalized. Thus in this experimental model, the restoration of the cardiac ATP pool was clearly associated with the complete normalization of left ventricular function.[59]

Intermittent Asphyxic Periods

In this next experimental model, the ATP content was reduced after application of five asphyxic periods separated by recovery periods of 3-minutes duration (see Figure 9-2). During the subsequent recovery period of

TABLE 9-2

Effect of Continuous IV Infusion of Ribose (450 mg/kg/hour) for 5 Hours on LV dP/dt_{max}, the Level of 3′,5′-cyclic AMP (cAMP), the Biosynthesis of Cardiac Adenine Nucleotides and on the ATP Content in the Myocardium of Rats Treated with Isoproterenol (24 mg/kg, s.c.)[a]

	LV dP/dt_{max} (mm Hg/sec in Percentages)	cAMP (pmol/g)	Biosynthesis of Adenine Nucleotides (nmol/g/hour)	ATP (μmol/g)
Control	100 $\pm$ 3.2 (12)	578 $\pm$ 23 (43)	6.0 $\pm$ 0.7 (25)	4.4 $\pm$ 0.1 (38)
Isoproterenol + NaCl infusion	171 $\pm$ 4.0 (14)	1393 $\pm$ 86 (8)	39.4 $\pm$ 2.3 (14)	3.1 $\pm$ 0.2 (4)
Isoproterenol + Ribose infusion	209 $\pm$ 2.2 (17)**	1409 $\pm$ 86 (5)	126.8 $\pm$ 1.8 (4)**	4.0 $\pm$ 0.2 (6)*

*$p < 0.025$

**$p < 0.0005$ versus isoproterenol + NaCl infusion

[a]Mean values $\pm$ SEM. Number of experiments in parentheses.

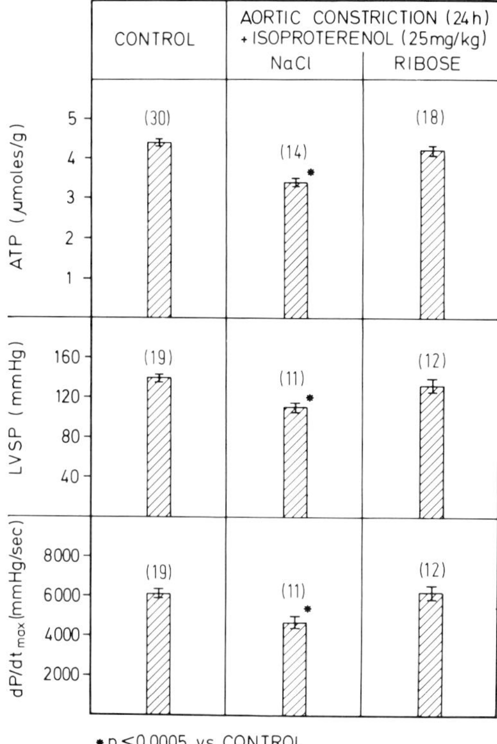

Figure 9-9. Content of myocardial ATP and left ventricular hemodynamic parameters (LSVP: left ventricular systolic pressure; dP/dt_{max}: maximal rate of rise in left ventricular pressure) in control rats and 24 hours after aortic constriction in combination with s.c. administration of isoproterenol. Continuous IV infusion of 0.9% NaCl or ribose (200 mg/kg/hour). Mean values ± SEM; number of experiments in parentheses.

60 minutes, adenine nucleotide biosynthesis was found to be enhanced by about 100% when only 0.9% NaCl was infused (Table 9-3). When ribose had been administered, cardiac adenine nucleotide *de novo* synthesis was further stimulated. Because of the inherent limitations of this model, the dose of ribose had to be increased to 450 mg/kg/hour. The relatively short postasphyxic recovery time of 60 minutes was not sufficient to achieve a substantial restitution of the cardiac adenine nucleotide pool. It is for this reason that another experimental model was selected in which the recovery of adenine nucleotides could be monitored for a longer period of time.

Experimental Myocardial Infarction

Temporary Regional Ischemia

Ligation of the left anterior descending (LAD) coronary artery was done for 15 minutes to induce a decline in the ATP content. During the postischemic recovery period that followed, both adenine nucleotide biosynthesis and the ATP pool were measured for up to several days. Table 9-3 shows that adenine nucleotide biosynthesis was increased during the fifth hour of postischemic recovery when the rats had received

TABLE 9-3

Changes in Myocardial Adenine Nucleotide Biosynthesis in Rats That Had Received Continuous IV Infusion of 0.9% NaCl or Ribose in Control and Several Pathophysiologic Conditions[a]

	NaCl Infusion (nmol/g/hour)	Ribose Infusion (nmol/g/hour)
Control	6.0 ± 0.7 (25)	26.7 ± 5.0 (7)
Recovery from intermittent asphyxic periods (1 hour)	12.6 ± 0.5 (6)	20.5 ± 2.2 (5)
Recovery from temporary regional ischemia of 15 minutes (5 hours)	14.4 ± 3.2 (4)	60.5 ± 11.2 (4)
Nonischemic myocardium after LAD ligation (48 hours)	32.2 ± 5.5 (8)	48.2 ± 8.0 (6)

[a]The dose of ribose was 500 mg/kg/hour during recovery from the intermittent asphyxic periods, and 200 mg/kg/hour in the other two experimental situations. Exposure time to 1-^{14}C-glycine was 60 minutes in every experimental model. The point in time of cardiac adenine nucleotide synthesis measurement is indicated in parentheses. Values are means ± SEM. Number of experiments in parentheses.

continuous IV infusion of 0.9% NaCl. When ribose had been administered for the same period of time, adenine nucleotide *de novo* synthesis was further enhanced to reach 60 nmol/g/hour. This value is of such a magnitude that it should have an effect on the actual cardiac ATP pool. This is, in fact, the case.

As the data in Figure 9-10 indicate, regional ischemia of 15 minutes caused a decline in ATP. There was a spontaneous increase when 0.9% NaCl was given, most likely due to the rephosphorylation of AMP and ADP. Ribose had no significant effect after this period of time (5 hours). After 12 hours, the ATP level was still significantly lower; there was no change compared to the 5-hour value. However, when ribose had been administered, the ATP content was entirely normal. The same is true for the 24 hour recovery period. There was no further increase in ATP in animals with no metabolic support, and the normalized ATP content was maintained with ribose infusion. It was only after 72 hours that the ATP level had achieved almost the control value without any metabolic intervention. Thus, the restoration in metabolic terms was acceler-

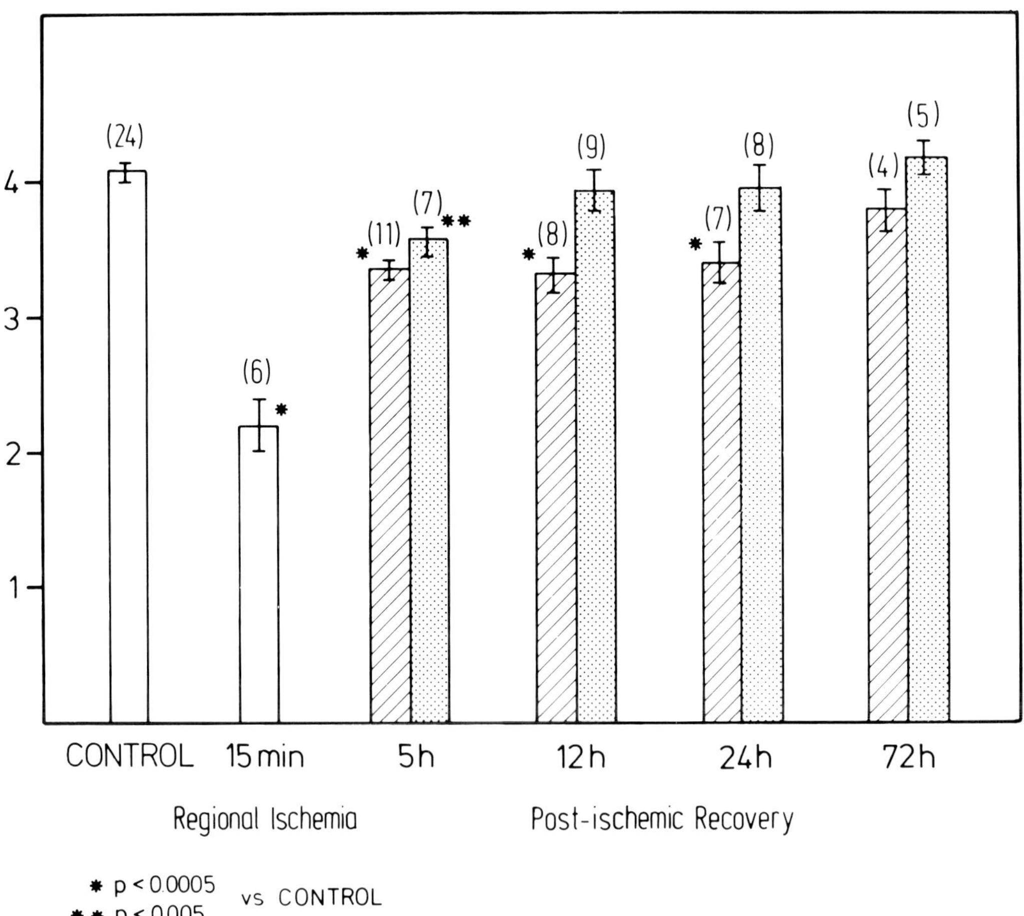

Figure 9-10. Alterations in the cardiac ATP pool (μmol/g) in controls, at the end of a 15-minute regional ischemia and during the postischemic recovery period in rats that received continuous IV infusion of 0.9% NaCl (hatched bars) or ribose (200 mg/kg/hour; stippled bars). Data are means ± SEM; number of experiments in parentheses.

ated considerably by ribose, at least by 60 hours.[21,63]

These studies demonstrate that the postischemic restitution of the cardiac adenine nucleotide pool is dependent almost exclusively on the *de novo* synthesis of adenine nucleotides. This seems to be the main reason for the slow metabolic recovery process. The heart in this particular situation urgently needs an efficient metabolic support, and this may also be true for a human with acute myocardial infarction and successful recanalization.[64,65] The beneficial effect of ribose during postischemic recovery has been confirmed by other research groups both in the isolated perfused rat heart and in the dog heart.[124–127]

Permanent Coronary Artery Ligation

Several adaptive processes take place in the nonischemic myocardium after the LAD is permanently ligated. One of these is the enhancement of adenine nucleotide biosynthesis, which is observed after only 24 hours and may be regarded as one of the first metabolic alterations to indicate the development of cardiac hypertrophy of the nonischemic portion of the heart.[62] In Table 9-3, the increase in adenine nucleotide biosynthesis is shown as measured 48 hours after coronary artery ligation in rats. When ribose had been given for this period of time, adenine nucleotide *de novo* synthesis was further enhanced. This clearly has consequences for the ATP pool in the nonischemic heart. As depicted in Figure 9-11, there is quite a marked drop in the ATP level after 1 day following coronary artery ligation. During the subsequent days, ATP recovers to some extent. This demonstrates that there is an excess functional load on the nonischemic myocardium that is represented by an ATP decline. When ribose had been ad-

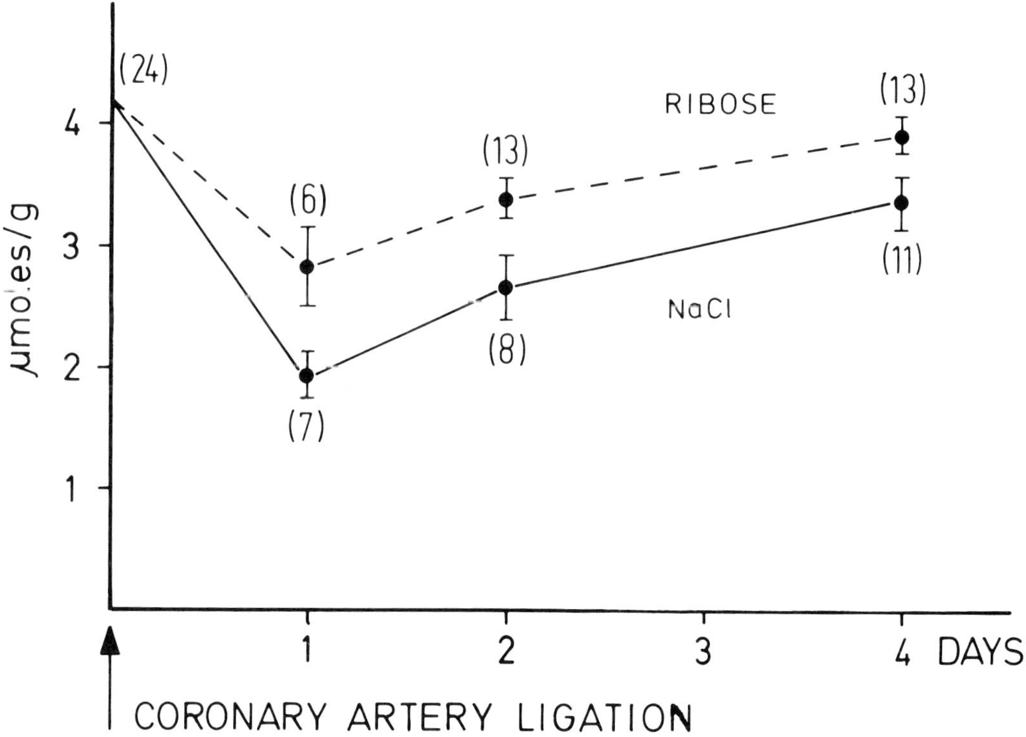

Figure 9-11. Changes in myocardial ATP level in the nonischemic area of hearts after permanent coronary artery ligation in rats with continuous IV infusion of 0.9% NaCl (continuous line) or ribose (200 mg/kg/hour; broken line). Data are mean values ± SEM; number of experiments in parentheses. (Data from reference 62.)

ministered, the fall in ATP was attenuated, and the restitution occurred faster.

As mentioned earlier (see "Experimental Myocardial Infarction"), there are pronounced changes in global heart function after the left coronary artery is ligated. These include a progressive decline in LSVP and in LV dp/dt_{max}. A characteristic feature is the elevation in LVEDP[62]. In Figure 9-12, the increase the LVEDP after 2 and 4 days is shown. This increase was attenuated when ribose was supplied as continuous IV infusion for the periods of time indicated. It thus appears that attenuation of the fall in ATP is associated with an improvement in global heart function, in this particular instance, in diastolic properties.

CRITERIA FOR SELECTION OF A CARDIOPROTECTIVE SUBSTRATE

From the previous discussion (see "Pharmacologic and Metabolic Interventions for Cardioprotection") it is obvious that there are numerous metabolic substrates that would qualify as cardioprotective. It would therefore be useful to have some criteria that would allow us to differentiate among the various interventions. In the following, seven criteria will be given and discussed mainly in relation to ribose:

1. The mechanism by which the substrate acts up on cardiac metabolism should be known. This is a prerequisite for any approach developed on a scientific basis. In the case of ribose, the mechanism of action is known. It is based on the fact that the capacity of the oxidative pentose phosphate pathway in the myocardium is very low. Ribose bypasses the critical step and stimulates the distal portion of this pathway, leading to an expansion of the 5-phosphoribosyl-1-pyrophosphate pool and to stimulation of purine nucleotide biosynthesis (see "The Oxidative Pentose Phosphate Pathway").

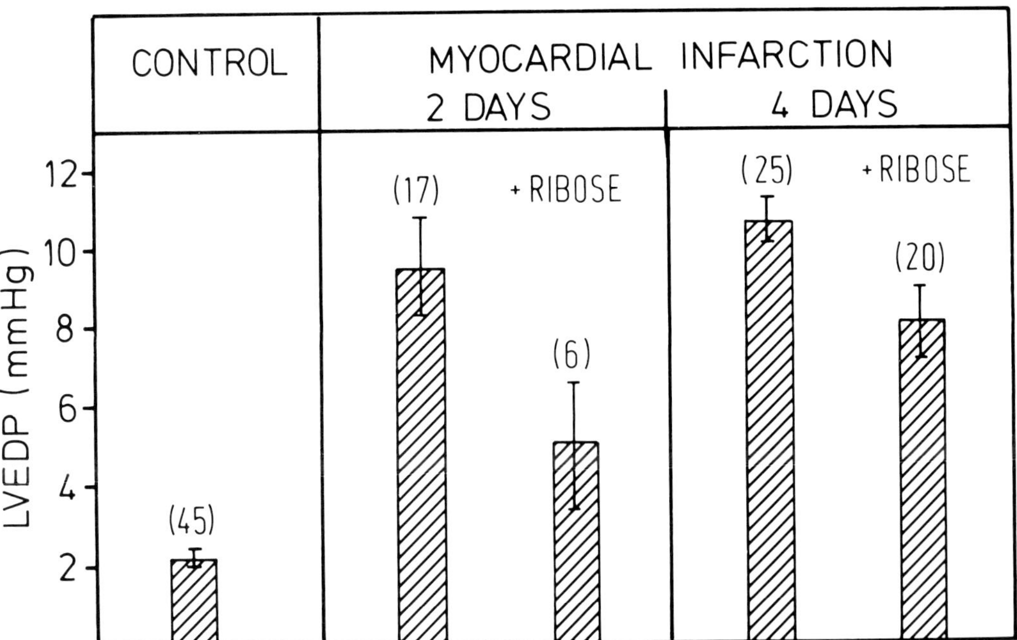

Figure 9-12. Elevation of left ventricular end diastolic pressure (LVEDP) in rat hearts 2 and 4 days after experimental myocardial infarction, and the effect of ribose (continuous IV infusion of 200 mg/kg/hour). Mean values ± SEM; number of experiments in parentheses. (Data from reference 62.)

2. The metabolic substrate should exhibit specificity for muscular organs, in particular, for the myocardium. Because of the low activity of glucose-6-phosphate dehydrogenase in muscular organs, the metabolic effects of ribose are confined to heart and skeletal muscle.

3. There should be marked and long-term effects on cardiac metabolism, ultimately leading to improvement of function. That ribose fulfills this specification has been demonstrated in long-term experiments (see "Effect of Ribose on Cardiac ATP Content and Function in Several Pathophysiologic Conditions").

4. The metabolic substrate should not have any hemodynamic or vasoactive properties of its own. This has been examined recently in extensive functional studies in which ribose was administered as continuous IV infusion for 5 and 24 hours in conscious, unrestrained rats. After the respective periods of infusion time, the left ventricular hemodynamic parameters were measured in Inactin[R] anesthesia. The data in Figure 9-13 indicate that ribose has no effect on heart rate, LVSP, or LV dp/dt_{max}, after either 5 or 24 hours. Also, as might be expected, the pressure-rate-product turned out to be essentially normal (Figure 9-14). When the measurements of left ventricular parameters had been taken, the Millar ultraminiature catheter pressure transducer was withdrawn from the left ventricle and positioned in the aorta so that the diastolic aortic pressure could be obtained. It also was not altered by ribose (Figure 9-14). Thus, ribose is entirely neutral in hemodynamic terms, both in acute experiments, as has been shown previously, and in long-term experiments, as demonstrated in Figure 9-13 and 9-14. This is in contrast to inosine, which was administered in combination with adenosine deaminase to exclude any contaminating or endogenous adenosine. When a cardioprotective dose of inosine was applied in acute experiments in anesthe-

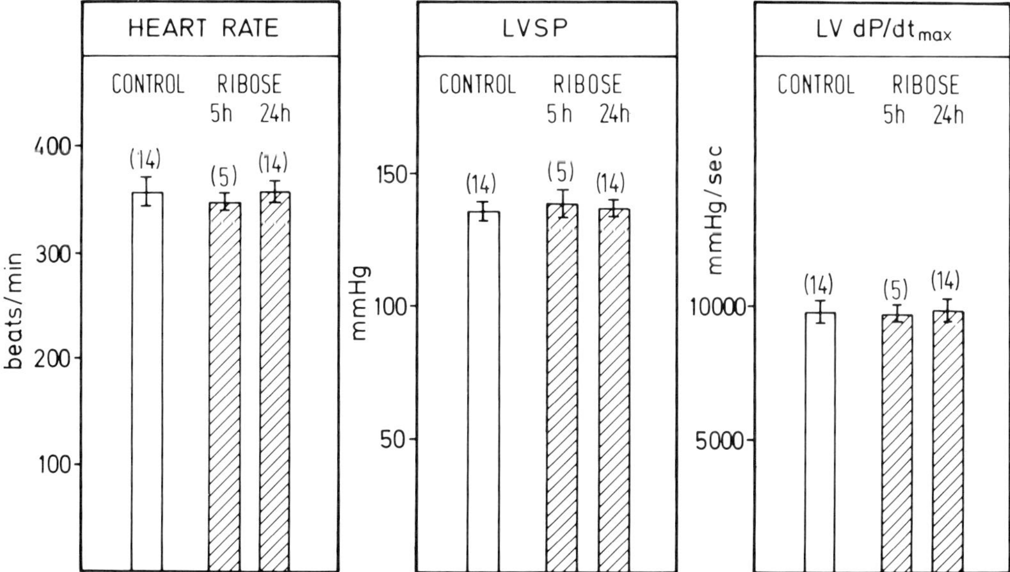

Figure 9-13. Effect of continuous IV infusion of ribose (200 mg/kg/hour) for 5 and 24 hours on left ventricular hemodynamic parameters. At the end of the respective infusion periods, the closed-chest rats were catheterized in Inactin[R]-anesthesia. Mean values ± SEM; number of experiments in parentheses.

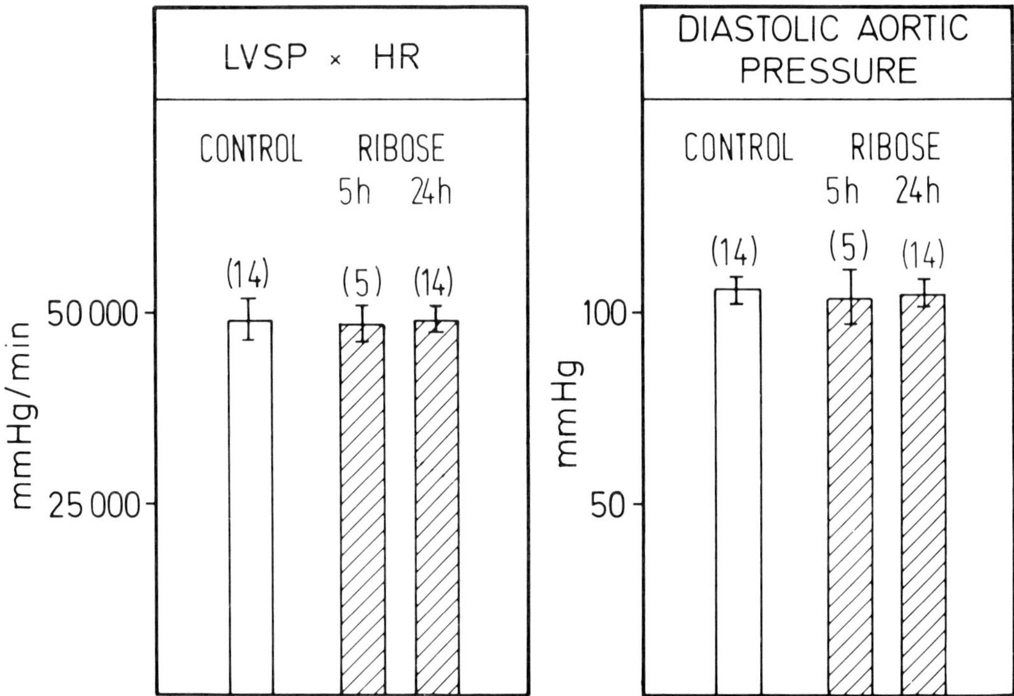

Figure 9-14. Effect of continuous IV infusion of ribose (200 mg/kg/hour) for 5 and 24 hours on the pressure-rate-product and on diastolic aortic pressure in rats. Mean values ± SEM; number of experiments in parentheses.

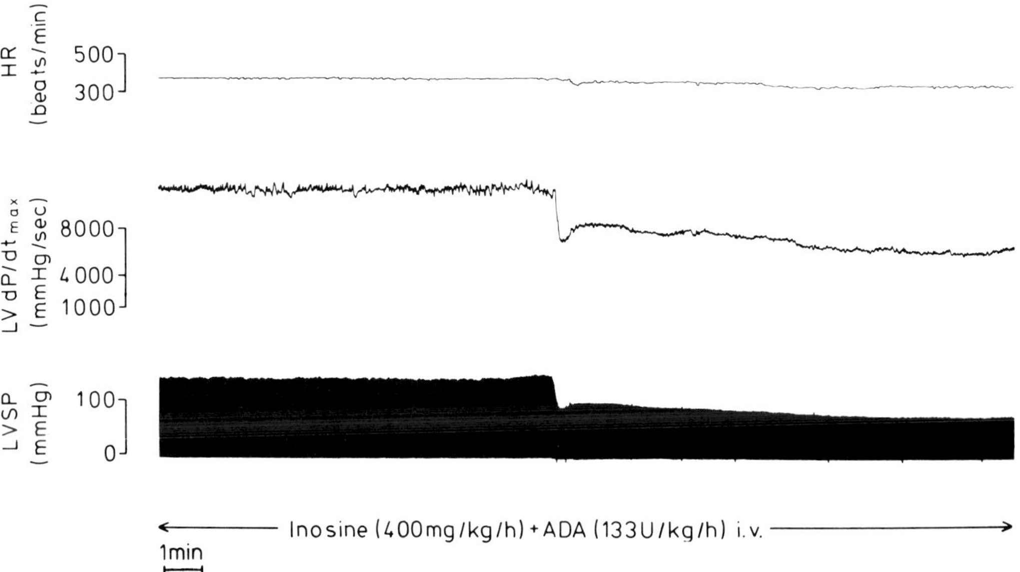

Figure 9-15. Acute effects of IV administration of inosine in combination with adenosine deaminase (ADA) on left ventricular functional parameters in closed-chest, Inactin[R]-anesthetized rats. Original recording of a typical experiment.

tized rats, there was a precipitous fall in LV dp/dt_{max} and in LVSP, with almost no change in heart rate (Figure 9-15). Thus inosine, which has metabolic effects similar to those of ribose, has pronounced acute negative hemodynamic effects.[112]

5. To be cardioprotective, an intervention should be compatible with conventional cardiac therapy. As far as ribose is concerned, this has been demonstrated recently for prazosin, for the calcium antagonist verapamil and for the beta$_1$-selective adrenergic blocker, metoprolol.[128-130] In these studies, the drugs were administered as continuous IV infusion in deliberately high doses so as to affect left ventricular hemodynamic parameters. Figure 9-16 shows that verapamil at the end of a 5-hour infusion period had depressed heart rate, LVSP, and LV dp/dt_{max}, and ribose had no effect on this depression. Likewise, the verapamil-induced decline of the pressure-rate-product and of diastolic aortic pressure was not influenced by concomitant ribose application (Figure 9-17). On the other hand, the stimulation of cardiac adenine nucleotide biosynthesis characteristic for ribose was not affected by verapamil (Figure 9-18). Thus, ribose did not interfere with the negative hemodynamic effects of verapamil, and verapamil did not impede the metabolic stimulation of ribose. Essentially similar results were obtained when ribose was combined with metoprolol.[129,130]

The situation is different when inosine is combined with these drugs. Figure 9-19 shows the depressive effects of verapamil on the hemodynamic parameters after 60 minutes of continuous infusion. Concomitant application of inosine in a cardioprotective dose had an additional deleterious effect on all functional parameters. Similarly,

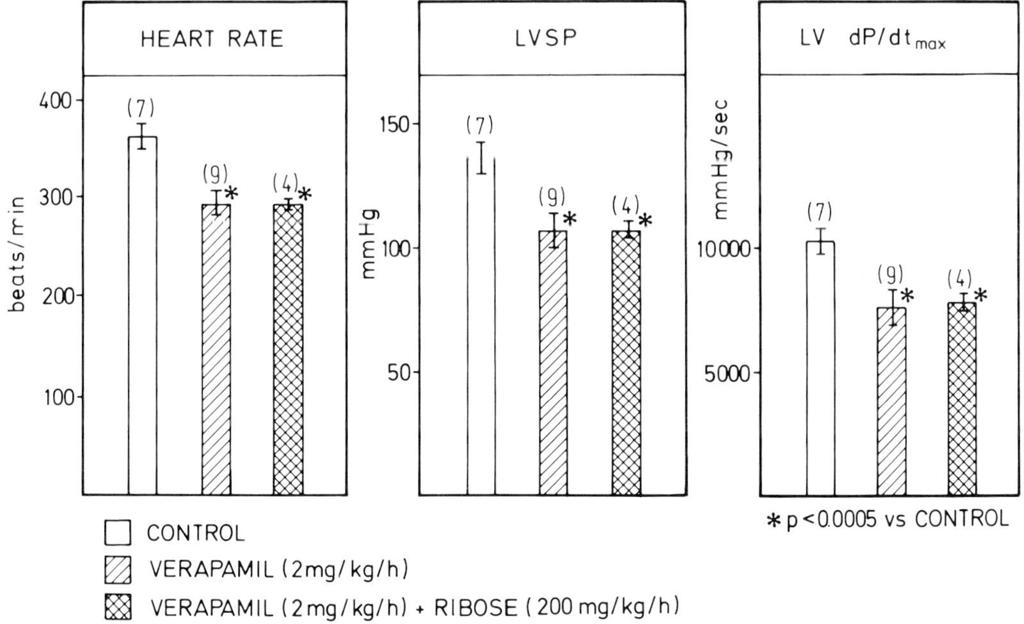

Figure 9-16. Effect of continuous IV infusion of verapamil, alone and in combination with ribose, for 5 hours on left ventricular functional parameters. Mean values ± SEM; number of experiments in parentheses.

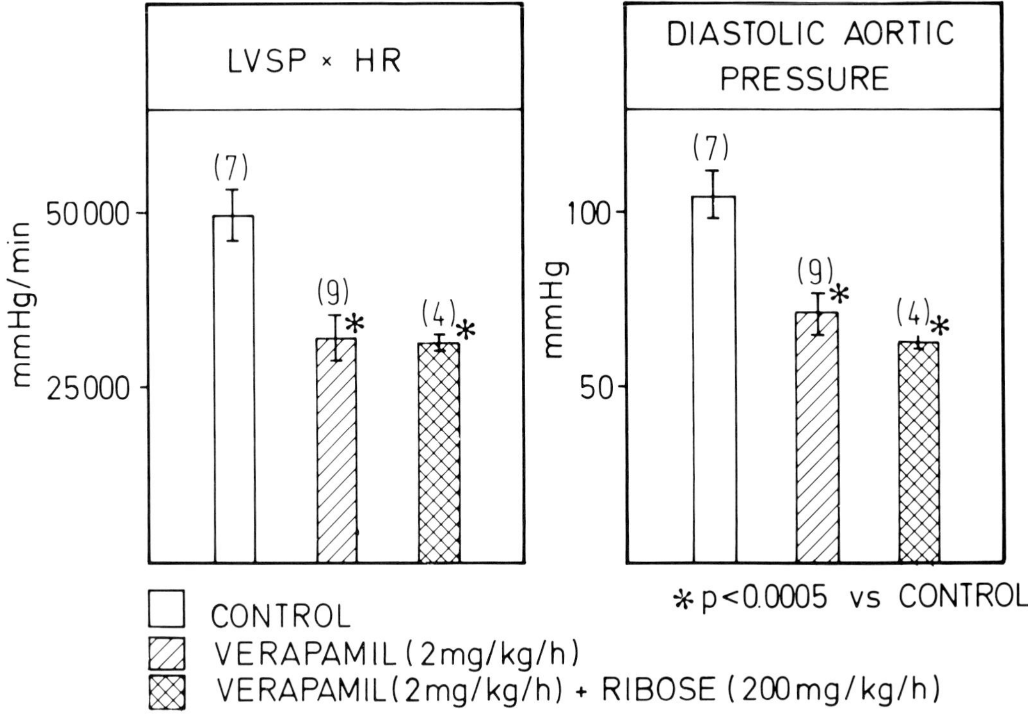

Figure 9-17. Pressure-rate-product and diastolic aortic pressure as depressed by continuous IV infusion of verapamil alone and in combination with ribose for 5 hours. Mean values ± SEM; number of experiments in parentheses.

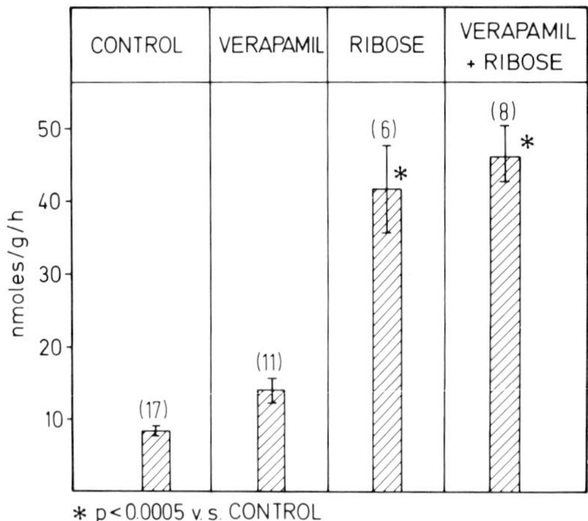

Figure 9-18. Effects of verapamil (2 mg/kg/hour), ribose (200 mg/kg/hour), and of the combination on the rates of cardiac adenine nucleotide biosynthesis. Measurements were done after continuous IV infusion for 5 hours. Mean values ± SEM; number of experiments in parentheses.

the pronounced negative hemodynamic effects of metoprolol were more pronounced when inosine was co-infused (Figure 9-20). Thus, the combination of inosine with verapamil or metoprolol is associated with a further decline in left ventricular hemodynamic parameters. When extrapolating these results (obtained in acute experiments on rats) to humans, this particular combination seems not to be advisable.

6. The metabolic substrate should be effective in the human heart (and skeletal muscle). The beneficial effects of ribose described in this chapter were all obtained in rats. It is therefore of interest whether ribose may also be cardioprotective in other species. For this, it

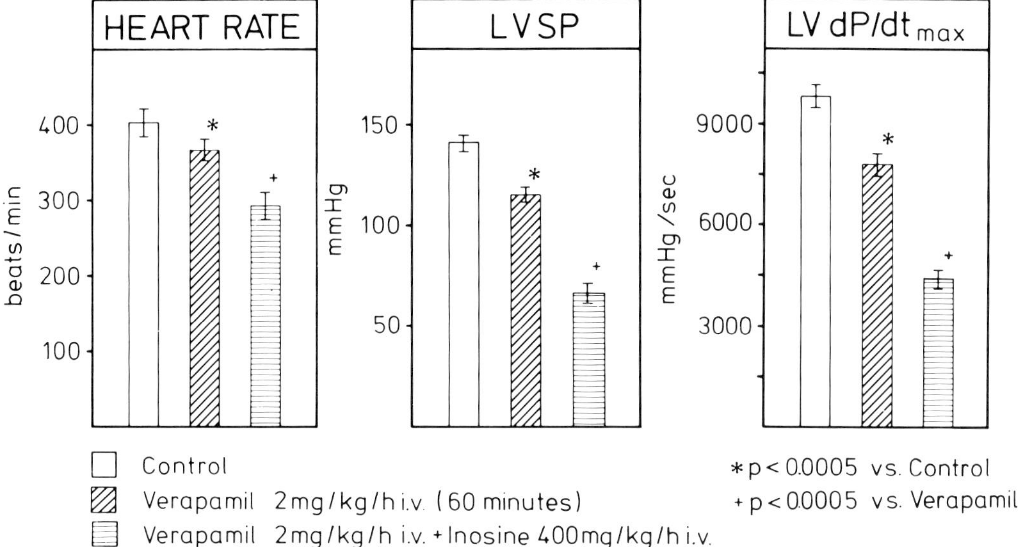

Figure 9-19. Heart rate, left ventricular systolic pressure, and LV dP/dt_{max} as influenced by verapamil, alone and in combination with inosine. Measurements were done after 60 minutes of continuous IV infusion. Mean values ± SEM.

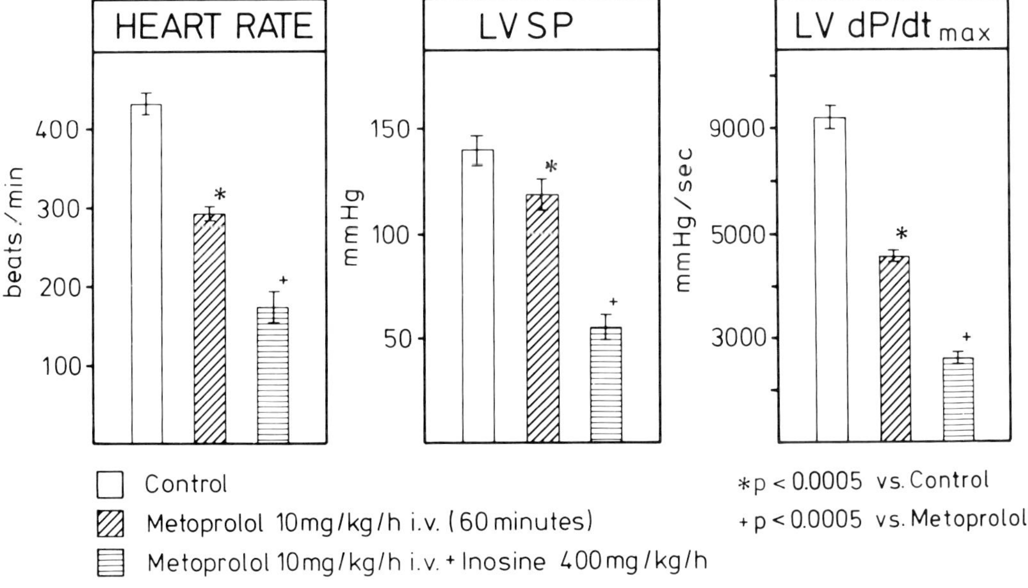

Figure 9-20. Effect of metroprolol, alone and in combination with inosine, on left ventricular function after 60 minutes of continuous IV infusion. Mean values ± SEM.

would be sufficient just to measure the activity of glucose-6-phosphate dehydrogenase, the critical enzyme of the oxidative pentose phosphate pathway. If its activity is low, then ribose can be expected to have the same effects as have been observed in rats.

In a comparative study, the activities of glucose-6-phosphate dehydrogenase and of 6-phosphogluconate dehydrogenase were measured in heart samples of a variety of animal species as well as as in human papillary muscles that were obtained during cardiac surgery.[131] Although there was some variation in glucose-6-phosphate dehydrogenase activity, it was a general feature that the activity of this enzyme was of the same order of magnitude in all animal species, including humans (Figure 9-21). Moreover, the activity of glucose-6-phosphate dehydrogenase was always lower than the activity of 6-phosphogluconate dehydrogenase. This is further evidence for the critical role of the first enzyme in the oxidative pentose phosphate pathway. That the activity of glucose-6-phosphate dehy-

drogenase closely parallels the rate of cardiac adenine nucleotide biosynthesis is evident for three animal species. The rate of adenine nucleotide synthesis in the guinea pig heart was 7.1 ± 4.4 nmol/g/hour, in the rat heart 6.0 ± 0.7 nmol/g/g (Tables 9-1 to 9-3), and in the dog heart 1.5 ± 0.3.[114,131] When extrapolating these data to the human heart, one can expect it to be lower than in the rat and higher than in the dog heart.

From all these results it appears that the metabolic basis for ribose seems to be the same in all species examined so far, and that ribose clearly has a chance to work in the human heart. That ribose can be effective in humans has been shown recently in a patient with myoadenylate deaminase deficiency in skeletal muscle.[132] In this disorder, there is a disruption of the purine nucleotide cycle, with resulting degradation of adenine nucleotides to adenosine similar to the situation in the heart.[133–135] The symptoms of this disease, such as muscle cramps and aches, can be prevented by ribose.[132]

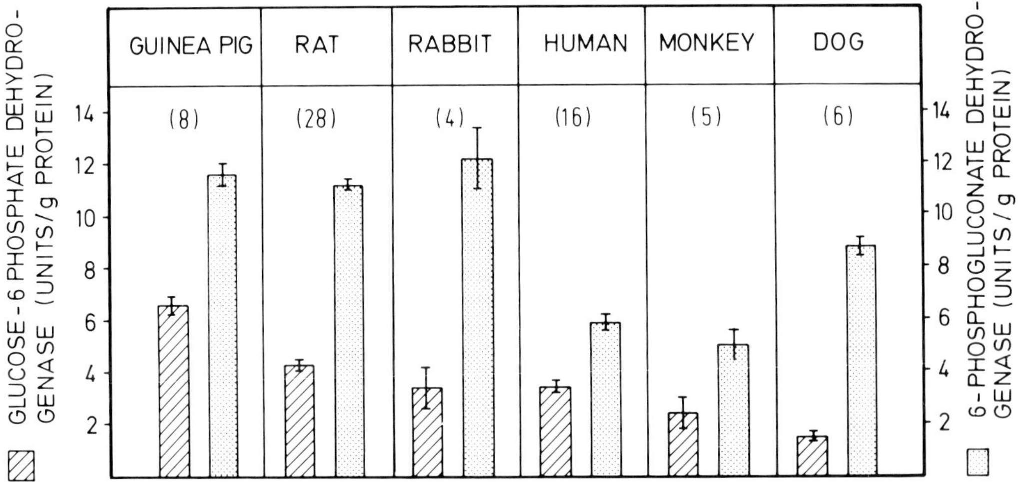

Figure 9-21. Activities of glucose-6-phosphate dehydrogenase (G-6-PDH) and of 6-phosphogluconate dehydrogenase (6-P-GDH) in the myocardium of several species. Mean values ± SEM; number of experiments in parentheses.

7. The side effects of the metabolic substrate should be known and manageable. There is perhaps no therapeutic intervention without any side effect. As regards ribose, the only side effect now known is hypoglycemia.[136-138]

SUMMARY, CONCLUSION, AND PROSPECTS

The studies presented in this chapter demonstrate that it is possible to stimulate the *de novo* synthesis of cardiac and skeletal muscle adenine nucleotides to such a degree that the decline in the adenine nucleotide pool in several pathophysiologic situations of the heart is attenuated or even entirely prevented. It is then possible to measure global heart function and to correlate it with the metabolic alterations. Using this experimental approach in the rat, it was found that ribose, by normalizing the myocardial ATP level, further enhances the isoproterenol-induced increase in contractility. On the other hand, a reduced heart function can also be normalized when the ATP pool is restored, as in the case of aortic constriction in combination with isoproterenol administration. Or diastolic heart function can be improved, as in the case of experimental myocardial infarction. Thus, there are experimental *in vivo* situations of cell injury in which one can demonstrate a good association between the restitution of the cardiac ATP pool and the improvement of heart function. When extrapolating these results to the clinical situation, one must stress that every possible effort should be made to support the heart metabolically. Numerous interventions have been developed to achieve that; however, only few of these have received general or partial acceptance.

The metabolic approach presented in this chapter is based on the fact that the oxidative pentose phosphate pathway that provides 5-phosphoribosyl-1-pyrophosphate, an essential substrate for pyrimidine and pu-rine nucleotide synthesis, is poorly developed in muscular organs. Ribose bypasses the critical and limiting step in the pathway, elevates the available pool of 5-phosphoribosyl-1-pyrophosphate, and stimulates the *de novo* synthesis of adenine nucleotides. This intervention is successful since ribose affects the myocyte directly, obviously in the appropriate adenine nucleotide compartment.[122] There is no direct information on this, so one can only speculate that ribose may affect the contractile machinery or, more likely, the membrane systems involved in the handling of Ca^{++} ions, such as the sarcolemma and the sarcoplasmic reticulum.

In addition to having appropriate effects on the subcellular level, ribose meets the main criteria of a good cardioprotective substrate (see "Criteria for Selection of a Cardioprotective Substrate"). There is thus the potential for therapeutic application in certain clinical states. Successful recanalization of an occluded coronary artery with streptokinase, with or without subsequent percutaneous transluminal coronary angioplasty, is a situation that may be particularly appropriate for this metabolic intervention. Since even brief periods of ischemia lead to a long-term depression of the myocardial ATP pool that may last for days (see Figure 9-10), the previously ischemic area of the heart urgently needs a metabolic support. Another situation could be heart failure due to myocardial energy deficiency. Also, patients with profound ventricular failure after open-heart surgery who need temporary circulatory support with a ventricular assist pump may benefit from ribose administration.[139] As far as skeletal muscle is concerned, the successful application of ribose in a patient with myoadenylate deaminase deficiency has already been demonstrated.[132] Excess purine degradation has also been shown recently in patients with glycogen storage disease types V and VII.[140] Thus, there are many prospects for this promising metabolic intervention.

ACKNOWLEDGMENTS

This research has been supported by the Deutsche Forschungsgemeinschaft (Zi 199/4-4, Zi 199/4-5) and by the Wilhelm Sander-Stiftung (No. 83.013.2).

REFERENCES

1. Koretsky AP, Wang S, Murphy-Boesch J, Klein MP, James TL, Weiner MW. ^{31}P NMR spectroscopy of rat organs, *in situ*, using chronically implanted radiofrequency coils. *Proc Natl Acad Sci USA*. 1983;80:7491.
2. Balaban RS, Kantor HL, Katz LA, Briggs RW. Relation between work and phosphate metabolite in the *in vivo* paced mammalian heart. *Science*. 1986;232:1121.
3. Wollenberger A. Relation between work and labile phosphate content in the isolated dog heart. *Circ Res*. 1957;5:175.
4. Hochrein H, Döring HJ. Die energiereichen Phosphate des Myokards bei Variation der Belastungsbedingungen. *Pflügers Arch*. 1960;271:548.
5. Gerlach E, Bader W, Schwoerer W. Über den Stoffwechsel säurelöslicher Phosphor-Verbindungen in der Rattenniere. *Pflügers Arch*. 1961;272:407.
6. Deuticke B, Gerlach E, Dierckesmann R. Abbau freier Nucleotide in Herz, Skeletmuskel, Gehirn und Leber der Ratte bei Sauerstoffmangel. *Pflügers Arch*. 1966;292:239.
7. Braasch W, Gudbjarnason S, Purin SP, Ravens KG, Bing R. Early changes in energy metabolism in the myocardium following acute coronary occlusion in anesthetized dogs. *Circ Res*. 1968;23:429.
8. Neely JR, Grotyohann LW. Role of glycolytic products in damage to ischemic myocardium: dissociation of adenosine triphosphate levels and recovery of function of reperfused ischemic hearts. *Circ Res*. 1984;55:816.
9. Neely JR, Rovetto MJ, Whitmer JT, Morgan HE. Effects of ischemia on function and metabolism of the isolated working rat heart. *Am J Physiol*. 1973;225:651.
10. Tennant R, Wiggers CJ. The effect of coronary occlusion on myocardial contraction. *Am J Physiol*. 1935;112:351.
11. Tennant R. Factors concerned with the arrest of contraction in an ischemic myocardial area. *Am J Physiol*. 1935;113:677.
12. Hearse DJ. Oxygen deprivation and early myocardial contractile failure: a reassessment of the possible role of adenosine triphosphate. *Am J Cardiol*. 1979;44:1115.
13. Reibel DK, Rovetto MJ. Myocardial ATP synthesis and mechanical function following oxygen deficiency. *Am J Physiol*. 1978;234:H620.
14. Watts JA, Koch CD, LaNoue KF. Effects of Ca^{2+} antagonism on energy metabolism: Ca^{2+} and heart function after ischemia. *Am J Physiol*. 1980;238:H909.
15. Nishioka K, Jarmakani JM. Effect of ischemia on mechanical function and high-energy phosphates in rabbit myocardium. *Am J Physiol*. 1982;242:H1077.
16. Gudbjarnason S, Puri PS, Mathes P. Biochemical changes in non-infarcted heart muscle following myocardial infarction. *J Mol Cell Cardiol*. 1971;2:253.
17. Heyndrickx GR, Millard RW, McRitchie RJ, Maroko PR, Vatner SF. Regional myocardial functional and electrophysiological alterations after brief coronary artery occlusion in conscious dogs. *J Clin Invest*. 1975;56:978.
18. Heyndrickx GR, Baig H, Nellens P, Leusen I, Fishbein MC, Vatner SF. Depression of regional blood flow and wall thickening after brief coronary occlusions. *Am J Physiol*. 1978;234:H653.
19. DeBoer KWV, Ingwall JS, Kloner RA, Braunwald E. Prolonged derangement of canine myocardial purine metabolism after a brief coronary artery occlusion not associated with anatomic evidence of necrosis. *Proc Natl Acad Sci USA*. 1980;77:5471.
20. Reimer KA, Hill ML, Jennings RB. Prolonged depletion of ATP and of the adenine nucleotide pool due to delayed resynthesis of adenine nucleotides following reversible myocardial ischemic injury in dogs. *J Mol Cell Cardiol*. 1981;13:229.
21. Zimmer H-G, Ibel H. Ribose accelerates the repletion of the ATP pool during recovery from reversible ischemia of the rat myocardium. *J Mol Cell Cardiol*. 1984;16:863.
22. Swain JL, Sabina RL, Hines JJ, Greenfield JC, Holmes EW. Repetitive episodes of brief ischemia (12 min) do not produce a cumulative depletion of high energy phosphate compounds. *Cardiovasc Res*. 1984;18:264.
23. Gudbjarnason S, Mathes P, Ravens KG. Functional compartmentation of ATP and creatine phosphate in heart muscle. *J Mol Cell Cardiol*. 1970;1:235.
24. Schrader J, Gerlach E. Compartmentation of cardiac adenine nucleotides and formation of adenosine. *Pflügers Arch*. 1976;367:129.
25. Soboll S, Bünger R. Compartmentation of adenine nucleotides in the isolated working guinea pig heart stimulated by noradrenaline. *Hoppe-Seyler's Z Physiol Chem*. 1981;362:125.
26. Bricknell OL, Davies PS, Opie LH. A relationship between adenosine triphosphate, glycolysis and ischaemic contracture in the isolated rat heart. *J Mol Cell Cardiol*. 1981;13:941.
27. Jennings RB, Hawkins HK, Lowe JE, Hill ML, Klotman S, Reimer KA. Relation between high energy phosphates and lethal injury in myocardial ischemia in the dog. *Am L Pathol*. 1978;92:187.
28. Cobbe SM, Poole-Wilson PA. The time of onset and severity of acidosis in myocardial ischaemia. *J Mol Cell Cardiol*. 1980;12:745.
29. Jacobus WE, Pores IH, Lucas SK, Weisfeldt ML,

Flaherty JT. Intracellular acidosis and contractility in the normal and ischemic heart as examined by ^{31}P NMR. *J Mol Cell Cardiol.* 1982;14(suppl. 3):13.

30. Steenbergen C, Deleuw G, Rick T, Williamson JR. Effects of acidosis and ischemia on contractility and intracellular pH of rat heart. *Circ Res.* 1977;41:849.

31. Kübler W, Katz AM. Mechanism of early "pump" failure of the ischemic heart: possible role of adenosine triphosphate depletion and inorganic phosphate accumulation. *Am J Cardiol.* 1977;40:467.

32. Fleckenstein A. *Calcium Antagonism in Heart and Smooth Muscle: Experimental Facts and Therapeutic Prospects.* New York: John Wiley, Ch. 3.

33. Rona G, Chappel CI, Balazs T, Gaudry R. An infarct-linke myocardial lesion and other toxic manifestations produced by isoproterenol in the rat. *Arch Pathol.* 1959;67:443.

34. Shen AC, Jennings RB. Kinetics of calcium accumulation in acute myocardial ischemic injury. *Am J Pathol.* 1972;67:441.

35. Bourdillon PD, Poole-Wilson PA. The effects of verapamil, quiescence, and cardioplegia on calcium exchange and mechanical function in ischemic rabbit myocardium. *Circ Res.* 1982;50:360.

36. Zimmermann ANE, Hülsmann WC. Paradoxical influence of calcium ions on the permeability of the cell membranes of the isolated rat heart. *Nature.* 1966;211:646.

37. Crevey BJ, Langer GA, Frank JS. Role of Ca^{2+} in maintenance of rabbit myocardial cell membrane structural and functional integrity. *J Mol Cell Cardiol.* 1978;10:1081.

38. Alto LE, Dhalla NS. Role of changes in microsomal calcium uptake in the effects of reperfusion of Ca^{2+}-deprived rat hearts. *Circ Res.* 1981;48:17.

39. Katz AM, Messineo FC. Lipid–membrane interactions and the pathogenesis of ischemic damage in the myocardium. *Circ Res.* 1981;48:1.

40. Chien KR, Reeves JP, Buja M, Bonte F, Parkey RW, Willerson JT. Phospholipid alterations in canine ischemic myocardium: temporal and topographical correlations with Tc-99m-PPi accumulation and an *in vitro* sarcolemmal Ca^{2+} permeability defect. *Circ Res.* 1981;48:711.

41. Steenbergen C, Jennings RB. Relationship between lysophospholipid accumulation and plasma membrane injury during total *in vitro* ischemia in dog hearts. *J Mol Cell Cardiol.* 1984;16:605.

42. McCord JM, Roy RS. The pathophysiology of superoxide: roles in inflammation and ischemia. *Can J Physiol Pharmacol.* 1982;60:1346.

43. Chambers JM, Yoshida S, Parmley LF, Downey JM. Xanthine oxidase as a source of free radical damage in myocardial ischemia. *J Mol Cell Cardiol.* 1985;17:145.

44. Jolly SR, Kane WJ, Bailie MB, Abrams GD, Lucchesi BR. Canine myocardial reperfusion injury: its reduction by the combined administration of superoxide dismutase and catalase. *Circ Res.* 1984;54:277.

45. Reimer KA, Jennings RB. Failure of the xanthine oxidase inhibitor allopurinol to limit infarct size after ischemia and reperfusion in dogs. *Circulation.* 1985;71:1069.

46. Romson J, Hook B, Rigot V, Schork A, Swanson D, Lucchesi B. The effect of ibuprofen on accumulation of indium-111 labeled platelets and leucocytes in experimental myocardial infarction. *Circulation.* 1982;66:1002.

47. Romson J, Hook B, Kunkel S, Abrams G, Schork A, Lucchesi B. Reduction in the extent of ischemic myocardial injury by neutrophil depletion in the dog. *Circulation.* 1983;67:1016.

48. Zimmer H-G, Ibel H. Effects of isoproterenol and dopamine on the myocardial hexose monophosphate shunt. *Experientia.* 1979;35:510.

49. Zimmer H-G. Measurement of left ventricular hemodynamic parameters in closed-chest rats under control and various pathophysiologic conditions. *Basic Res Cardiol.* 1983;78:77.

50. Zimmer H-G, Ibel H. Effects of ribose on cardiac metabolism and function in isoproterenol-treated rats. *Am J Physiol.* 1983;245:H880.

51. Glock GE, McLean P. Further studies on the properties and assay of glucose-6-phosphate dehydrogenase and 6-phosphogluconate dehydrogenase of rat liver. *Biochem J.* 1953;55:400.

52. Glock GE, McLean P. Levels of enzymes of the direct oxidative pathway of carbohydrate metabolism in mammalian tissues and tumours. *Biochem J.* 1954;56:171.

53. Zimmer H-G, Gerlach E. Stimulation of myocardial adenine nucleotide biosynthesis by pentose and pentitols. *Pflügers Arch.* 1978;376:223.

54. Zimmer H-G, Trendelenburg C, Kammermeier H, Gerlach E. *De novo* synthesis of myocardial adenine nucleotides in the rat: acceleration during recovery from oxygen deficiency. *Circ Res.* 1973;32:635.

55. Gerlach E, Deuticke B, Dreisbach RH. Zum Verhalten von Nucleotiden und ihren dephosphorylierten Abbauprodukten in der Niere bei Ischämie und kurzzeitiger post-ischämischer Wiederdurchblutung. *Pflügers Arch.* 1963;278:296.

56. Zimmer H-G, Steinkopff G, Gerlach E. Changes of protein synthesis in the hypertrophying rat heart. *Pflügers Arch.* 1972;336:311.

57. Zimmer H-G. Correlation between haemodynamic and metabolic changes in three models of experimental cardiac hypertrophy. *Eur Heart J.* 1984;5(suppl F):171.

58. Zimmer H-G, Gerlach E. Early metabolic alterations during the development of experimentally induced cardiac hypertrophy. *Arzneim Forsch /Drug Res.* 1980;30(II, 11a):2001.

59. Zimmer H-G. Normalization of depressed heart function in rats by ribose. *Science.* 1983;220:81.

60. Zimmer H-G. Restitution of myocardial adenine nucleotides: acceleration by administration of ribose. *J Physiol (Paris)* 1980;76:769.

61. Bechtelsheimer H, Zimmer H-G, Gedigk P. Histologische und histochemische Untersuchungen am nicht infarzierten Myokardanteil beim experimentellen Herzinfarkt der Ratte. *Vichows Arch Abt B Zellpath.* 1971;9:115.

62. Zimmer H-G, Martius PA, Marschner G. Myocardial infarction in rats: effects of metabolic and pharmacologic interventions. *Basic Res Cardiol.* 1989;84:332.

63. Ibel H, Zimmer H-G. Metabolic recovery following temporary regional myocardial ischemia in the rat. *J Mol Cell Cardiol.* 1986;18(suppl. 4):61.

64. Rentrop P, Blanke H, Köstering H, Karsch KR. Intrakoronare streptokinase-applikation bei akutem infarkt und instabiler angina pectoris. *Dtsch Med Wschr.* 1980;105:221.

65. Grüntzig AR. Transluminal dilatation of coronary-artery stenosis. *Lancet.* 1978;1:263.

66. Berg R, Kendall RW, Duvoisin GE, Ganji JH, Rudy LW, Everhart FJ. Acute myocardial infarction: a surgical emergency. *J Thorac Cardiovasc Surg.* 1975;70:432.

67. Bolooki H, Kotler MD, Lottenberg L, et al. Myocardial revascularization after acute infarction. *Am J Cardiol.* 1975;36:395.

68. Mueller HS, Ayres SM, Religa A, Evans RG. Propranolol in the treatment of acute myocardial infarction: effect on myocardial oxygenation and hemodynamics. *Circulation.* 1974;49:1078.

69. MecLean D, Fishbein MC, Braunwald E, Maroko PR. Long-term preservation of ischemic myocardium after experimental coronary artery stenosis. *J Clin Invest.* 1978;61:541.

70. Reimer KA, Lowe JE, Jennings RB. Effect of the calcium antagonist verapamil on necrosis following temporary coronary artery occlusion in dogs. *Circulation.* 1977;55:581.

71. DeJong JW, Harmsen E, DeTombe PP, Keijzer E. Nifedipine reduces adenine nucleotide breakdown in ischemic rat heart. *Eur J Pharmacol.* 1982;81:89.

72. Lange R, Ingwall JS, Hale SL, Alker KJ, Braunwald E, Kloner RA. Preservation of high-energy phosphates by verapamil in reperfused myocardium. *Circulation.* 1984;70:734.

73. Kübler W, Spieckermann PG. Regulation of glycolysis in the ischemic and anoxic myocardium. *J Mol Cell Cardiol.* 1970;1:351.

74. Rovetto MJ, Whitmer JT, Neely JR. Comparison of the effects of anoxia and whole heart ischemia on carbohydrate utilization in isolated working rat hearts. *Circ Res.* 1973;32:699.

75. Rovetto MJ, Lamberton WF, Neely JR. Mechanisms of glycolytic inhibition in ischemic rat hearts. *Circ Res.* 1975;37:742.

76. Calva E, Mújica A, Bisteni A, Sodi-Pallares D. Oxidative phosphorylation in cardiac infarct: effect of glucose-KC1-insulin solution. *Am J Physiol.* 1965;209:371.

77. Maroko PR, Libby P, Sobel BE et al. Effect of glucose-insulin-potassium infusion on myocardial infarction following experimental coronary artery occlusion. *Circulation.* 1972;44:1160.

78. Apstein CS, Gravino FN, Haudenschild CC. Determinants of a protective effect of glucose and insulin on the ischemic myocardium: effects on contractile function, diastolic compliance, metabolism, and ultrastructure during ischemia and reperfusion. *Circ Res.* 1983;52:515.

79. Sundstet CD, Sylvrén C, Mogensen L. Glucose-insulin-potassium-albumin infusion in the early phase of acute myocardial infarction: a controlled study. *Acta Med Scand.* 1981;210:67.

80. Liedtke AJ, Nellis SH, Neely JR, Hughes HC. Effects of treatment with pyruvate and tromethamine in experimental myocardial ischemia. *Circ Res.* 1976;39:378.

81. Gutterman DD, Chilian WM, Eastham CL, Inon T, White CW, Marcus ML. Failure of pyruvate to salvage myocardium after prolonged ischemia. *Am J Physiol.* 1986;250:H114.

82. Mjøs OD, Miller NE, Riemersma RA, Oliver MF. Effects of dichloracetate on myocardial substrate extraction, epicardial ST-segment elevation and ventricular blood flow following coronary occlusion in dogs. *Cardiovasc Res.* 1976;10:427.

83. Kjekshus JK, Mjøs OD. Effect of free fatty acids on myocardial function and metabolism in the ischemic dog heart. *J Clin Invest.* 1972;51:1767.

84. Kjekshus JK, Mjøs OD. Effect of inhibition of lipolysis on infarct size after experimental coronary artery occlusion. *J Clin Invest.* 1973;52:1170.

85. Liedtke AJ, Nellis SH. Effects of carnitine in ischemic and fatty acid supplemented swine hearts. *J Clin Invest.* 1979;64:440.

86. Shug AL, Shrago E, Bittar N, Folts JD, Koke JR. Acyl-CoA inhibition of adenine nucleotide translocation in ischemic myocardium. *Am J Physiol.* 1975;228:689.

87. Shug AL, Thomsen JH, Folts JD, et al. Changes in tissue levels of carnitine and other metabolites during myocardial ischemia and anoxia. *Arch Biochem Biophys.* 1978;187:25.

88. Gerlach E, Deuticke B, Dreisbach RH. Der Nucleotid-Abbau im Herzmuskel bei Sauerstoffmangel und seine mögliche Bedeutung für die Coronardurchblutung. *Naturwiss.* 1963;50:228.

89. Berne RM. Cardiac nucleotides in hypoxia: possible role in regulation of coronary blood flow. *Am J Physiol.* 1963;204:317.

90. Ronca-Testoni S, Borghini F. Degradation of perfused adenine compounds up to uric acid in isolated rat heart. *J Mol Cell Cardiol.* 1982;14:177.

91. Gerlach E, Nees S, Becker BF. The vascular endothelium: a survey of some newly evolving biochemical and physiological features. *Basic Res Cardiol.* 1985;80:459.

92. Wiedmeier VT, Rubio R, Berne RM. Inosine incorporation into myocardial nucleotides. *J Mol Cell Cardiol.* 1972;4:445.

93. Henderson JF. *Regulation of Purine Biosynthesis.* Washington, DC: American Chemical Society; 1972. ACS Monograph 170.

94. Gerlach E, Deuticke B. Bildung und Bedeutung von Adenosin in dem durch Sauerstoffmangel geschädigten Herzmuskel unter dem Einfluss von 2,6-Bis-(diaethanolamino-)4,8-dipideridino-pyrimido (5,4-d)pyrimidin. *Arzneim Forsch/Drug Res.* 1963;13:48.

95. Kübler W, Spieckermann PG, Bretschneider HJ. Influence of dipyridamol (Persantin) on myocardial adenosine metabolism. *J Mol Cell Cardiol.* 1970;1:23.

96. Feinberg H, Levitsky S, Coughlin TR, Merchant

F, Holland CH, O'Donaghue M. Effect of dipyridamole on ischemia-induced changes in cardiac performance and metabolism. *J Cardiovasc Pharmacol.* 1980;2:299.

97. Blumenthal DS, Hutchins GM, Jugdutt BI, Becker LC. Salvage of ischemic myocardium by dipyridamole on the conscious dog. *Circulation.* 1981;64:915.

98. Foker JE, Einzig S, Wang T. Adenosine metabolism and myocardial preservation: consequences of adenosine catabolism on myocardial high-energy compounds and tissue blood flow. *J Thorac Cardiovasc Surg.* 1980;80:506.

99. Humphrey SM, Seelye RN. Improved functional recovery of ischemic myocardium by suppression of adenosine catabolism. *J Thorac Cardiovasc Surg.* 1982;84:16.

100. DeWall RA, Vasko KA, Stanley EL, Kezdi P. Responses of the ischemic myocardium to allopurinol. *Am Heart J.* 1971;82:362.

101. Arnold WL, DeWall RA, Kezdi P, Zwart HH. The effect of allopurinol on the degree of early myocardial ischemia. *Am Heart J.* 1980;99:614.

102. Crowell JW, Jones CE, Smith EE. Effect of allopurinol on hemorrhagic shock. *Am J Physiol.* 1969;216:744.

103. Isselhard W, Hinzen DH, Geppert E, Mäurer W. Beeinflussung des post-asphyktischen Wiederaufbaues der Adeninnucleotide im Kaninchenherzen *in vivo* durch Substratangebot. *Pflügers Arch.* 1970;320:195.

104. Isselhard W, Eitenmüller J, Mäurer W, et al. Increase in myocardial adenine nucleotides induced by adenosine: dosage, mode of application and duration, species differences. *J Mol Cell Cardiol.* 1980;12:619.

105. Isselhard W, Hamaji M, Mäuer W, Erkens H, Welter H. Adenosine-induced increase in myocardial adenine nucleotides without adenosine-induced systemic hypotension. *Basic res. cardiol.* 1985;80:47.

106. Reibel DK, Rovetto MJ. Myocardial adenosine salvage rates and restoration of ATP content following ischemia. *Am J Physiol.* 1979;237:H247.

107. Seifart HI, Delabar U, Siess M. The influence of various precursors on the concentration of energy-rich phosphates and pyridine nucleotides in cardiac tissues and its meaning for anoxic survival. *Basic Res Cardiol.* 1980;75:57.

108. Siess M, Seifart HI. Anoxic energy production and contractile activity in mammalian cardiac muscle. In: Tajuddin M, Bhatia B, Siddiqui HH, Rona G, eds. *Advances in Myocardiology,* Vol. 2. Baltimore: University Park Press; 1980:295.

109. Liu MS, Feinberg H. Incorporation of adenosine-8-^{14}C and inosine-8-^{14}C into rabbit heart adenine nucleotides. *Am J Physiol.* 1971;220:1242.

110. Goldhaber SZ, Pohost GM, Kloner RA, Andrews E, Newell JB, Ingwall JS. Inosine: a protective agent in an organ culture model of myocardial ischemia. *Circ Res.* 1982;51:181.

111. Ingwall JS. Phosphorus nuclear magnetic resonance spectroscopy of cardiac and skeletal muscle. *Am J Physiol.* 1982;242:H729.

112. Zimmer H-G. Effects of inosine on cardiac adenine nucleotide metabolism in rats. In: Dhalla NS, Hearse DJ, eds. *Advances in Myocardiology,* Vol. 6. New York: Plenum; 1985:173.

113. Swain J, Hines JJ, Sabina RL, Holmes EW. Accelerated repletion of ATP and GTP pools in postischemic canine myocardium using a precursor of purine *de novo* synthesis. *Circ Res.* 1982;51:102.

114. Mauser M, Hoffmeister HM, Nienhaber C, Schaper W. Influence of ribose, adenosine, and "AICAR" on the rate of myocardial adenosine triphosphate synthesis during reperfusion after coronary artery occlusion in the dog. *Circ Res.* 1985;56:220.

115. Verdetti J, Aussedat J, Rossi A. Effets de l'administration *in vivo* de divers précurseurs sur le teneurs en nucléotides adényliques et uridyliques du coeur de rat. *J Physiol (Paris).* 1980;76:693.

116. Olivares J, Rossi A. Incorporation de l'acid orotique dans les nucléotides uridyliques du tissu myocardique: effet de l'isoprénaline et du ribose. *J Physiol (Paris).* 1982;78:175.

117. Eggleston LV, Krebs HA. Regulation of the pentose phosphate cycle. *Biochem J.* 1974;138:425.

118. Löhr GW, Waller HD. Zur Biochemie angeborener hämolytischer Anämien. *Folia Haematol.* 1963;8:377.

119. Zimmer H-G, Ibel H. Studies on the mechanism for the isoproterenol-induced stimulation of cardiac glucose-6-phosphate dehydrogenase. *FEBS Lett.* 1979;106:335.

120. Harmsen E, DeTombe PP, DeJong JW, Achterberg PW. Enhanced ATP and GTP synthesis from hypoxanthine or inosine after myocardial ischemia. *Am J Physiol.* 1984;246:H37.

121. Brown AK, Raeside DL, Bowditch J, Dow JW. Metabolism and salvage of adenine and hypoxanthine by myocytes isolated from mature rat heart. *Biochim Biophys Acta.* 1985;845:469.

122. Dow JW, Nigdikar S, Bowditch J. Adenine nucleotide synthesis *de novo* in mature rat cardiac myocytes. *Biochim Biophys Acta.* 1985;847:223.

123. Zimmer H-G, Ibel H, Steinkopff G, Korb G. Reduction of the isoproterenol-induced alterations in cardiac adenine nucleotides and morphology by ribose. *Science.* 1980;207:319.

124. Pasque MK, Spray TL, Pellom GL, et al. Ribose-enhanced myocardial recovery following ischemia in the isolated working rat heart. *J Thorac Cardiovasc Surgery.* 1982;83:390.

125. Haas G, DeBoer LV, O'Keefe DD, et al. Reduction of postischemic myocardial dysfunction by substrate repletion during reperfusion. *Circulation.* 1984;70(suppl. I):65.

126. Ward HB, Kriett J, St. Cyr JA, et al. Relationship between recovery of myocardial ATP levels and cardiac function following ischemia. *J Am Coll Cardiol.* 1984;3(2):544.

127. Foker JE, Ward HB, St. Cyr JA, Alyono D, Bianco RW, Kriett JM. Enhanced ATP return after global ischemia. *Eur Heart J.* 1984;5(suppl. 1):377.

128. Lortet S, Zimmer H-G. Functional and metabolic effects of ribose in combination with prazosin,

verapamil and metoprolol in rats in vivo. *Cardiovasc Res.* 1989;23:702.

129. Zimmer H-G, Zierhut W, Marschner G. Combination of ribose with calcium antagonist and β-blocker treatment in closed-chest rats. *J Mol Cell Cardiol.* 1987;19:635.

130. Zimmer H-G, Seesko RC, Zierhut W, Pechan I. Cardiac effects of β-adrenergic blockers in combination with metabolic interventions. In: Beamish RE, Panagia V, Dhalla NS, eds. *Pharmacological Aspects of Heart Disease.* Boston: Martinus Nijhoff; 1987:233.

131. Zimmer H-G, Ibel H, Suchner U, Schad H. Ribose intervention in the cardiac pentose phosphate pathway is not species-specific. *Science.* 1984;223:712.

132. Zöllner N, Reiter S, Gross M, et al. Myoadenylate deaminase deficiency: successful symptomatic therapy by high dose oral administration of ribose. *Klin Wschr.* 1986;64:1281.

133. Goodman MN, Lowenstein JM. The purine nucleotide cycle: studies of ammonia production by skeletal muscle *in situ* and in perfused preparations. *J Biol Chem.* 1977;252:5054.

134. Sabina RL, Swain JL, Patten BM, et al. Disruption of the purine nucleotide cycle: a potential explanation for muscle dysfunction in myoadenylate deaminase deficiency. *J Clin Invest.* 1980;66:1419.

135. Sabina RL, Swain JL, Olanow CW, et al. Myoadenylate deaminase deficiency: functional and metabolic abnormalities associated with disruption of the purine nucleotide cycle. *J Clin Invest.* 1984;73:720.

136. Bierman EL, Baker EM, Plough IC, Hall WH. Metabolism of D-ribose in diabetes. *Diabetes.* 1959;8:453.

137. Larson HW, Bradshaw PJ, Ewing ME, Sawyer SD, Blatherwick NR. The metabolism of ribose. *Fed Proc.* 1943;2:66.

138. Segal S, Foley J, Wyngaarden JB. Hypoglycemic effect of D-ribose in man. *Proc Soc Exp Biol.* 1957;95:551.

139. Pierce WS. Artificial heart and blood pumps in the treatment of profound heart failure. *Circulation.* 1983;68:883.

140. Mineo I, Kono N, Shimizu T, et al. Excess purine degradation in exercising muscles of patients with glycogen storage disease types V and VII. *J Clin Invest.* 1985;76:556.

Index

Numerals in *italics* indicate a figure, "t" following a page number indicates tabular matter.

1027718

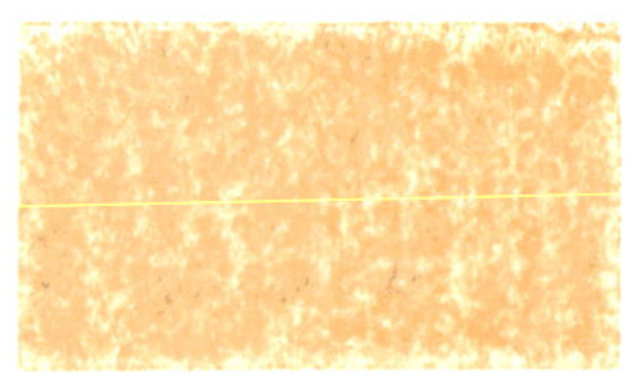

DATE DUE

FEB 1 1 1991		
FEB 2 6 1996		
MAY 0 9 1997		
NOV 2 3 1998		
DISCARD		
GAYLORD		PRINTED IN U.S.A.